LINEAR PROGRAMMING

An-min Chung

Professor of Economics,
Drexel Institute of Technology

CHARLES E. MERRILL BOOKS, INC.,
COLUMBUS, OHIO

Library of Congress Catalog Card Number: 63-11327

PRINTED IN THE UNITED STATES OF AMERICA

To my wife, Tsu-shan Mei,

and my mother, Alice S. F. Sun,

whose silent contributions to the undertaking

have been invaluable.

PREFACE

The increased use of mathematics and statistics in solving business and industrial problems focuses attention on linear programming as one of the few sophisticated analytical devices finding wide acceptance in industry. A number of universities have introduced courses in linear programming, but, until recently, there was no suitable textbook. Most of them either avoided entanglement with the mathematical aspects of linear programming completely, or else discussed them at a level comprehensible only to the mathematically initiated. To a growing group of non-mathematically-oriented people, both on and off campus, who heed the trend and therefore wish to understand the theory of linear programming, the lack of an effective vehicle of learning remains a continued source of frustration.

This book serves as an introduction to the subtle mathematical reasoning that underlies and gives strength to linear programming as an analytical tool. That the reader should emerge with a somewhat enlarged stock of mathematical knowledge, as a result of having studied this book, would gratify, rather than surprise, the author.

Assuming only a knowledge of elementary algebra, the book begins by explaining the mathematical characteristics of a linear programming problem. This is followed by a careful explanation of the rudiments of matrix algebra and determinants, and such basic concepts as linear independence, vector space, and basis. In Chapter III the mathematical properties of linear programming solutions are discussed so as to pave the way for the development of the simplex method. The method itself is treated in Chapters IV and V which deal with its theoretical and computational aspects respectively. Two important variations of the simplex method, the revised simplex and the dual simplex, are introduced in Chapter VI. Sensitivity analysis is taken up in Chapter VII and is illustrated with practical examples. The theory of the dual is developed in Chapter VIII by employing the novel concept of a vector solution pair.

With much of the basic theory out of the way, the transportation problem is introduced as a special case of the general linear programming problem. Finally, in Chapter X, a number of the more advanced linear programming topics are described.

Numerical examples throughout the text illustrate important concepts and theorems. Examples with business or industrial content are used whenever possible so that the reader will constantly be reminded of the practical relevance of linear programming. No attempt is made to circumvent concepts and theorems just because they are mathematical. On the contrary, continuous effort is made to deal with them simply and clearly, but with increasingly liberal use of mathematics consistent with the expected development of the reader's mathematical ability. Mathematics is basically an essay in logic, and what is logical should not be difficult to understand once its baffling symbols and fundamental precepts are firmly grasped. As intimated earlier, one of the principal aims of this book is to sharpen the reader's mathematical facilities to a point where, upon completion of this volume, he may venture into technical journals in this field.

Students whose major interests lie in areas such as operations research, industrial engineering, management science, and applied economics, whether they be undergraduate or graduate students, will find this book especially designed for use as a text. Exercises of a varied nature, such as those that illuminate theory and those that develop skill, are included in each chapter, thus making this book suitable also for managerial and staff personnel in industry as well as for anyone else who is willing to use the self-study method. Tested in a quarterly course at Drexel for two years, it has been found to contain more material than can be handled comfortably in one quarter. If time is a limiting factor, then one may omit, without loss of continuity, the following material: Sections 6–5, 6–6 of Chapter VI, Chapter VII, Section 8–4 of Chapter VIII, and Chapter X. Students with adequate background knowledge of matrix algebra, however, should have no difficulty in covering the whole book simply by substituting Chapter II for the omissions suggested above. One advantage of this book is that it permits individual instructors to add to it in whatever way they deem desirable. For instance, those who favor theory may introduce additional materials on network flow theory, whereas others who favor applications may expose the students to more industrial examples found in current literatures.

Thanks are due to my colleagues at Drexel, particularly Samuel S. McNeary who read and offered important advices on the mathematical parts of the original manuscript. All errors are, of course, responsibilities of my own.

A. M. Chung

May, 1962

TABLE OF CONTENTS

INTRODUCTION

1.1. General.

In the history of management technology, there is perhaps no other period more comparable in importance to the era of Frederick Taylor than the decade immediately following the Second World War. During this decade two important developments—of breakthrough proportions—paved the way for important advances in the essential art of management planning and control. One was the advent of the digital computer and the other was the application of higher mathematics and statistics to problems of industrial management. The former made possible rapid and economical manipulation of massive data, while the latter provided the necessary theoretical framework for the organization and analysis of these data. The result is that business problems of unprecedented complexity can now be solved and made a part of the rational decisions on which the success of businesses so importantly depends.

A vital part of these advances in management technology is the increasing application of linear programming to solve a wide range of managerial problems. This analytical tool has found ready adoption in many industries, including petroleum, chemicals, steel, and agriculture. Exploratory studies using the linear programming approach have also been made by airlines, railroads, utilities and financial institutions.

Moreover, interest in linear programming extends beyond the

world of business. Economists, for instance, have found it helpful in attempting a re-appraisal of the theory of the firm and the theory of resource-allocation in a free-price economy. Mathematicians have through it discovered new avenues of research and investigation. In short, it is a subject that will be of considerable interest and importance to the business and academic worlds for some time to come. Our approach here, therefore, is to emphasize the fundamentals of the theory of linear programming, which, however, will be illustrated from time to time by problems relevant to the practical business world.

1.2. Characteristics of Linear Programming Problems.

Generally speaking, linear programming is a mathematical optimizing technique applicable to a class of problems having certain characteristics in common. The interpretation of these characteristics varies in accordance with the specific content of the problem. For this reason, it is best to describe these common characteristics in mathematical terms so as to retain their generality.

(1) A linear objective function—All linear programming problems have as their objective the optimization of some explicit linear function of many variables. If we denote this linear objective function by $f(X)$ and the appropriate variables of the problem by $x_1, x_2, \ldots, x_n$, then the goal of a linear programming problem is always to maximize (or minimize)

$$f(X) = c_1x_1 + c_2x_2 + \ldots + c_nx_n^*, \tag{1.2.1}$$

where $c_1, \ldots, c_n$ represent the parameters of the problem. In other words, the goal is not just to accomplish something but to accomplish it in the "best" possible manner. The fact that business operations quite often are conducted with this type of goal in mind, e.g., to maximize profit, or to minimize cost, substantially explains why linear programming is found useful in business.

However, not all problems with optimizing objectives call for linear programming solutions. The objective must also be capable of being stated as a linear function of the variables of the problem. What do we mean by a linear function then? In mathematics, a function,

---

* Obviously the coefficients $c_1, \ldots, c_n$ are not restricted as to signs either individually or collectively.

say $f(X)$, is described as linear if and only if the following two conditions are satisfied:*

$a.\ f(kX) = kf(X),$
$b.\ f(X_1 + X_2) = f(X_1) + f(X_2),$

where k is a constant coefficient; X, the set of variables, $x_1, x_2, \ldots, x_n$; and X_1, X_2, two different sets of values of $x_1, \ldots, x_n$. Stated in words, the first condition simply stipulates that multiplying the variables by a certain constant k should result in the same functional value as multiplying the functional value of X by the same constant k. The second condition is a little more difficult to explain. Essentially it requires that, given two sets of values of the variables, X_1 and X_2, the functional value of $(X_1 + X_2)$ must be the same as the sum of the two functional values obtained from X_1 and X_2 separatively.

Based on the above definition, we can easily verify that a first degree polynomial such as the one shown in (1.2.1) is a linear function. For if

$$f(X) = c_1x_1 + c_2x_2 + \ldots + c_nx_n,$$

then

$$\begin{aligned} f(kX) &= c_1(kx_1) + c_2(kx_2) + \ldots + c_n(kx_n) \\ &= k(c_1x_1 + c_2x_2 + \ldots + c_nx_n) \\ &= kf(X). \end{aligned}$$

Moreover, letting X_1 represent the set of values, $x_{11}, \ldots, x_{1n}$, and X_2 represent the other set of values, $x_{21}, \ldots, x_{2n}$, we have

$$\begin{aligned} f(X_1 + X_2) &= c_1(x_{11} + x_{21}) + c_2(x_{12} + x_{22}) + \ldots + c_n(x_{1n} + x_{2n}) \\ &= (c_1x_{11} + c_2x_{12} + \ldots + c_nx_{1n}) \\ &\qquad + (c_1x_{21} + c_2x_{22} + \ldots + c_nx_{2n}) \\ &= f(X_1) + f(X_2). \end{aligned}$$

On the other hand, such functions as $\sin x_1 + \sin x_2$, $x_1^2 + x_2^2$, or in general, all trigonometric functions, exponential functions and polynomials of higher degree than 1, are not linear functions because both conditions cannot be satisfied. Some functions may satisfy one but not the other condition. For instance, $f(X) = \dfrac{x_1^3 + x_2^3}{x_1^2 + x_2^2}$ satisfies the

* Throughout this text, the domain of real numbers is always assumed implicitly. In other words, functions and variables are always defined over the domain of real numbers, which is all that is necessary for the usual applications of linear programming.

first but not the second condition. Consequently, it is not a linear function.

(2) A set of linear constraints—Suppose we were to maximize a linear function such as $2x_1 - 3x_2$. Since x_1 is associated with a positive coefficient and x_2 with a negative coefficient, obviously we could increase the value of this function as long as we increase x_1 and decrease x_2. This could be continued indefinitely until x_1 is increased to positive infinity and x_2 decreased to negative infinity with the result that the value of the function would be positive infinity. As a matter of fact, all linear functions by themselves have a maximal value of positive infinity and a minimal value of negative infinity if no restrictions are placed on the values of their variables. This means that they all have the same maximum and minimum. Consequently, the problem of optimizing a linear function is not a mathematically meaningful problem unless the variables are constrained as to the ranges of values they may assume. In linear programming such restrictions are contained in a set of linear inequalities as follows:

$$\begin{aligned} a_{11}x_1 + a_{12}x_2 + \dots + a_{1n}x_n &\leq b_1, \\ a_{21}x_1 + a_{22}x_2 + \dots + a_{2n}x_n &\leq b_2, \\ \vdots \qquad \vdots \qquad\quad & \quad \vdots \\ a_{m1}x_1 + a_{m2}x_2 + \dots + a_{mn}x_n &\leq b_m, \end{aligned} \tag{1.2.2}$$

where $a_{11}, a_{12}, \dots, a_{1n}, \dots, a_{m1}, a_{m2}, \dots, a_{mn}$ are constant coefficients* and $b_1, b_2, \dots, b_m$ simply constants. There are three points that should be borne in mind in connection with these constraints (called *structural* constraints in linear programming terminology). First, they are linear in the same sense as the linear function is defined above. Second, they are characteristically inequalities although equality-constraints are not categorically excluded. Finally, the number of such constraints, i.e., m, is not restricted in any way except as it affects the practical problem of computation. Together these constraints define a region of acceptable values of the variables. Since this region may or may not include values up to infinity, linear programming becomes a problem of selecting a set of values (of the

* Since there are m inequalities and n variables, it is necessary to differentiate among the coefficients by double subscripts with the first referring to the inequality and the second to the variable. Thus, a_{1n} is the coefficient of the nth variable (x_n) in the first inequality.

variables) within this region, which will yield the maximal (or minimal) value of the objective function.

The relevance of these mathematical expressions to the real world is not far to see. For example, if we are describing a management problem of deciding on how much of each of a group of n products to produce, it is reasonable to expect that some of the products may require the same type of steel, or the same kind of welding operation, or the same assembling facilities. To the extent that the supply of raw material, or of machine time, or of assembling facilities is fixed, which is again reasonable to expect, there is an effective constraint on the relative production levels of these products. In addition, there may be time constraints imposed by delivery schedules, financial constraints imposed by budgets, technical constraints imposed by engineering considerations, and so on. As the next section will show, the question is not so much whether there are constraints or not, but whether they lend themselves to unambiguous mathematical statements and, if they do, whether the resulting statements are linear inequalities. If they are not, then the problem is not a linear programming problem.*

(3) The non-negativity constraints—As a direct outgrowth of its application in business and industry, all linear programming problems require that the solution values of their variables be non-negative, i.e., either zero or positive but not negative. This is necessary because the variables in business and industry, e.g., level of production, amount in inventory, etc. usually have no meaningful negative counterparts. Mathematically these non-negativity constraints are written as follows:

(1.2.3) $$x_1, x_2, \ldots, x_n \geq 0.$$

Although at first glance the inclusion of these constraints seems to compound the difficulty of solving the problem, actually—as we will see in Chapter IV—they can be taken care of by a simple rule of transformation which automatically excludes all unacceptable negative solutions. Their presence in the problem has no more operational significance than the secondary advantage of providing an additional

* Current efforts to extend the simplex method of solving linear programming problems to programming problems with non-linear constraints are evidenced by Peter Wegner's article, "A Non-linear Extension of the Simplex Method," *Management Science*, VII (1960), pp. 43–56.

check on the accuracy of computations—a point which will become clear as we complete Chapter V.

In summary then, a linear programming problem is recognized by the following three parts:

$$\text{Maximize } f(X) = c_1x_1 + c_2x_2 + \ldots + c_nx_n, \tag{1.2.1}$$

$$(1.2.4)\quad \begin{aligned} \text{subject to:} \quad & a_{11}x_1 + a_{12}x_2 + \ldots + a_{1n}x_n \leq b_1, \\ & a_{21}x_1 + a_{22}x_2 + \ldots + a_{2n}x_n \leq b_2, \quad (1.2.2) \\ & \quad \cdot \qquad\quad \cdot \qquad\qquad\qquad \cdot \qquad \cdot \\ & \quad \cdot \qquad\quad \cdot \qquad\qquad\qquad \cdot \qquad \cdot \\ & \quad \cdot \qquad\quad \cdot \qquad\qquad\qquad \cdot \qquad \cdot \\ & a_{m1}x_1 + a_{m2}x_2 + \ldots + a_{mn}x_n \leq b_m, \\ \text{and} \quad & x_1, x_2, \ldots, x_n \geq 0. \quad (1.2.3) \end{aligned}$$

It may be well to point out that a "$\leq$" inequality may be converted into a "$\geq$" one if multiplied on both sides by -1. Hence, the expression of the structural constraints in (1.2.4) is perfectly general. Furthermore, since all inequalities can be converted into equalities by means of adding or subtracting non-negative variables, some prefer to write the constraints in the form of equations. In that event, however, it must be stipulated that $n \geq m$, or the problem may very well have no solution.

A casual survey of (1.2.4) would probably suggest that the problem could be solved by first finding all possible solutions to (1.2.2); then eliminating those which violate (1.2.3), and finally, selecting among the remaining solutions the one which optimizes $f(X)$ in (1.2.1). We soon realize, however, that (1.2.2) admits of an infinite number of solutions if any solution exists at all; therefore a complete enumeration of all of them is quite impracticable if not impossible. Accordingly, a more efficient method based on careful analysis of the theoretical properties of the problem is necessary. Such a method is the simplex method, first proposed by George B. Dantzig in 1947 in collaboration with Marshall Wood, Alex Orden and others while working on research projects in the U.S. Department of Air Force. This method together with its subsequent modifications, extensions and applications will constitute the subject matter of the ensuing chapters of this text.

1.3. Illustrative Linear Programming Problems.

(1) The product-mix problem—Let us assume that a certain manufacturing concern regularly produces three different products, called A, B, and C, which are sold for \$5, \$10 and \$20 per unit respectively. It is further assumed that these prices are constant irrespective of the firm's output, i.e., the market can absorb any amount of the products it may turn out without adverse effects on their prices. A total of four input factors is required for their production, and the amounts of each input factor required for each unit of the three products, together with the unit costs and the available supplies of these inputs are given in the table below:

Input	Unit cost (\$)	Product A	Product B	Product C	Supply of Input
1	2	0	1	2	100
2	1	1	2	1	200
3	0.5	4	6	10	400
4	2	0	0	2	100

Assuming that there are no other restrictions on the firm's production decisions, how much of each product should be produced in order to maximize profit?

A moment's reflection should reveal the general nature of this problem. Three products, all profitable, are competing for a fixed supply of resources. If the objective is to maximize profit, it would seem intuitively logical that the product with the highest profit margin should be given the highest priority in the use of resources. However, further considerations soon reveal the fallacious nature of this approach. For instance, the supply of resources may be badly balanced so far as the requirements of this product are concerned, thus seriously restricting its output potential. Or, its production may exhaust the supply of a particular input factor critically needed for the production of others, causing large quantities of the other inputs to be left unused (i.e., idle capacity). These preliminary considerations should be sufficient to indicate that the output decisions on the individual products are interdependent. Consequently, a "simultaneous" approach—such as that represented by linear programming—is needed.

Let us now formulate the problem systematically. First, consider the objective of the problem. Since we are trying to maximize the total profit, naturally we have to know the unit profit of each of the three products.

Since each unit of A requires 1 unit of input 2 and 4 units of input 3 at unit costs of \$1 and \$0.5 respectively, the total cost of producing one unit of A is $1 \times 1 + 4 \times 0.5 = \$3$. As the price of A is \$5, the unit profit of A is $\$5 - \$3 = \$2$. Similarly, we find that the unit profit of B and the unit profit of C are respectively \$3 and \$6. Now, let the unknown outputs of A, B and C be the variables x_1, x_2 and x_3; the total profit is then represented by $f(X) = 2x_1 + 3x_2 + 6x_3$, which is, of course, the objective function we wish to maximize.

However, production is restricted by the supply of resources available. Since A requires no input factor 1, whereas B and C require 1 and 2 units respectively, the total amount of input 1 needed for x_1 units of A, x_2 units of B and x_3 units of C is $0 \cdot x_1 + 1 \cdot x_2 + 2 \cdot x_3$, which must be no greater than the amount of input 1 available. Therefore, $x_2 + 2x_3 \leq 100$. Similarly, we can write $x_1 + 2x_2 + x_3 \leq 200$ for input 2, $4x_1 + 6x_2 + 10x_3 \leq 400$ for input 3, and finally $2x_3 \leq 100$ for input 4. Moreover, as production cannot be negative, we require $x_1 \geq 0$, $x_2 \geq 0$ and $x_3 \geq 0$. The whole problem then becomes

$$\begin{aligned}
\text{Maximize } f(X) &= 2x_1 + 3x_2 + 6x_3, \\
\text{subject to:} \quad & \phantom{2x_1 + {}} x_2 + 2x_3 \leq 100, \\
& x_1 + 2x_2 + x_3 \leq 200, \\
& 4x_1 + 6x_2 + 10x_3 \leq 400, \\
& 2x_3 \leq 100, \\
\text{and} \quad & x_1, x_2, x_3 \geq 0,
\end{aligned}$$

which, according to (1.2.4), is clearly a linear programming problem.

(2) The feed-mix problem—A problem closely related to the above is the so-called feed-mix problem. It is assumed that the farmer is interested in finding the most inexpensive formula for mixing a number of ingredients, e.g., corn, wheat, soybean and flour middlings, to make the feed for his hogs that will satisfy the minimal nutrition requirements in fat, protein, vitamins, minerals, etc. Presumably, of course, he is informed about the costs and nutrient contents of

these ingredients. The question is: What proportions should he use to produce a satisfactory feed at least cost?

Clearly the objective of this problem is to minimize a cost function rather than to maximize a profit function as in the previous example. Furthermore, the constraints are in the form of the lower limits that must be met, and hence are of the "$\geq$" rather than the "$\leq$" type. The variables, which are the various portions of the ingredients in the desired formula, are certainly subject to the non-negativity constraints. Hence, again we have a linear programming problem the details of which, however, need not be spelled out here. It is interesting to note that some years ago, linear programming was actually employed to solve a similar human diet problem. It consisted in the selection from a total of 77 varieties of food, of those items which would satisfy minimal nutritive requirements at the lowest possible cost. It was found that this "health" diet would include only 9 items, viz., wheat flour, corn meal, evaporated milk, peanut butter, lard, beef liver, cabbage, potatoes and spinach, and would cost a mere $39.67 per person at 1939 prices for the whole year.* However, it is safe to assume that most people are willing to "squander" considerably more than that to get variety in addition to nutrition.

(3) The transportation problem—The transportation problem is actually a sub-class of linear programming problems that can be adapted to a wide variety of situations. In its simplest version, the problem describes a situation wherein a firm desires to find the cheapest way of shipping its product from its warehouses with known stocks to its markets with known demands.

Specifically let us assume that this firm has 3 warehouses, called W_1, W_2 and W_3, which regularly supply its product to 5 markets, designated $M_1, M_2, \ldots, M_5$. The available stocks at these warehouses are b_1, b_2, b_3 and the demands at the markets are $a_1, a_2, \ldots, a_5$. The transportation cost from W_1 to M_1 is c_{11} per unit; from W_1 to M_2 it is c_{12} per unit, etc. In general, it is c_{ij} from W_i, $i = 1, 2, 3$ to M_j, $j = 1, 2, \ldots, 5$. Since the total shipping costs depend upon the various amounts shipped it seems natural that we should let the latter be the variables of the problem. Designating by x_{ij} the unknown amount that should be shipped from W_i to M_j, $i = 1, 2, 3$, and $j = 1, 2, \ldots, 5$. We have then the following objective function to be minimized:

* S. I. Gass, *Linear Programming* (New York: McGraw-Hill, 1958), p. 171.

$$\begin{aligned}\text{Minimize } f(X) = {} & c_{11}x_{11} + c_{12}x_{12} + \ldots + c_{15}x_{15} \\ & + c_{21}x_{21} + c_{22}x_{22} + \ldots + c_{25}x_{25} \\ & + c_{31}x_{31} + c_{32}x_{32} + \ldots + c_{35}x_{35}.\end{aligned}$$

However, as each warehouse has only a limited stock, the total amount shipped from each warehouse to all markets must be no more than its stock. Hence, the variables are restricted by the following three "capacity" constraints:

$$\begin{aligned} x_{11} + x_{12} + \ldots + x_{15} &\leq b_1 \text{ (capacity of } W_1), \\ x_{21} + x_{22} + \ldots + x_{25} &\leq b_2 \text{ (capacity of } W_2), \\ x_{31} + x_{32} + \ldots + x_{35} &\leq b_3 \text{ (capacity of } W_3). \end{aligned}$$

Additionally, since each market has a fixed demand that must be met, the total amount shipped to that market from all warehouses must be no less than the demand. Consequently, the variables are further restricted by the following five "requirement" constraints:

$$\begin{aligned} x_{11} + x_{21} + x_{31} &\geq a_1 \text{ (requirement of } M_1), \\ x_{12} + x_{22} + x_{32} &\geq a_2 \text{ (requirement of } M_2), \\ x_{13} + x_{23} + x_{33} &\geq a_3 \text{ (requirement of } M_3), \\ x_{14} + x_{24} + x_{34} &\geq a_4 \text{ (requirement of } M_4), \\ x_{15} + x_{25} + x_{35} &\geq a_5 \text{ (requirement of } M_5). \end{aligned}$$

Lastly, since the shipments cannot be negative, all variables are subject to the non-negativity constraints.

Noting that the objective function and the constraints are all linear, we conclude that the transportation problem is a linear programming problem. The transportation problem is of special interest because its constraints contain special structural features which make possible the use of a very simple method of solution. The simplicity of the method, in turn, enhances its value as a practical tool of analysis. Consequently, we will devote an entire chapter (Chapter IX) to the solution and application of the transportation problem.

(4) The rail-freight movement problem*—Whereas the foregoing illustrations are more or less "natural" linear programming problems in the sense that the formulation is relatively straightforward, the remaining two examples are more typical of the tasks involved in actual applications. More often than not, industrial problems do not

---

* Excerpted from A. Charnes and M. H. Miller, "A Model for the Optimal Programming of Railway Freight Train Movements," *Management Science*, III (1956), pp. 74–93.

display their linear programming characteristics in the most obvious manner. One has to tax his brain, coax his ingenuity and call on his experience—if he has any of these—to determine how to define the variables, describe their relationships and make the necessary assumptions so that the problem can be structured as a linear programming problem. The ability to do this, of course, is largely a matter of art rather than science, and so falls outside the scope of this text. It is hoped however that through the following two examples, one might at least develop a "feel" for the task involved in the discovery and formulation of linear programming problems.

The rail-freight movement problem described is the end-product of a research study of an actual managerial problem. How practicable the result is as a solution to the problem is not known. But for that matter, practicability is not important here.

Consider a railroad having a given number of loaded freight trains to switch between 3 terminals, say A, B and C. These terminals are located from 10 to 15 miles apart and are linked by direct lines. The movement of a freight train calls for the assignment of an engine and engine crew to the proper terminal. Although the engine crew (henceforth implying also the engine) can be assigned to any terminal, it must be paid on a portal-to-portal basis, i.e., from starting point back to starting point. For instance, if a crew is assigned to move a train from A to B, it will be paid also for the trip back to A whether the return trip is a freight move or a light move (i.e., the engine by itself). Furthermore, the crews work in shifts, and normally each crew is given a schedule which brings it back to the starting point within the same shift. This, in effect, limits the maximum number of terminals a one-shift assignment can cover to four. For instance, assignments such as A–B–A (from A to B then back to A) and A–B–C–A (from A to B, then to C, then back to A) are feasible. But A–B–C–B–A and A–B–A–B–A (two round trips between A and B) are not. Assuming that the engine-and-crew cost is the predominant element in the cost of switching operations and that the number of trains to be moved from A to B is t_{ab}, from B to A is t_{ba}, from A to C is t_{ac}, from C to A is t_{ca}, from B to C is t_{bc}, and from C to B is t_{cb}, how should crews be assigned to move the freight at minimal cost?

Since the crew is paid portal-to-portal in one shift, one is driven (after some hard thinking) to determine, first of all, what route-assignments are possible. These turn out to be five, viz., A–B–A,

B–C–B, C–A–C, A–B–C–A, and A–C–B–A. Due to the fact that a crew can be started at any terminal, all other route-assignments, such as B–A–B, C–B–C, B–A–C–B, etc. are no longer distinctly different assignments but are equivalent to, in order, A–B–A, B–C–B, A–C–B–A, etc. which have already been considered. The engine-and-crew cost of each of these assignments can be, of course, readily determined. The total cost then, is the sum of the number of crews assigned to each route multiplied by crew cost of that route. This enables us to define the following variables:

x_1 = number of engine-crews assigned to route A–B–A,
x_2 = number of engine-crews assigned to route B–C–B,
x_3 = number of engine-crews assigned to route C–A–C,
x_4 = number of engine-crews assigned to route A–B–C–A,
x_5 = number of engine-crews assigned to route A–C–B–A.

Letting the crew costs of the above routes be c_1, c_2, c_3, c_4 and c_5 respectively, the objective of the problem can now be stated explicitly as:

$$\text{Minimize } f(X) = c_1x_1 + c_2x_2 + c_3x_3 + c_4x_4 + c_5x_5.$$

Now, the next question is: Can the constraints be stated in terms of these variables? We observe that one assignment to route A–B–A means that one freight train could be moved from A to B and one from B to A. Similar analysis of the other routes leads to the following table showing the two-point freight-movement components of all routes:

Freight-movement component	Route				
	A–B–A	B–C–B	C–A–C	A–B–C–A	A–C–B–A
A – B	1			1	
A – C			1		1
B – A	1				1
B – C		1		1	
C – A			1	1	
C – B		1			1

It should now become clear that the constraints can indeed be stated in terms of the variables defined above. For instance, since x_1 and x_4 represent the numbers of crews assigned to route A–B–A and route

A–B–C–A respectively, and since according to the above table each assignment to these two routes results in one freight train moved from A to B, therefore, the total number of trains that could be moved from A to B is $x_1 + x_4$, which must be no less than t_{ab}, or the number that has to be moved. Consequently, we obtain the following six constraints, which are linear, to insure that all required freight movements be accomplished:

$$\begin{aligned}
x_1 \quad\quad\quad\quad &+ x_4 &\geq t_{ab},\\
x_3 \quad &+ x_5 \geq t_{ac},\\
x_1 \quad\quad\quad\quad\quad\quad &+ x_5 \geq t_{ba},\\
x_2 \quad\quad &+ x_4 &\geq t_{bc},\\
x_3 &+ x_4 &\geq t_{ca},\\
x_2 \quad\quad\quad\quad &+ x_5 \geq t_{cb}.
\end{aligned}$$

Needless to say, all these variables have to be prevented from becoming negative as negative freight movement is meaningless within the context of this problem. Thus, we must add the non-negativity constraints, and thereby complete the linear-programming formulation of the problem.

(5) The stock-slitting problem*—In many manufacturing processes, raw material such as sheet steel, paper board, or fabric, is purchased in rolls of standard width, and is then slit into narrower strips on a slitting machine in accordance with fabrication needs. The slitting machine is generally equipped with a fixed number of cutters, all or some of which may be employed simultaneously. Since the standard width of the raw material is seldom the exact equal of a certain combination of desired widths, waste results. The question, therefore, is: How can the slitting be programmed to satisfy the raw material requirements with minimal waste?

Let us assume that the slitting machine is equipped with 6 cutters and the desired slit-widths are $1\frac{1}{2}'$, $1'$ and $\frac{1}{3}'$. The raw material comes in 100′ rolls which are 4′ wide and the total requirements are 1,000′ of $1\frac{1}{2}'$ strip, 2,000′ of 1′ strip and 1,000′ of $\frac{1}{3}'$ strip.

First of all, let us consider how the full width of the raw material can be best utilized. Bearing in mind that the maximum number of cuts allowed by the machine is 6, we obtain the following combina-

* R. W. Metzger, *Elementary Mathematical Programming* (New York: John Wiley & Sons, Inc., 1958), Chapter VIII. In the present text, the problem is trimmed of many features not essential to linear programming, but considered for various objective functions.

tions of desired slit-widths together with their corresponding waste factors:

Table 1.1

Width of strip desired	Combination of desired strip-widths						
	C_1	C_2	C_3	C_4	C_5	C_6	C_7
$1\frac{1}{2}'$	2	2	1	1	0	0	0
$1'$	1	0	2	1	4	3	2
$\frac{1}{3}'$	0	3	1	4	0	3	4
Total cuts	2	4	4	6	3	5	6
Total width utilized	$4'$	$4'$	$3\frac{5}{6}'$	$3\frac{5}{6}'$	$4'$	$4'$	$3\frac{1}{3}'$
Width of waste	0	0	$\frac{1}{6}'$	$\frac{1}{6}'$	0	0	$\frac{2}{3}'$

For example, combination C_1 will yield two 100′ rolls of $1\frac{1}{2}'$ strip and one 100′ roll of 1′ strip with no waste. Combination C_3, on the other hand, will yield one 100′ roll of $1\frac{1}{2}'$ strip, two 100′ rolls of 1′ strip and one 100′ roll of $\frac{1}{3}'$ strip with a waste of $\frac{1}{6}'$ of the 4′ original width There are, of course, other combinations that are technically possible, such as slitting the roll into six $\frac{1}{3}'$ strips (limited by the number of cutters). But the waste would be so obviously excessive that it could hardly be expected to figure in the best slitting program. These combinations are accordingly not included in the problem. This element of human judgment is, of course, subject to error, which could be costly. But the cost of insuring against this error, represented by the enlargement of the problem and the lengthening of computations, could be even higher. In actual practice, careful judgment is often both desirable and necessary, but here the question is quite immaterial because the structure of the problem is not affected by the addition of more feasible combinations.

Before we proceed further, let us make the assumption that changing the cutting combination during the slitting of a roll is impractical. This means that the required strips must be obtained in 100′ rolls. Converting the previously stated requirements into rolls, we have the demands for $1\frac{1}{2}'$, $1'$ and $\frac{1}{3}'$ strips as 10, 20 and 10 rolls respectively. Now, if we let x_1 be the number of original rolls slit according to C_1, x_2 the number slit according to C_2, etc., the total number of $1\frac{1}{2}'$ rolls obtained must be $2x_1 + 2x_2 + x_3 + x_4$ based on the data shown in Table 1.1. Consequently, following a line of reasoning reminiscent of the rail-freight movement problem, we obtain three structural constraints as follows:

$$\begin{aligned} 2x_1 + 2x_2 + x_3 + x_4 &\geq 10, \\ x_1 + 2x_3 + x_4 + 4x_5 + 3x_6 + 2x_7 &\geq 20, \\ 3x_2 + x_3 + 4x_4 + 3x_6 + 4x_7 &\geq 10. \end{aligned}$$

These insure that the requirements for all three types of strip will be met. We now turn our attention to the objective function. Since it has been clearly stated that the objective is to minimize total waste, and the waste factor of each combination is given in Table 1.1, we arrive at the following simple statement of purpose (no waste in C_1, C_2, C_5, and C_6):

$$\text{Minimize } f(X) = \tfrac{1}{6}x_3 + \tfrac{1}{6}x_4 + \tfrac{2}{3}x_7.$$

It should be noticed that the value of the function above is a measure of waste in terms of $100'$ rolls rather than actual length. But the coefficients $(\frac{1}{6}, \frac{1}{6}, \frac{2}{3})$ do reflect correctly the relative wastefulness of the various combinations of cuts. In linear programming, the variables are always cast in competing roles in some relevant respects, hence it is the *relative* desirability or undesirability of these variables that is the important measurement rather than the absolute.

With the constraints and the objective function identified as linear, and with the non-negativity constraints obviously relevant, we may conclude that the stock-slitting problem can be formulated and solved as a linear programming problem.

Actually, the constraints above are not sufficiently "tight" to guarantee a meaningful optimal solution as desired. We will, however, leave it to the reader as an exercise to discover the flaw and the appropriate remedy.

This concludes our discussion on the identification and formulation of linear programming problems. The rest of the text will be concerned almost exclusively with the solution of these problems and some non-linear versions thereof. To do this, however, we will need some basic mathematical knowledge. This will be introduced in the next chapter.

EXERCISES

1. A vendor customarily mixes three types of nuts in different proportions for sale at different prices. The specifications of two nut mixtures and the available supply of ingredients are given below.

Mixture	Peanut	Cashew	Hazel
A	$\frac{1}{4}$	$\frac{1}{2}$	$\frac{1}{4}$
B	0	$\frac{1}{2}$	$\frac{1}{2}$
Supply (lbs)	1,000	400	400

The costs of peanuts, cashews and hazelnuts are \$0.60, \$1.20 and \$1.00 per pound respectively. The mixtures are sold for \$1.50 per pound of A and \$1.80 per pound of B. Any ingredient that is left over after the mixtures are prepared can also be sold at cost. The vendor wishes to know how much of A and B should be prepared. Formulate this as a linear programming problem.

2. A trucking firm has received an order to move 3,000 tons of industrial materials to a destination 1,000 miles away. The firm has available at the moment a fleet of 150 Class A 15-ton trailer trucks and another fleet of 100 Class B 10-ton trailer trucks. The operating costs of these trucks are 3¢ and 4¢ per ton-mile respectively. Based on past experience, the firm has a policy of retaining at least one Class A truck with every two Class B trucks in reserve. It desires to know how many of the two classes of vehicles should be dispatched to move the materials at minimal total operating costs. Formulate this as a linear programming problem.

3. Two ingredients, A and B, are mixed in varying proportions to make two kinds of salad dressing called Oriental and Eskimo which are sold at wholesale for \$3.25 and \$2.75 per quart respectively. The cost of A is \$1.50 per quart and that of B is \$2.00 per quart. Although there is no strict formula to follow in mixing the ingredients, two rules are generally observed. These are (1) Oriental dressing may contain no less than 25% of A and no less than 50% of B; and (2) Eskimo dressing, on the other hand, may contain no more than 75% of A. There are 30 quarts of A and 20 quarts of B available. Assuming that the objective is to determine the most profitable quantities of the two salad dressings to produce, formulate the above as a linear programming problem.

ELEMENTS OF MATRIX ALGEBRA AND DETERMINANTS

2.1. Matrix—Definition and General Properties.

A matrix is a rectangular array of numbers. For our purposes we are interested only in matrices that are rectangular arrays of real numbers—called real matrices. Each number in a matrix is called an element, and as such has no implied mathematical relationship with other elements in the same matrix. In this respect it is decidedly different from a determinant (cf. Section 1.14 of this chapter), which is a single number and implies that certain arithmetical operations are to be performed in a specified manner. To clearly distinguish between matrices and determinants, it is now the accepted practice to use large parentheses, square brackets or double vertical bars to denote the former while reserving single vertical bars for the latter. Some examples of matrices are shown in Figure 2.1.

The number of rows and the number of columns contained in a matrix are its row and column dimensions. A matrix is often described by these dimensions. For instance, (a) in Figure 2.1 is a three-by-three (3×3) matrix and (b) is a 2×3 matrix, etc. A matrix is called a square matrix if its row and column dimensions are the same.

We describe one matrix as the *transpose* of another if the rows of the former are identical with the columns of the latter in exactly the same order. For instance, (b) in Figure 2.1 is the transpose of (c) and vice versa. If (b) is denoted by A, then (c), its transpose, is customarily denoted by the symbol A^T.

$$\begin{bmatrix} 2 & 0 & 1 \\ 0 & -1 & 0 \\ 0 & 2 & -1 \end{bmatrix} \qquad \begin{bmatrix} 7 & -3 & 1 \\ -4 & 0 & 0 \end{bmatrix} \qquad \begin{bmatrix} 7 & -4 \\ -3 & 0 \\ 1 & 0 \end{bmatrix}$$
$$\text{(a)} \qquad\qquad\qquad \text{(b)} \qquad\qquad\qquad \text{(c)}$$

$$\begin{bmatrix} 1 & 0 & 0 \\ 0 & 1 & 0 \\ 0 & 0 & 1 \end{bmatrix} \qquad \begin{bmatrix} 0 & 0 & 0 \\ 0 & 0 & 0 \\ 0 & 0 & 0 \end{bmatrix}$$
$$\text{(d)} \qquad\qquad\qquad \text{(e)}$$

FIGURE 2.1

We speak of *equality* of two matrices if and only if (1) the row and column dimensions of the two matrices are identical and (2) the corresponding elements in the two matrices are the same. If both conditions are not completely satisfied, then no equality exists.

Some matrices are distinguished by the characteristics of their contents. For instance, the matrix in Figure 2.1 (d) is a square matrix with 1 on its main diagonal (from upper left to lower right) and 0 elsewhere. It is known as the *unit* or *identity* matrix and usually designated I. Another special matrix is the matrix (e) in Figure 2.1 whose elements are all zero. It is known as the *zero* matrix and designated O. In many respects, the I matrix and the O matrix in matrix algebra correspond closely to the numbers 1 and 0 in elementary algebra. This will be made clear later in the sections dealing with matrix operations.

2.2. Row and Column Vectors.

A special type of matrix includes those which have only one row or column. They are frequently referred to as row or column vectors.* Any $m \times n$ matrix may be viewed as an ordered arrangement of m row or n column vectors. For instance, matrix (b) in Figure 2.1 is composed of 2 row vectors or 3 column vectors. The dimension of a vector

---

* Readers who have had a course in vector analysis must pay especial attention to this new definition of vector in order to avoid confusion.

is the number of elements in the vector. Two vectors are equal if and only if (1) they are of the same type, i.e., either both row or both column; (2) they have the same number of elements; and (3) their corresponding elements are the same. A unit vector is defined as a vector which has one of its elements equal to 1, and all other elements equal to 0. For instance, $\begin{bmatrix} 0 \\ 1 \\ 0 \end{bmatrix}$ is a three-dimensional column unit vector. Thus, an $n \times n$ identity matrix may be regarded as a collection of n n-dimensional row or column unit vectors in a specific order. A zero or null vector is one whose elements are all 0.

2.3. Index Notations of Matrices.

All elements in a matrix are identified by their row and column positions. Thus the element in the third row and second column of a matrix is indicated by the use of the double subscript 3,2. If the elements of the matrix A are denoted by the general symbol a then the element in the ith row and the jth column may be designated specifically by a_{ij}. In general, an $m \times n$ matrix A may be written as follows:

$$A = [a_{ij}], \qquad i = 1, 2, \ldots, m, \text{ and } j = 1, 2, \ldots, n.$$

Or alternatively,

$A = [A_j]$, $j = 1, 2, \ldots, n$, where A_j is the jth column vector of A.

$A = [\overline{A}_i]$, $i = 1, 2, \ldots, m$, where $\overline{A}_i$ is the ith row vector of A.

With the use of double subscript or index notations, many matrices and matrix operations can be conveniently defined or described without resorting to the lengthy process of writing down the full details.

Example 1. The identity matrix, I, shown in Fig. 1 (d), may be defined as follows:

$$I = [a_{ij}], \quad a_{ij} \begin{cases} = 1 & \text{for } i = j, \\ = 0 & \text{for } i \neq j, \end{cases}$$
$$i, j = 1, 2, \ldots, m.$$

Example 2. The following matrix,

$$B = [b_{ij}], \quad \text{where } b_{ij} = i - j, \quad \text{with } i = 1, 2, \ldots, 6, \quad j = 1, 2, \ldots, 5,$$

is the 6×5 matrix shown below when written out in full:

$$B = \begin{bmatrix} 0 & -1 & -2 & -3 & -4 \\ 1 & 0 & -1 & -2 & -3 \\ 2 & 1 & 0 & -1 & -2 \\ 3 & 2 & 1 & 0 & -1 \\ 4 & 3 & 2 & 1 & 0 \\ 5 & 4 & 3 & 2 & 1 \end{bmatrix}.$$

2.4. Matrix Addition and Subtraction.

The first rule of matrix addition and subtraction is that the matrices must have identical dimensions. Assuming the dimensions are identical, matrix addition is defined as follows:

Given two $m \times n$ matrices, $A = [a_{ij}]$ and $B = [b_{ij}]$, the addition of A and B determines an $m \times n$ matrix C whose elements $c_{ij} = a_{ij} + b_{ij}$ for all i and j.

Matrix subtraction is defined similarly. Given two matrices, $A = [a_{ij}]$ and $B = [b_{ij}]$ having identical dimensions, the subtraction of B from A results in a third matrix, $C = [c_{ij}]$ of the same dimensions, whose elements $c_{ij} = a_{ij} - b_{ij}$, for all i and j.

Example 1. Let

$$A = \begin{bmatrix} 2 & -1 & 3 \\ 0 & 1 & -3 \end{bmatrix} \quad \text{and} \quad B = \begin{bmatrix} 0 & -4 & 1 \\ 2 & 1 & 0 \end{bmatrix}.$$

Then

$$A + B = \begin{bmatrix} 2 & -1 & 3 \\ 0 & 1 & -3 \end{bmatrix} + \begin{bmatrix} 0 & -4 & 1 \\ 2 & 1 & 0 \end{bmatrix}$$

$$= \begin{bmatrix} 2+0 & -1-4 & 3+1 \\ 0+2 & 1+1 & -3+0 \end{bmatrix} = \begin{bmatrix} 2 & -5 & 4 \\ 2 & 2 & -3 \end{bmatrix}.$$

Example 2. Using the same matrices as above

$$A - B = \begin{bmatrix} 2 & -1 & 3 \\ 0 & 1 & -3 \end{bmatrix} - \begin{bmatrix} 0 & -4 & 1 \\ 2 & 1 & 0 \end{bmatrix}$$

$$= \begin{bmatrix} 2-0 & -1-(-4) & 3-1 \\ 0-2 & 1-1 & -3-0 \end{bmatrix}$$

$$= \begin{bmatrix} 2 & 3 & 2 \\ -2 & 0 & -3 \end{bmatrix}.$$

Since vectors have been defined here as special forms of matrices, their addition and subtraction follow the rules described above.

2.5. Matrix Multiplication.

Like addition and subtraction, multiplication of matrices basically involves two matrices. But unlike the two preceding operations, the order in which the two matrices are written in multiplication is extremely important. When two matrices A and B are multiplied together and written as AB, the result in general is different from BA. For this reason, we distinguish between the two matrices related by multiplication by referring to the first matrix as the lead-matrix and the second one as the lag-matrix. Hence, in AB, A is the lead- and B the lag-matrix.

The first rule of matrix multiplication is that the column dimension of the lead-matrix must be equal to the row dimension of the lag-matrix. When this condition is satisfied, we say the matrices are compatible. Otherwise, they are incompatible and multiplication cannot be performed. Note that the row dimension of the lead-matrix and the column dimension of the lag-matrix do not impose any restriction on the feasibility of matrix multiplication.

Example. Given the following matrices

$$A = \begin{bmatrix} a_{11} & a_{12} & a_{13} \\ a_{21} & a_{22} & a_{23} \end{bmatrix}, \quad B = \begin{bmatrix} b_{11} & b_{12} & b_{13} \\ b_{21} & b_{22} & b_{23} \\ b_{31} & b_{32} & b_{33} \end{bmatrix}, \quad C = \begin{bmatrix} c_1 \\ c_2 \\ c_3 \end{bmatrix}.$$

By checking their respective row and column dimensions, we know that the only multiplications possible are: AB, AC, BC. The others, i.e., BA, CA, CB are all impossible.

Using the matrices A and B defined above, we may now describe specifically the meaning of the multiplication, AB.

$$AB = \begin{bmatrix} a_{11} & a_{12} & a_{13} \\ a_{21} & a_{22} & a_{23} \end{bmatrix} \begin{bmatrix} b_{11} & b_{12} & b_{13} \\ b_{21} & b_{22} & b_{23} \\ b_{31} & b_{32} & b_{33} \end{bmatrix}$$

$$= \begin{bmatrix} a_{11}b_{11}+a_{12}b_{21}+a_{13}b_{31} & a_{11}b_{12}+a_{12}b_{22}+a_{13}b_{32} & a_{11}b_{13}+a_{12}b_{23}+a_{13}b_{33} \\ a_{21}b_{11}+a_{22}b_{21}+a_{23}b_{31} & a_{21}b_{12}+a_{22}b_{22}+a_{23}b_{32} & a_{21}b_{13}+a_{22}b_{23}+a_{23}b_{33} \end{bmatrix}.$$

Although the above expression appears to be lengthy and complicated, the actual arithmetical operations involved are quite systematic and easy to grasp. First, take the first row of the lead-matrix and operate on it with the first column of the lag-matrix—the operations being (1) form the products of the corresponding elements of the row and column, e.g., $a_{11}b_{11}$, $a_{12}b_{21}$, $a_{13}b_{31}$, and (2) sum up all these products. The resulting total represents the element in the first row and the first column of the product matrix. Do the same to the first row of the lead-matrix using successively the second and third columns of the lag-matrix to obtain respectively the second and third elements in the first row of the product matrix. Then, repeat the same operation on the second row of the lead-matrix with the first, second and third columns of the lag-matrix to obtain respectively the first, second and third elements in the second row of the product matrix. The multiplication, AB, is now complete and the final result is a 2×3 matrix.

In general, then, matrix multiplication is defined as follows: given an $m \times p$ matrix $A = [a_{ik}]$ and a $p \times n$ matrix $B = [b_{kj}]$,

$$\text{(2.5.1)} \qquad AB = C = [c_{ij}], \quad \text{where } c_{ij} = \sum_{k=1}^{p} a_{ik}b_{kj}, \qquad \text{for } i = 1, 2, \ldots, m, \; j = 1, 2, \ldots, n.$$

In other words, the product AB is a third matrix C whose row dimension is that of the lead-matrix, A; whose column dimension is that of the lag-matrix, B; and whose elements c_{ij} is the sum of the products of corresponding elements in the ith row of A and the jth column of B.

Example. Let

$$A = \begin{bmatrix} 7 & -1 & 0 \\ 0 & -1 & -2 \end{bmatrix}, \qquad B = \begin{bmatrix} 1 & 2 \\ 0 & -2 \\ 4 & 0 \end{bmatrix}.$$

Then

$$AB = \begin{bmatrix} 7 & -1 & 0 \\ 0 & -1 & -2 \end{bmatrix} \begin{bmatrix} 1 & 2 \\ 0 & -2 \\ 4 & 0 \end{bmatrix}$$

$$= \begin{bmatrix} 7+0+0 & 14+2+0 \\ 0+0-8 & 0+2+0 \end{bmatrix} = \begin{bmatrix} 7 & 16 \\ -8 & 2 \end{bmatrix},$$

and

$$BA = \begin{bmatrix} 1 & 2 \\ 0 & -2 \\ 4 & 0 \end{bmatrix} \begin{bmatrix} 7 & -1 & 0 \\ 0 & -1 & -2 \end{bmatrix}$$

$$= \begin{bmatrix} 7+0 & -1-2 & 0-4 \\ 0+0 & 0+2 & 0+4 \\ 28+0 & -4+0 & 0+0 \end{bmatrix} = \begin{bmatrix} 7 & -3 & -4 \\ 0 & 2 & 4 \\ 28 & -4 & 0 \end{bmatrix}.$$

Note that AB is not equal to BA even though both multiplications can be performed.

2.6. Multiplication of Matrix by Vector.

Since a vector is regarded here as a special form of matrix, the multiplication of matrix by vector or of vector by matrix follows the same general definition presented in the preceding section. Dimensional compatibility must be observed. For example, if A is an $m \times m$ matrix, U, an m-component row vector and V, an m-component column vector, then, UA and AV are permissible. On the other hand, AU is not permissible since there are m columns in the lead-matrix, A, and only one row in the lag-matrix, U. Similarly, VA is not permissible.

Example. Let

$$A = \begin{bmatrix} 7 & -1 & 0 \\ 0 & -1 & -2 \end{bmatrix}, \qquad U = [-2 \quad 3], \qquad V = \begin{bmatrix} 1 \\ -1 \\ 2 \end{bmatrix}.$$

Then

$$UA = [-2 \quad 3] \begin{bmatrix} 7 & -1 & 0 \\ 0 & -1 & -2 \end{bmatrix}$$

$$= [-14+0 \quad 2-3 \quad 0-6] = [-14 \quad -1 \quad -6]$$

$$AV = \begin{bmatrix} 7 & -1 & 0 \\ 0 & -1 & 2 \end{bmatrix} \begin{bmatrix} 1 \\ -1 \\ 2 \end{bmatrix} = \begin{bmatrix} 7+1+0 \\ 0+1+4 \end{bmatrix} = \begin{bmatrix} 8 \\ 5 \end{bmatrix}.$$

2.7. Multiplication of Vector by Vector.

There are two types of vector multiplication which have to be distinguished clearly. One is the multiplication of a row vector by a column vector or vice versa. The other is the multiplication of a row vector by a row vector or a column vector by a column vector. The former falls within the scope of regular matrix multiplication (since vectors are regarded as special matrices), and is governed by the rule of dimensional compatibility. The product is always a matrix, though it may consist of only one number. For instance, let U be the row vector, $[2 \quad -1 \quad 3]$ and V, the column vector, $\begin{bmatrix} 5 \\ 0 \\ -1 \end{bmatrix}$

$$UV = [2 \quad -1 \quad 3]\begin{bmatrix} 5 \\ 0 \\ -1 \end{bmatrix} = [10 + 0 - 3] = 7 \qquad \text{a } 1 \times 1 \text{ matrix.}$$

But

$$VU = \begin{bmatrix} 5 \\ 0 \\ -1 \end{bmatrix}[2 \quad -1 \quad 3]$$

$$= \begin{bmatrix} 5\cdot 2 & 5(-1) & 5\cdot 3 \\ 0\cdot 2 & 0(-1) & 0\cdot 3 \\ -1\cdot 2 & (-1)\cdot(-1) & (-1)\cdot 3 \end{bmatrix} = \begin{bmatrix} 10 & -5 & 15 \\ 0 & 0 & 3 \\ -2 & 1 & -3 \end{bmatrix}.$$

The second type of vector multiplication, involving vectors of the same kind, i.e., row by row or column by column, produces what is called the inner product of vectors.* The inner product is simply a number called a *scalar*** in matrix algebra, and represents the sum of the products of corresponding elements of two n-component row (or column) vectors. The symbol of inner product is a dot between the vectors as illustrated below: Let $U = [u_1, u_2, \ldots, u_n]$ and $W = [w_1, w_2, \ldots, w_n]$, the inner product,

$$U \cdot W = u_1w_1 + u_2w_2 + \ldots + u_nw_n = \sum_{i=1}^{n} u_iw_i.$$

---

* In vector analysis, two types of vector multiplication are recognized: the vector product and the scalar product (or dot product). Although scalar product corresponds to inner product discussed here, no further parallelism should be attempted beyond this. There is no such concept as "vector product" in matrix algebra. In this respect, it is of particular advantage for a person who has studied vector analysis to think of vectors as special forms of matrices.

** A scalar is not to be confused with a 1×1 matrix. They are completely different mathematical concepts and follow different rules of operations as will be made clear in Section 2.8.

Obviously, inner product is not a regular matrix multiplication. Its use in linear programming is usually a matter of notational expediency only.

2.8. Multiplication of Matrix or Vector by Scalar.

The multiplication of a matrix (or a vector) by a scalar simply means the simultaneous multiplication of all the elements of the matrix (or the vector) by the same scalar. For instance, let A be an $m \times n$ matrix with elements a_{ij}, and let k be a scalar quantity, then kA is another $m \times n$ matrix with elements equal to ka_{ij}. In other words, if

$$A = [a_{ij}] = \begin{bmatrix} a_{11} & \cdots & a_{1n} \\ \cdot & & \cdot \\ \cdot & & \cdot \\ \cdot & & \cdot \\ a_{m1} & \cdots & a_{mn} \end{bmatrix},$$

then

$$kA = [ka_{ij}] = \begin{bmatrix} ka_{11} & \cdots & ka_{1n} \\ \cdot & & \cdot \\ \cdot & & \cdot \\ \cdot & & \cdot \\ ka_{m1} & \cdots & ka_{mn} \end{bmatrix}.$$

Example. Given

$$A = \begin{bmatrix} -2 & 3 \\ 1 & 0 \\ -5 & 1 \end{bmatrix} \quad \text{and} \quad k = 0.5,$$

$$kA = \begin{bmatrix} 0.5\cdot(-2) & 0.5\cdot 3 \\ 0.5\cdot 1 & 0.5\cdot 0 \\ 0.5\cdot(-5) & 0.5\cdot 1 \end{bmatrix} = \begin{bmatrix} -1 & 1.5 \\ 0.5 & 0 \\ -2.5 & 0.5 \end{bmatrix}.$$

2.9. Commutative, Associative, Distributive Laws and Matrix Operations.

In the algebra of numbers, the addition of two numbers, a and b, can be carried out as either a added to b, or b added to a, without affecting the result. Similarly, $a\cdot b$ yields the same result as $b\cdot a$. In short, when the result is independent of the order in which the objects are given, we say that the operation obeys the *commutative law* of addition or multiplication.

Now, consider a series of similar operations, such as the addition (or multiplication) of three numbers a, b, and c. If the result is invariant whether (1) a and b are added first, then c is added to the sum of a and b, or (2) b and c are added first, then a is added to the sum of b and c, we say that the operation of addition obeys the *associative law*. Symbolically, if $(a + b) + c = a + (b + c)$, then the associative law holds. A little reflection will reveal that the associative law also holds for algebraic multiplication.

Finally, we say that the *distributive law* holds if $a(b + c) = ab + ac$, i.e., if the product of a and the sum of b and c is the same as the sum of the products ab and ac.

Referring to matrix algebra, it can easily be shown that matrix addition and subtraction obey the commutative and the associative laws. However, matrix multiplication (excepting multiplication of matrix by scalar) obeys only the associative law but not the commutative law, i.e., if A, B, and C stand for three different matrices, $(AB)C = A(BC)$ but, $AB \neq BA$ unless either A or B is an identity matrix. The examples in Section 2.5 serve to illustrate that matrix multiplication is not commutable. Finally, the distributive law holds for matrix operations, i.e., $A(B + C) = AB + AC$, and $(B + C)A = BA + CA$.

2.10. Partitioning of a Matrix.

A matrix may be partitioned vertically and/or horizontally into a number of submatrices. Each submatrix may then be viewed as an "element" of the partitioned matrix and treated accordingly in matrix operations, particularly multiplication. Practically speaking, the advantage of partitioning a matrix is that it simplifies multiplication when the matrix contains special submatrices such as the identity matrix and the zero matrix. However, to carry out such a multiplication, the lead-matrix and the lag-matrix must be conformably partitioned, i.e., the lead-matrix must be partitioned vertically in exactly the same way as the lag-matrix is partitioned horizontally. The following example will make the concept clear.

Example. Find the product, AB, where:

$$A = \left[\begin{array}{c:cc:cc} 3 & 2 & -1 & 4 & -7 \\ 1 & -9 & 0 & 1 & 0 \\ -5 & 0 & 3 & 0 & -2 \\ 4 & 7 & -6 & 0 & -8 \end{array}\right], \qquad B = \left[\begin{array}{cc:c} 0 & 0 & 1 \\ \hdashline 1 & 7 & 0 \\ 2 & 0 & 0 \\ \hdashline 1 & 0 & 0 \\ 0 & 1 & 0 \end{array}\right]$$

Since B contains a large number of "0"s and "1"s, we will first partition B into six submatrices as shown to create as many identity and zero submatrices as possible. The matrix A will then have to be partitioned conformably as shown. We may now regard A as a 1×3 vector and B as a 3×2 matrix:

$$A = [A_1 \quad A_2 \quad A_3], \quad \text{where,} \quad A_1 = \begin{bmatrix} 3 \\ 1 \\ -5 \\ 4 \end{bmatrix}, \; A_2 = \begin{bmatrix} 2 & -1 \\ -9 & 0 \\ 0 & 3 \\ 7 & -6 \end{bmatrix},$$

$$\text{and} \qquad A_3 = \begin{bmatrix} 4 & -7 \\ 1 & 0 \\ 0 & -2 \\ 0 & -8 \end{bmatrix};$$

$$\text{and} \quad B = \begin{bmatrix} B_{11} & B_{12} \\ B_{21} & B_{22} \\ B_{31} & B_{32} \end{bmatrix} \qquad B_{11} = [0 \quad 0], \quad B_{12} = [1]$$

$$B_{21} = \begin{bmatrix} 1 & 7 \\ 2 & 0 \end{bmatrix}, \quad B_{22} = \begin{bmatrix} 0 \\ 0 \end{bmatrix},$$

$$B_{31} = \begin{bmatrix} 1 & 0 \\ 0 & 1 \end{bmatrix}, \quad B_{32} = \begin{bmatrix} 0 \\ 0 \end{bmatrix},$$

and $AB = [A_1B_{11} + A_2B_{21} + A_3B_{31} \qquad A_1B_{12} + A_2B_{22} + A_3B_{32}]$.

Since B_{11}, B_{22}, and B_{32} are zero matrices, A_1B_{11}, A_2B_{22}, and A_3B_{32} are also zero matrices. Furthermore, since B_{12} and B_{31} are both identity matrices, A_1B_{12} is simply A_1, and A_3B_{31} is A_3. Therefore

$$AB = (A_2B_{21} + A_3A_1) = \left[\begin{bmatrix} 0 & 14 \\ -9 & -63 \\ 6 & 0 \\ -5 & 49 \end{bmatrix} + \begin{bmatrix} 4 & -7 \\ 1 & 0 \\ 0 & -2 \\ 0 & -8 \end{bmatrix} \quad \begin{matrix} 3 \\ 1 \\ -5 \\ 4 \end{matrix}\right]$$

$$= \begin{bmatrix} 4 & 7 & 3 \\ -8 & -63 & 1 \\ 6 & -2 & -5 \\ -5 & 41 & 4 \end{bmatrix}.$$

Partitioning a matrix into zero and identity submatrices whenever possible may also facilitate the determination of its inverse. This will be made clear in Exercise 14 that follows the conclusion of the present chapter.

In addition to the advantage of simplifying multiplication, the partitioning of a matrix permits a greater insight into the nature of matrices and matrix operations. For instance, consider the problem of multiplying an $m \times n$ matrix, A, by an n-element column vector, U. If A is partitioned into n submatrices, $A_1, A_2, \ldots, A_n$, each a column vector, one may then regard A as a row vector consisting of elements, $A_1, A_2, \ldots, A_n$, and the multiplication of U by A becomes

$$AU = [A_1, A_2, \ldots, A_n]\begin{bmatrix} u_1 \\ u_2 \\ \cdot \\ \cdot \\ \cdot \\ u_n \end{bmatrix}$$

$$= u_1A_1 + u_2A_2 + \ldots + u_nA_n = \sum_{i=1}^{n} u_iA_i.$$

This expression has a geometrical interpretation. Since A_1, A_2, etc. are m-component vectors which can be erected in m-dimensional space, and $u_1, u_2, \ldots, u_n$ are scalars, the result of U multiplied by A is simply the vector sum of A_1, A_2, etc. after having been scaled by u_1, u_2, etc. respectively.

Example. Let $A = \begin{bmatrix} 2 & 3 & -1 \\ 1 & -1 & 2 \end{bmatrix}$ and $U = \begin{bmatrix} \frac{1}{2} \\ -1 \\ 2 \end{bmatrix}$.

Then, $$AU = \begin{bmatrix} 2 & 3 & -1 \\ 1 & -1 & 2 \end{bmatrix}\begin{bmatrix} \frac{1}{2} \\ -1 \\ 2 \end{bmatrix}$$

$$= \tfrac{1}{2}\begin{bmatrix} 2 \\ 1 \end{bmatrix} - 1\begin{bmatrix} 3 \\ -1 \end{bmatrix} + 2\begin{bmatrix} -1 \\ 2 \end{bmatrix} = \begin{bmatrix} -4 \\ 5\frac{1}{2} \end{bmatrix}.$$

2.11. Inverse and Inversion of a Matrix.

In the algebra of numbers, division is defined in terms of multiplication. The division of a (real) number a by another (real) number b, written as $\frac{a}{b}$, is simply an accustomed way of writing the multiplication of the reciprocal of b, (b^{-1}), by a. In other words, $\frac{a}{b} = a(b^{-1})$. Since the multiplication of numbers is commutable, the multiplier and the multiplicand can switch roles without affecting the product, i.e., $\frac{a}{b} = a(b^{-1}) = (b^{-1})a$.

In matrix algebra, however, because multiplication is not commutable, this familiar notion of division cannot be applied. Given two matrices, A and B, the expression $\frac{A}{B}$ is ill-defined for it could mean either $A(B^{-1})$ or $(B^{-1})A$ and the two are generally not the same. For this reason, we avoid using the term "division" and the expression $\frac{A}{B}$ in matrix algebra. Instead, we resort to the fundamental operation of multiplication and perform $A(B^{-1})$ or $(B^{-1})A$ as the case may be.

There is, however, a correspondence between the reciprocal of a number and the "reciprocal" of a matrix. Just as the reciprocal of a number d is defined as another number d^{-1} such that $(d^{-1})d = 1$, the "reciprocal" of a matrix A (called the *inverse* of A) is another matrix, denoted A^{-1} such that $A(A^{-1}) = I$. Although matrix multiplication is not, in general, commutable, in this case, $A^{-1}A = AA^{-1} = I$.

Not every matrix has an inverse. As a matter of fact, only a square matrix whose row (or columns) are independent of each other has an inverse. Such a matrix is called a non-singular matrix and its relationship with invertibility will be discussed further in Section 2.16. In the meantime, we will illustrate the concept of an inverse with a numerical example.

Example. Find the inverse of

$$A = \begin{bmatrix} 2 & -1 \\ 0 & 2 \end{bmatrix}.$$

Let the inverse be

$$A^{-1} = \begin{bmatrix} b_{11} & b_{12} \\ b_{21} & b_{22} \end{bmatrix},$$

then by definition of the inverse,

$$A^{-1}A = \begin{bmatrix} b_{11} & b_{12} \\ b_{21} & b_{22} \end{bmatrix} \begin{bmatrix} 2 & -1 \\ 0 & 2 \end{bmatrix} = \begin{bmatrix} 1 & 0 \\ 0 & 1 \end{bmatrix}.$$

After carrying out the multiplication fully, we obtain the following linear equation systems:

$$\begin{cases} 2b_{11} + 0 \cdot b_{12} = 1 \\ -b_{11} + \quad 2b_{12} = 0 \end{cases} \quad \text{and} \quad \begin{cases} 2b_{21} + 0 \cdot b_{22} = 0 \\ -b_{21} + \quad 2b_{22} = 1 \end{cases}$$

By solving the first pair of equations, we obtain: $b_{11} = \frac{1}{2}$, $b_{12} = \frac{1}{4}$; by solving the second pair, we have: $b_{21} = 0$, and $b_{22} = \frac{1}{2}$. Therefore the inverse is

$$A^{-1} = \begin{bmatrix} \frac{1}{2} & \frac{1}{4} \\ 0 & \frac{1}{2} \end{bmatrix}.$$

Of course, the validity of this answer can be quickly checked by determining $A^{-1}A$ to see if the result is actually a 2×2 identity matrix. If it is not, then some mistake has been made, and a recalculation is in order.

The above example is aimed at clarifying the concept of an inverse rather than illustrating the method of determining one. There are many other general methods of finding an inverse. One of these will be explained in Section 2.17. Special methods also exist which are more efficient for inverting special matrices. These may be found in the appropriate mathematical and computer literatures and will not be discussed here.

2.12. Linear Independence.

To capitalize on the reader's knowledge of the algebra of numbers, we will approach the concept of linear independence via a discussion of its antithesis, viz., linear dependence. When two variables x and y are related in a certain way such that given the value of x (or y), the value of y (or x) is immediately determined, we say that x and y are dependent on each other. The dependent relationship is said to be linear if one can be expressed as a linear function of the other, such as $y = ax$.*

The same concept of linear dependence may be extended to matrix algebra. An $m \times m$ matrix, A, may be regarded as composed of m row vectors, $\overline{A}_1, \overline{A}_2, \ldots, \overline{A}_i, \ldots, \overline{A}_m$, each having m elements. If one of

* See definition of linear function in Chapter I, Section 1.2.

these row vectors, say $\overline{A}_1$, can be expressed as a linear function (or linear combination) of the other row vectors, such as

$$c_1\overline{A}_1 = c_2\overline{A}_2 + c_3\overline{A}_3 \ldots + c_i\overline{A}_i \ldots + c_m\overline{A}_m, \tag{2.12.1}$$

where $c_1, c_2, c_3, \ldots, c_m$ are not all zero,

then, we say that the row vectors are linearly dependent. Contrarily, if no row can be expressed as a linear combination of other rows unless all coefficients are zero, then the rows are said to be linearly independent. It is important to note that linear dependence or independence is a mutual relationship among the vectors concerned. Whereas it may appear in (2.12.1) that $\overline{A}_1$ is dependent on $\overline{A}_2, \overline{A}_3, \ldots, \overline{A}_m$, it can, nevertheless, be shown that some (at least one) of the latter, say $\overline{A}_2$, whose coefficient c_2 is not zero, can be similarly expressed as a linear function of the rest merely by a suitable transposition of terms. Thus, the correct interpretation of linear dependence of vectors is that they are mutually dependent as a group. This explains the mathematicians' preference for the following definition of linear dependence and independence: a set of vectors, $\overline{A}_1, \overline{A}_2, \ldots, \overline{A}_m$, is said to be linearly dependent if a corresponding set of coefficients, $c_1, c_2, \ldots, c_m$, not all zeroes, can be found such that

$$c_1\overline{A}_1 + c_2\overline{A}_2 + \ldots + c_m\overline{A}_m = \bar{O}, \tag{2.12.2}$$

where $\bar{O}$ is zero row vector. If (2.12.2) can be satisfied *only* when $c_1, c_2, \ldots, c_m$ are all zero, then the set of vectors is said to be linearly independent.

Example 1. Determine if the rows in the matrix A below are linearly dependent.

$$A = \begin{bmatrix} 7 & 0 & 1 \\ -3 & 2 & 0 \\ 1 & 4 & 1 \end{bmatrix}.$$

Designating the first, second and third rows of A by $\overline{A}_1$, $\overline{A}_2$, and $\overline{A}_3$, we may follow (2.12.1) and write

$$c_1\overline{A}_1 = c_2\overline{A}_2 + c_3\overline{A}_3$$

$$c_1[7 \quad 0 \quad 1] = c_2[-3 \quad 2 \quad 0] + c_3[1 \quad 4 \quad 1]$$

which can be translated into the following three simultaneous equations:

$$\begin{cases} 7c_1 = -3c_2 + c_3 \\ 0 = 2c_2 + 4c_3 \\ c_1 = 0c_2 + c_3 \end{cases}$$

From the third equation, we obtain, $c_3 = c_1$. Substituting this value into the first and second equations, we get identical results, viz. $c_2 = -2c_1$, which indicate that the equations are not contradictory. Hence, by letting $c_1 = 1$, we obtain $\bar{A}_1 = -2\bar{A}_2 + \bar{A}_3$, indicating that the rows are dependent. Parenthetically, we may add that since the rows are dependent, A is singular and has no inverse (cf. Section 2.11).

Example 2. Show that the matrix A below has an inverse:

$$A = \begin{bmatrix} 7 & 0 & 1 \\ -3 & 2 & 0 \\ 1 & 0 & 2 \end{bmatrix}.$$

Assume that the first row, $\bar{A}_1$ can be expressed as a linear combination of the second and third rows, $\bar{A}_2$ and $\bar{A}_3$, so that:

$$c_1\bar{A}_1 = c_2\bar{A}_2 + c_3\bar{A}_3.$$

Following the procedure employed in the previous example, three linear simultaneous equations may be set forth:

$$\begin{cases} 7c_1 = -3c_2 + c_3 \\ 0 = 2c_2 \\ c_1 = 2c_3 \end{cases}$$

Clearly, then, $c_2 = 0$ from the second equation. But if $c_2 = 0$, then, $c_3 = 7c_1$ from the first equation and $c_3 = \frac{1}{2}c_1$ from the third. These contradictory results indicate that no solution of c_2 and c_3 is possible, or equivalently, (2.12.2) can be satisfied only when c_1, c_2 and c_3 are all zero. Consequently, we assert that $\bar{A}_1$ cannot be expressed as a linear combination of $\bar{A}_2$ and $\bar{A}_3$, or $\bar{A}_1$, $\bar{A}_2$ and $\bar{A}_3$ are linearly independent; therefore, A has an inverse.

To conclude this section, we will introduce the following notions without extensive discussion or proof: (1) the *rank* of a matrix is the number of independent rows (or columns) in the matrix, (2) for a square matrix, linear independence among the rows implies linear independence among the columns.

2.13. Basis and Vector Space.

We know intuitively that a two-component vector can be represented by a point in a plane with reference to a pair of axes at right angles to each other. For instance, the vector, $U = \begin{bmatrix} 3 \\ 2 \end{bmatrix}$, may be represented as shown in Figure 2.2, i.e., by a point located 3 unit-lengths horizontally from axis X_2 and 2 unit-lengths vertically from axis X_1.

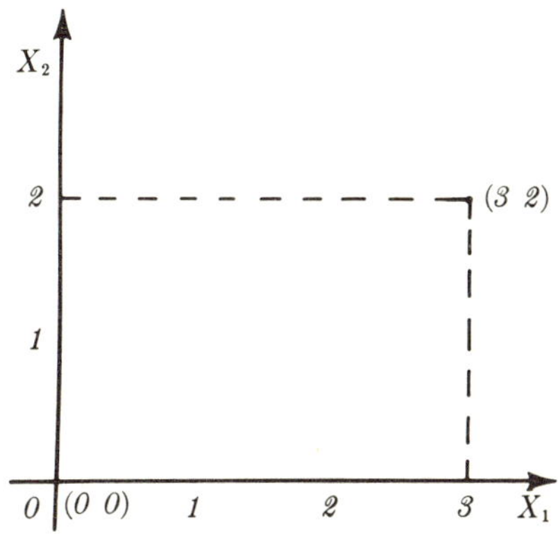

FIGURE 2.2

Provided we consistently use horizontal distance from the axis X_2 to represent the magnitude of first component of a two-dimensional vector, and vertical distance from X_1 to represent the second component, we have a geometrical scheme to represent all two-component real vectors uniquely in the same plane. This plane represents the totality of all two-component real vectors and is known as a two-dimensional *vector space.* The fact that a pair of perpendicular axes has been chosen as bases of reference is incidental. Similarly, unique representation of vectors in a plane can be made by reference to a pair of axes intersecting at any oblique angle. What is significant is that in all such schemes of unique representation of two-dimensional vectors a pair of intersecting axes is required. One is not enough and three become redundant. The reader may wish to verify this himself quickly on a piece of paper.

It is clear upon brief reflection that the pair of axes has to be intersecting. The reason is simply that we need two different directions to represent the two distinctly separate components of the vector. If we construct two unit vectors $U_{(1)} = \begin{bmatrix} 1 \\ 0 \end{bmatrix}$ and $U_{(2)} = \begin{bmatrix} 0 \\ 1 \end{bmatrix}$, they will fall right on the axes as shown in Figure 2.3. Since they represent unit length in their respective directions, greater length

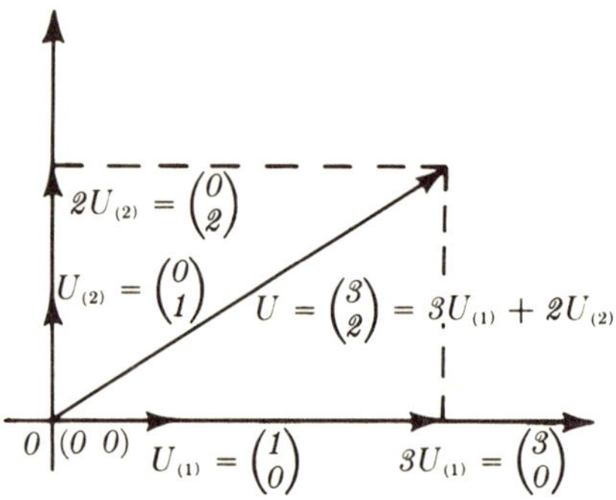

FIGURE 2.3

along these directions can simply be represented by the scalar multiples of these unit vectors. Any two-dimensional vector may be represented in a plane by a unique point, its position determined by the magnitude of its components expressed in distance from the origin along the axes. It can also be represented by a combination of the scalar multiples of the two unit vectors $U_{(1)}$ and $U_{(2)}$. For example, the vector $U = \begin{bmatrix} 3 \\ 2 \end{bmatrix}$ can be regarded as the diagonal of a parallelogram with sides which are the 3-multiple of $U_{(1)}$ along one axis and the 2-multiple of $U_{(2)}$ along the other. Algebraically,

$$\begin{aligned} U = \begin{bmatrix} 3 \\ 2 \end{bmatrix} = \begin{bmatrix} 3 \\ 0 \end{bmatrix} + \begin{bmatrix} 0 \\ 2 \end{bmatrix} = 3\begin{bmatrix} 1 \\ 0 \end{bmatrix} + 2\begin{bmatrix} 0 \\ 1 \end{bmatrix} \\ = 3U_{(1)} + 2U_{(2)}. \end{aligned}$$

Since this is true of *any* two-dimensional vector, it is true of *all* two-dimensional vectors. Now, according to our discussion of linear independence in Section 2.12, $U_{(1)}$ and $U_{(2)}$ are clearly independent.

Therefore, we conclude that all two-dimensional vectors can be expressed as a linear combination of two independent vectors. Such a set of linearly independent vectors is called a *basis*, and the vectors that comprise a basis are known as *basis vectors.*

As the dimension of vectors is increased beyond three, graphical representation becomes impracticable. But the concepts remain true. In general, we define an n-dimensional vector space as the totality of all n-component real-number vectors and an n-dimensional basis as any set of n linearly independent vectors in that space.* Such a basis has the property that any vector U in that space can be expressed as a linear combination of the basis vectors, $U_{(1)}$, $U_{(2)}$, $U_{(3)}$ . . . , $U_{(n)}$. Or, equivalently,

$$U = c_1U_{(1)} + c_2U_{(2)} \ldots + c_nU_{(n)} = \sum_{i=1}^{n} c_iU_{(i)}. \tag{2.13.1}$$

In linear programming, basis is one of the most important concepts and will be encountered frequently in the chapters to follow. The most natural basis of an n-dimensional vector space is the most obvious set of n independent vectors, viz. the n unit vectors,

$$U_{(1)} = \begin{bmatrix} 1 \\ 0 \\ \cdot \\ \cdot \\ \cdot \\ 0 \end{bmatrix}, \qquad U_{(2)} = \begin{bmatrix} 0 \\ 1 \\ \cdot \\ \cdot \\ \cdot \\ 0 \end{bmatrix}, \qquad \ldots \qquad U_{(n)} = \begin{bmatrix} 0 \\ 0 \\ \cdot \\ \cdot \\ \cdot \\ 1 \end{bmatrix}.$$

It is customary to express a basis in the form of a square matrix with its columns (or rows) representing the set of independent vectors comprising the basis. For instance, the above set of basis vectors $U_{(1)}, \ldots, U_{(n)}$ may be simply represented by the matrix B defined as follows:

$$B = \begin{bmatrix} 1 & 0 & 0 & \ldots & 0 \\ 0 & 1 & 0 & \ldots & 0 \\ 0 & 0 & 1 & \ldots & 0 \\ \cdot & \cdot & \cdot & & \cdot \\ \cdot & \cdot & \cdot & & \cdot \\ \cdot & \cdot & \cdot & & \cdot \\ 0 & 0 & 0 & \ldots & 1 \end{bmatrix}$$

* For a more rigorous definition of vector space and basis, refer to Birkoff and MacLane, *A Survey of Modern Algebra* (New York: The MacMillan Co., 1956) Chapter VII, Sections 3 and 8.

Referring to (2.13.1) it is quite possible that some of the coefficients may be zero. This indicates that the vector U is a linear combination of less than the complete set of n independent basis vectors. In linear programming this phenomenon has some special significance as it gives rise to what is known as the problem of *degeneracy*. We shall return to this problem in greater detail in Chapter IV Section 4.4. Meanwhile, we shall demonstrate in Section 2.18 of this chapter how the coefficients, $c_1, c_2, \ldots, c_n$, in (2.13.1) can be determined.

2.14. Determinant.

After an extended discussion of linear independence, it seems that the method of testing for linear independence by means of solving simultaneous equations is a rather cumbersome one. In many cases, we are more interested in knowing whether linear independence exists than in the exact relationship between the vectors should they prove to be dependent. For this kind of quick diagnosis, we now develop a simpler method which involves the notion of *determinant.*

Determinant, historically known as "eliminant," was developed as a by-product of mathematicians' primary concern with solving simultaneous equations. We will attempt to show how the latter leads to the former by solving a general two-equation two-unknown linear system using the standard elimination method.

$$a_{11}x_1 + a_{12}x_2 = b_1 \tag{2.14.1}$$

$$a_{21}x_1 + a_{22}x_2 = b_2 \tag{2.14.2}$$

Multiplying (2.14.1) by a_{21} and (2.14.2) by a_{11}, we obtain,

$$a_{11}a_{21}x_1 + a_{12}a_{21}x_2 = a_{21}b_1, \tag{2.14.3}$$

$$a_{11}a_{21}x_1 + a_{11}a_{22}x_2 = a_{11}b_2. \tag{2.14.4}$$

Subtracting (2.14.3) from (2.14.4) yields,

$$(a_{11}a_{22} - a_{12}a_{21})x_2 = a_{11}b_2 - a_{21}b_1. \tag{2.14.5}$$

Hence,

$$x_2 = \frac{a_{11}b_2 - a_{21}b_1}{a_{11}a_{22} - a_{12}a_{21}}. \tag{2.14.6}$$

Similarly, we can solve for x_1 by multiplying (2.14.1) by a_{22} and multiplying (2.14.2) by a_{12}, thus obtaining,

$$a_{11}a_{22}x_1 + a_{12}a_{22}x_2 = a_{22}b_1,$$
$$a_{12}a_{21}x_1 + a_{12}a_{22}x_2 = a_{12}b_2.$$

Subtracting the second equation from the first we have

$$(a_{11}a_{22} - a_{12}a_{21})x_1 = a_{22}b_1 - a_{12}b_2,$$

from which we obtain

$$x_1 = \frac{a_{22}b_1 - a_{12}b_2}{a_{11}a_{22} - a_{12}a_{21}}. \tag{2.14.7}$$

It is interesting to note that the denominators of the solutions (2.14.6) and (2.14.7) are the same and depend only on the coefficients of x_1 and x_2 in the original equations, (2.14.1) and (2.14.2). Let us, therefore, represent these coefficients by a matrix defined as follows:

$$A = \begin{bmatrix} a_{11} & a_{12} \\ a_{21} & a_{22} \end{bmatrix}.$$

We notice that the denominator, $a_{11}a_{22} - a_{12}a_{21}$, contains two terms, each of which is the product of two elements of A, selected in such a way that the following conditions are satisfied: (1) in each term, each element represents a different row and a different column so that all rows and all columns of A are represented; (2) the two terms collectively represent all possible ways of forming such terms without any duplication. Furthermore, each term is assigned a sign determined according to the following rule: arrange the elements in each product term in the numerical order of their first subscripts (row indices); if the order of their second subscripts (column indices) has an even number of inversions, the sign is positive; otherwise, the sign is negative. These are essentially the rules for forming all the terms in a determinant and the determinant itself is the algebraic sum of all these terms properly formed and signed. Checking the denominator, $a_{11}a_{22} - a_{12}a_{21}$, against these rules, we find that the first term, $a_{11}a_{22}$, contains a_{11} from the first row and first column of A, and a_{22} from the second row and second column of A. Therefore, the first condition is satisfied. Similarly, the second term, $a_{12}a_{21}$, satisfies the first condition. Since there are only two columns in A, these two terms exhaust the possibility of combining two elements according to condition (1) without duplication. Therefore, the second condition is also satisfied. Now, the sign of $a_{11}a_{22}$ is positive because the first subscripts are in numerical order and the second subscripts, in the order of (1, 2), have no inversion. However, $a_{12}a_{21}$ is negative because while the first subscripts are arranged in order, their second subscripts have 2 ahead of 1, thus registering one inversion. The term, $a_{11}a_{22} - a_{12}a_{21}$, then, is defined as the determinant of A, or,

$$|A| = \begin{vmatrix} a_{11} & a_{12} \\ a_{21} & a_{21} \end{vmatrix} = a_{11}a_{22} - a_{12}a_{21}. \tag{2.14.8}$$

(Note that single vertical bars are used exclusively for *determinant* as distinguished from parentheses and square brackets for *matrix*). Similarly, the numerator of (2.14.6), viz. $a_{11}b_2 - a_{21}b_1$, can be written as $\begin{vmatrix} a_{11} & b_1 \\ a_{21} & b_2 \end{vmatrix}$, and the numerator of (2.14.7), viz. $a_{22}b_1 - a_{12}b_2$, as $\begin{vmatrix} b_1 & a_{12} \\ b_2 & a_{22} \end{vmatrix}$. The solutions of x_1 and x_2 as shown in (2.14.6) and (2.14.7) now become

$$x_1 = \frac{\begin{vmatrix} b_1 & a_{12} \\ b_2 & a_{22} \end{vmatrix}}{|A|} \quad \text{and} \quad x_2 = \frac{\begin{vmatrix} a_{11} & b_1 \\ a_{21} & b_2 \end{vmatrix}}{|A|}.$$

These represent Cramer's Rule as applied to a 2-equation linear system in 2 unknowns.*

We may now attempt a general definition of determinant. A *determinant* of order n, $|A| = |a_{ij}|$ with $i, j = 1, 2, \ldots, n$, is the algebraic sum of $n!$ terms each representing the product of n elements selected one from each row and each column in all possible and distinct combinations, bearing either a positive or negative sign depending on whether the number of inversions in the order of j (after the order of i in the product term has been arranged in numerical ascendency) is even or odd.

Example. Write down in general terms the determinant of a 3×3 matrix.**

---

* A. Spitzbart and R. H. Bardell, *College Algebra and Plane Trigonometry* (Reading, Massachusetts: Addison-Wesley Publishing Co., 1955), pp. 312–316.

** In most elementary algebra, a third order determinant is usually evaluated by the simplified rule shown below. Positive product terms are formed of elements linked by solid lines; negative product terms, of elements linked by dotted lines. The reader must be cautioned against using a similar rule in evaluating determinants whose order is higher than three.

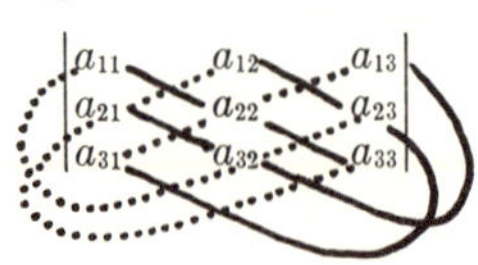

Let $$A = \begin{bmatrix} a_{11} & a_{12} & a_{13} \\ a_{21} & a_{22} & a_{23} \\ a_{31} & a_{32} & a_{33} \end{bmatrix}.$$

Since the order of the determinant is 3, i.e. $n = 3$, there will be $3! = 3 \times 2 \times 1 = 6$ terms. Let us form each term by selecting the first, second and third elements always from the first, second and third rows in that order so that all the rows will be covered and none covered more than once. Then, if a_{11} is the first element, the second element must be either a_{22} or a_{23}. If the second element is a_{22}, then the third can be only a_{33}. If the second element is a_{23}, then the third can only be a_{32}. This way, all the columns will be covered and covered only once. Therefore, we have two terms, viz. $a_{11}a_{22}a_{33}$ and $a_{11}a_{23}a_{32}$, which have a_{11} as the first element. Similarly, we have two terms in which a_{12} is the first element, and two more terms in which a_{13} is the first element. These six terms which exhaust all possible and distinct combinations (as evidenced by the permutation of the three column indices in a term) are listed below together with their signs.

Term	No. of inversions	Sign	Term	No. of inversions	Sign	Term	No. of inversions	Sign
$a_{11}a_{22}a_{33}$	0	+	$a_{12}a_{23}a_{31}$	2	+	$a_{13}a_{22}a_{31}$	3	−
$a_{11}a_{23}a_{32}$	1	−	$a_{12}a_{21}a_{33}$	1	−	$a_{13}a_{21}a_{32}$	2	+

Consequently,

(2.14.9) $$|A| = a_{11}a_{22}a_{33} - a_{11}a_{23}a_{32} + a_{12}a_{23}a_{31} - a_{12}a_{21}a_{33} + a_{13}a_{21}a_{32} - a_{13}a_{22}a_{31}.$$

2.15. Minors, Cofactors and Laplace's Expansion of a Determinant.

If the equation (2.14.9) is rearranged slightly we obtain

(2.15.1) $$|A| = a_{11}(a_{22}a_{33} - a_{23}a_{32}) + a_{12}(a_{23}a_{31} - a_{21}a_{33}) + a_{13}(a_{21}a_{32} - a_{22}a_{31}).$$

Now, by the definition of a determinant, $a_{22}a_{33} - a_{23}a_{32}$ is clearly the determinant of $\begin{bmatrix} a_{22} & a_{23} \\ a_{32} & a_{33} \end{bmatrix}$, and hence can be written as $\begin{vmatrix} a_{22} & a_{23} \\ a_{32} & a_{33} \end{vmatrix}$. Similarly, $a_{23}a_{31} - a_{21}a_{33}$ and $a_{21}a_{32} - a_{22}a_{31}$ can be written respectively as $-\begin{vmatrix} a_{21} & a_{23} \\ a_{31} & a_{33} \end{vmatrix}$ and $\begin{vmatrix} a_{21} & a_{22} \\ a_{31} & a_{32} \end{vmatrix}$. It will be noticed that these determinants are actually determinants of the submatrices of A obtained by deleting certain rows and columns in A. In other words, referring to (2.15.1), it is seen that a_{11} is associated with the determinant of a submatrix obtained by deleting the 1st row and 1st column of A. Similarly, a_{12} is associated with negative determinant of a submatrix obtained by deleting the 1st row and 2nd column of A. These determinants are called *minors.* A minor, $|M_{ij}|$, is defined as the determinant of the submatrix of a matrix, obtained by deleting its ith row and jth column. Equation (2.15.1) can now be written in terms of the minors, i.e.,

$$|A| = a_{11}|M_{11}| - a_{12}|M_{12}| + a_{13}|M_{13}|. \tag{2.15.2}$$

The evaluation of a third-order determinant is thus reduced to a problem of evaluating three second-order determinants. A closer look at (2.15.2) reveals that the signs of these minors alternate between positive and negative. To determine the sign of any particular minor, $|M_{ij}|$, we evaluate the function $(-1)^{i+j}$. If the sum of i and j (which are indices of the row and column deleted) is even, the sign of the minor is positive; otherwise, the sign is negative. A minor with its proper sign is called a *cofactor.* A cofactor C_{ij} may be defined as a signed minor. More precisely, the cofactor is $C_{ij} = (-1)^{i+j}|M_{ij}|$. In terms of cofactors, then, equation (2.15.2) becomes simply

$$|A| = a_{11}C_{11} + a_{12}C_{12} + a_{13}C_{13}. \tag{2.15.3}$$

In general, the minor, $|M_{ij}|$, of an $n \times n$ matrix A is the determinant of an $(n - 1) \times (n - 1)$ submatrix of A, obtained by deleting its ith row and jth column, and the cofactor, C_{ij}, of A is simply $(-1)^{i+j}|M_{ij}|$. The determinant of A can then be written as:

$$|A| = a_{i1}C_{i1} + a_{i2}C_{i2} \dots + a_{ij}C_{ij} \dots + a_{in}C_{in} = \sum_{j=1}^{n} a_{ij}C_{ij}. \tag{2.15.4}$$

Or equivalently

$$|A| = a_{1j}C_{1j} + a_{2j}C_{2j} \dots + a_{ij}C_{ij} \dots + a_{nj}C_{nj} = \sum_{i=1}^{n} a_{ij}C_{ij}. \tag{2.15.5}$$

Both (2.15.4) and (2.15.5) are special cases of Laplace's expansion of a determinant: the former by the cofactors of the ith row and the latter by the cofactors of the jth column. Since cofactors are basically also determinants, they can be evaluated by further expansion into determinants of still lower order. Thus, repeated expansion will eventually bring an nth-order determinant down to a series of second or third-order determinants which can be evaluated quickly by hand. In actual computation, much time can be saved by choosing for Laplace expansion that row or column which has the most zero elements. This is so because if a_{ij} is zero, then $a_{ij}C_{ij}$ is also zero; hence the evaluation of C_{ij} can be omitted.

Example. Evaluate the determinant of

$$A = \begin{bmatrix} 5 & 1 & -1 & 0 \\ 0 & 1 & 0 & 2 \\ 3 & 0 & 3 & -1 \\ 0 & -1 & 1 & 6 \end{bmatrix}.$$

Since the second row has the most zeroes (also the first column), we will expand $|A|$ by the cofactors of the second row. Applying (2.15.4), we have,

$$|A| = a_{21}C_{21} + a_{22}C_{22} + a_{23}C_{23} + a_{24}C_{24} = a_{22}C_{22} + a_{24}C_{24}$$

$$= (-1)^{2+2}\begin{vmatrix} 5 & -1 & 0 \\ 3 & 3 & -1 \\ 0 & 1 & 6 \end{vmatrix} + 2(-1)^{2+4}\begin{vmatrix} 5 & 1 & -1 \\ 3 & 0 & 3 \\ 0 & -1 & 1 \end{vmatrix}.$$

The two third-order determinants above can, in turn, be evaluated either by a second round of Laplace's expansion or by the use of (2.14.9). Following the latter approach we arrive at

$$|A| = (90 + 0 + 0 + 0 + 5 + 18) \\ + 2(0 + 3 + 0 + 0 + 15 - 3) = 143.$$

It will be noticed that in the Laplace expansion of a determinant, e.g. (2.15.4) and (2.15.5), the cofactor C_{i1} is multiplied by the element a_{i1}; C_{i2} by a_{i2}, etc. In short, all cofactors are multiplied by elements bearing the same row and column indices. A strange but interesting result takes place, however, if the cofactors of a row are multiplied by the corresponding elements of a different row. Invariably the sum of these terms reduces identically to zero. Symbolically,

(2.15.6) $\quad a_{l1}C_{i1} + a_{l2}C_{i2} + \ldots + a_{ln}C_{in} = 0 \quad \text{for all } l \neq i.$

We will not attempt to prove this relationship here but will merely illustrate it by the following example.

Example. Given the matrix

$$A = \begin{bmatrix} 2 & 0 & 1 \\ -1 & 0 & 3 \\ 0 & 2 & -1 \end{bmatrix},$$

evaluate the sum of the terms which are products of cofactors of row 1 and corresponding elements of row 2, i.e.

$$\begin{aligned} a_{21}C_{11} + a_{22}C_{12} + a_{23}C_{13} & \\ &= (-1)(-1)^{1+1}\begin{vmatrix} 0 & 3 \\ 2 & -1 \end{vmatrix} + 0 + 3(-1)^{1+3}\begin{vmatrix} -1 & 0 \\ 1 & 2 \end{vmatrix} \\ &= 6 - 6 = 0. \end{aligned}$$

Repeating the above using elements of row 3 we have

$$\begin{aligned} a_{31}C_{11} + a_{32}C_{12} + a_{33}C_{13} & \\ &= 0 + 2(-1)^{1+2}\begin{vmatrix} -1 & 3 \\ 1 & -1 \end{vmatrix} + (-1)(-1)^{1+3}\begin{vmatrix} -1 & 0 \\ 0 & 2 \end{vmatrix} \\ &= -2 + 2 = 0. \end{aligned}$$

In Section 2.17, we will use this relationship, called *expansion by alien cofactors* to develop a new method of finding the inverse of a non-singular square matrix.

2.16. Determinant and Linear Independence.

In this section, we will develop an expedient test for linear independence using determinants.

Before we proceed, however, we will discuss one of the fundamental theorems in the algebra of determinants. This theorem is as follows: if in the square matrix, A, the ith row is some multiple (say c) of the kth row, then the determinant of A is zero. We will not give the formal proof here, but merely indicate the reasoning behind it. Bear in mind that a determinant is the sum of all terms which are products of elements of the matrix selected in accordance with the rules described in Section 2.14. In an nth-order determinant, then, we may focus our attention on one general term, viz. $a_{1h} \dots a_{ij}a_{kl} \dots a_{np}$. For this term, there must be a counterpart, $a_{1h} \dots a_{il}a_{kj} \dots a_{np}$, made up of identical elements except two. These two elements are taken from ith row, lth column and kth row,

jth column, as distinguished from the two corresponding elements in the first term, which are selected from ith row, jth column and kth row, lth column respectively. Since we assume that the ith row is the c-multiple of the kth row, $a_{ij} = ca_{kj}$ and $a_{il} = ca_{kl}$. Therefore, $a_{ij}a_{kl} = a_{il}a_{kj}$, and $a_{1h} \ldots a_{ij}a_{kl} \ldots a_{np} = a_{1h} \ldots a_{il}a_{kj} \ldots a_{np}$. But the second term has l before j whereas the first term has j before l. Hence their signs must be opposite; consequently they cancel each other in the determinant. Similarly, all other terms in the determinant can be paired and cancelled with the result that the determinant is reduced identically to zero.

It takes only one more step to show that if an $n \times n$ matrix, A, contains a row $\overline{A}_i$ that is a linear combination of some other rows, say $\overline{A}_k$ and $\overline{A}_m$ such that $\overline{A}_i = c_k\overline{A}_k + c_m\overline{A}_m$, then the determinant of A is zero. The proof of this closely parallels the reasoning indicated above and may be left as an exercise. Since linear dependence is defined in terms of linear combination, (cf. Section 2.12), we conclude that the determinant of an $n \times n$ matrix whose rows are linearly dependent is zero. On the other hand, if the rows are linearly independent, the determinant is non-zero.

From the foregoing discussion it is clear that the determinant of a square matrix, A, may be used to ascertain whether the matrix contains linearly dependent rows or not. If it does, it is called a *singular* matrix and its determinant is zero. If it does not, it is called a *non-singular* matrix and its determinant is non-zero.* Such a matrix, then, forms a basis in terms of which all other vectors in the same vector space can be expressed (cf. Section 2.13).

2.17. Matrix Inversion Using Determinant and Adjoint.

Matrix inversion is the process by which the inverse of a non-singular square matrix is determined. The link between matrix inversion and determinant lies in Laplace's expansion formula for a determinant, viz. $|A| = a_{i1}C_{i1} + a_{i2}C_{i2} + \ldots + a_{in}C_{in}$, where i represents the index of any row and n, the dimension of the matrix. Bearing this in mind as well as the theorem that expansion of a determinant by alien cofactors results in zero (cf. Section 2.15), we

* For a rigorous proof of this relationship, read Birkoff and Maclane, *op. cit.*, pp. 304–5.

may construct a matrix whose columns are the cofactors of the first row, second row, etc. of the original matrix, A, and arrive at the following desired result. Let

$$A = \begin{bmatrix} a_{11} & a_{12} & \cdots & a_{1n} \\ a_{21} & a_{22} & \cdots & a_{2n} \\ \cdot & \cdot & & \cdot \\ \cdot & \cdot & & \cdot \\ \cdot & \cdot & & \cdot \\ a_{n1} & a_{n2} & \cdots & a_{nn} \end{bmatrix}, \quad B = \begin{bmatrix} C_{11} & C_{21} & \cdots & C_{n1} \\ C_{12} & C_{22} & \cdots & C_{n2} \\ \cdot & \cdot & & \cdot \\ \cdot & \cdot & & \cdot \\ \cdot & \cdot & & \cdot \\ C_{1n} & C_{2n} & \cdots & C_{nn} \end{bmatrix}.$$

Then

$$AB = \begin{bmatrix} \sum_{j=1}^{n} a_{1j}C_{1j} & \sum_{j=1}^{n} a_{1j}C_{2j} & \cdots & \sum_{j=1}^{n} a_{1j}C_{nj} \\ \sum_{j=1}^{n} a_{2j}C_{1j} & \sum_{j=1}^{n} a_{2j}C_{2j} & \cdots & \sum_{j=1}^{n} a_{2j}C_{nj} \\ \cdot & \cdot & & \cdot \\ \cdot & \cdot & & \cdot \\ \cdot & \cdot & & \cdot \\ \sum_{j=1}^{n} a_{nj}C_{1j} & \sum_{j=1}^{n} a_{nj}C_{2j} & \cdots & \sum_{j=1}^{n} a_{nj}C_{nj} \end{bmatrix}.$$

All the elements on the main diagonal of resulting matrix, i.e. $\sum_{j=1}^{n} a_{1j}C_{1j}, \sum_{j=1}^{n} a_{2j}C_{2j}, \ldots, \sum_{j=1}^{n} a_{nj}C_{nj}$, are Laplace's expansion of $|A|$, and therefore are equal to $|A|$. All other elements are expansion by alien cofactors and are equal to zero. Consequently

$$AB = \begin{bmatrix} |A| & 0 & \cdots & 0 \\ 0 & |A| & \cdots & 0 \\ \cdot & \cdot & & \cdot \\ \cdot & \cdot & & \cdot \\ \cdot & \cdot & & \cdot \\ 0 & 0 & \cdots & |A| \end{bmatrix} = |A| \begin{bmatrix} 1 & 0 & \cdots & 0 \\ 0 & 1 & \cdots & 0 \\ \cdot & \cdot & & \cdot \\ \cdot & \cdot & & \cdot \\ \cdot & \cdot & & \cdot \\ 0 & 0 & \cdots & 1 \end{bmatrix} = |A|\, I.$$

Dividing both sides by $|A|$

$$A \frac{B}{|A|} = I.$$

By definition of an inverse

$$A^{-1} = \frac{B}{|A|}.$$

The matrix B, which is the transpose of the matrix consisting of the cofactors of a_{ij}, is known as the *adjoint* of A and is designated A_{adj}.

(2.17.1) $$A_{adj} = [C_{ij}]^T \qquad \text{where } C_{ij} \text{ are the cofactors of } a_{ij} \text{ in } A.$$

(2.17.2) $$A^{-1} = \frac{1}{|A|} A_{adj}.$$

Matrix inversion, though useful, is often a tedious process. For small matrices, such as a 3×3 or a 4×4, (2.17.2) offers a convenient general procedure for inversion by hand as the determinants which have to be evaluated are all of second or third order. Inversion of large matrices, on the other hand, is usually a job for electronic computers.

Example. Find the inverse of $A = \begin{bmatrix} 1 & 1 & 1 \\ 2 & -1 & -1 \\ 0 & 2 & 3 \end{bmatrix}$ by the use of adjoint.

From (2.17.1),

$$A_{adj} = \begin{bmatrix} C_{11} & C_{21} & C_{31} \\ C_{12} & C_{22} & C_{32} \\ C_{13} & C_{23} & C_{33} \end{bmatrix}$$

$$= \begin{bmatrix} \begin{vmatrix} -1 & -1 \\ 2 & 3 \end{vmatrix} & -\begin{vmatrix} 1 & 1 \\ 2 & 3 \end{vmatrix} & \begin{vmatrix} 1 & 1 \\ -1 & -1 \end{vmatrix} \\ -\begin{vmatrix} 2 & -1 \\ 0 & 3 \end{vmatrix} & \begin{vmatrix} 1 & 1 \\ 0 & 3 \end{vmatrix} & -\begin{vmatrix} 1 & 1 \\ 2 & -1 \end{vmatrix} \\ \begin{vmatrix} 2 & -1 \\ 0 & 2 \end{vmatrix} & -\begin{vmatrix} 1 & 1 \\ 0 & 2 \end{vmatrix} & \begin{vmatrix} 1 & 1 \\ 2 & -1 \end{vmatrix} \end{bmatrix} = \begin{bmatrix} -1 & -1 & 0 \\ -6 & 3 & 3 \\ 4 & -2 & -3 \end{bmatrix}.$$

From (2.15.4),

$$|A| = \sum_{j=1}^{3} a_{3j}C_{3j} = 0 + 2(3) + 3(-3) = -3.$$

Hence, by (2.17.2),

$$A^{-1} = \frac{1}{|A|} A_{adj} = \frac{1}{-3} \begin{bmatrix} -1 & -1 & 0 \\ -6 & 3 & 3 \\ 4 & -2 & -3 \end{bmatrix} = \begin{bmatrix} \frac{1}{3} & \frac{1}{3} & 0 \\ 2 & -1 & -1 \\ -\frac{4}{3} & \frac{2}{3} & 1 \end{bmatrix}.$$

It is a good practice to check the validity of the inverse by forming the product AA^{-1} to see if the result is actually an identity matrix. The verification of A^{-1} in this example can be quickly established by such a matrix multiplication.

2.18. Simultaneous Linear Equations and Matrix Algebra.

By now, it should be clear that matrix operations in the main closely parallel ordinary algebraic operations. They can always be translated into the latter although in many cases it can be done only at the expense of considerable time and effort. This advantage of economy in time and effort is most evident in the solution of linear simultaneous equations. We will illustrate this point by solving a three-equation three-unknown linear system as follows:

$$\begin{aligned} a_{11}x_1 + a_{12}x_2 + a_{13}x_3 &= b_1, \\ a_{21}x_1 + a_{22}x_2 + a_{23}x_3 &= b_2, \\ a_{31}x_1 + a_{32}x_2 + a_{33}x_3 &= b_3. \end{aligned} \tag{2.18.1}$$

We will define a coefficient matrix A and two column vectors X and P as follows:

$$A = \begin{bmatrix} a_{11} & a_{12} & a_{13} \\ a_{21} & a_{22} & a_{23} \\ a_{31} & a_{32} & a_{33} \end{bmatrix}, \quad X = \begin{bmatrix} x_1 \\ x_2 \\ x_3 \end{bmatrix}, \quad \text{and} \quad P = \begin{bmatrix} b_1 \\ b_2 \\ b_3 \end{bmatrix}.$$

According to the rule of matrix multiplication, we can write (2.18.1) in the much simpler matrix form

$$AX = P. \tag{2.18.2}$$

Or equivalently, we can write in index form

$$A_1x_1 + A_2x_2 + A_3x_3 = P, \tag{2.18.3}$$

where A_1, A_2, and A_3 are the first, second and third columns of A.

Now, pre-multiplying both sides of (2.18.2) by the inverse of A, A^{-1}, we have,

$$A^{-1}(AX) = A^{-1}P. \tag{2.18.4}$$

Since matrix multiplication complies with the associative law,

$$A^{-1}(AX) = (A^{-1}A)X.$$

But, by the definition of an inverse, $A^{-1}A = I$. Therefore

$$(A^{-1}A)X = IX = X.$$

Hence, (2.18.4) becomes

$$X = A^{-1}P. \tag{2.18.5}$$

In other words, to solve the linear system by matrix algebra, we have first to invert the coefficient matrix A, then post-multiply the inverse A^{-1} by the column vector P. The result is a column vector whose first, second, etc. elements represent the values of the first, second, etc. unknowns. Let us now take up the following numerical illustrations.

Example 1. Solve the following linear equation system by the matrix method.

$$\begin{aligned} x_1 + x_2 + x_3 &= 6, \\ 2x_1 - x_2 - x_3 &= 9, \\ 2x_2 + 3x_3 &= 1. \end{aligned}$$

Let

$$A = \begin{bmatrix} 1 & 1 & 1 \\ 2 & -1 & -1 \\ 0 & 2 & 3 \end{bmatrix}, \quad X = \begin{bmatrix} x_1 \\ x_2 \\ x_3 \end{bmatrix}, \quad \text{and} \quad P = \begin{bmatrix} 6 \\ 9 \\ 1 \end{bmatrix}.$$

We first find the inverse of A. This has already been done in the example of Section 2.17, since the same matrix is used. So

$$A^{-1} = \begin{bmatrix} \frac{1}{3} & \frac{1}{3} & 0 \\ 2 & -1 & -1 \\ -\frac{4}{3} & \frac{2}{3} & 1 \end{bmatrix}.$$

Following (2.18.5), we obtain

$$X = A^{-1}P = \begin{bmatrix} \frac{1}{3} & \frac{1}{3} & 0 \\ 2 & -1 & -1 \\ -\frac{4}{3} & \frac{2}{3} & 1 \end{bmatrix} \begin{bmatrix} 6 \\ 9 \\ 1 \end{bmatrix} = \begin{bmatrix} 5 \\ 2 \\ -1 \end{bmatrix}.$$

Or, $x_1 = 5$, $x_2 = 2$ and $x_3 = -1$.

Example 2. Solve the following linear equation system by matrix algebra.

$$\begin{aligned} x_1 - x_2 &= 3, \\ 2x_2 + x_3 &= 1, \\ 2x_1 + x_3 &= 7. \end{aligned}$$

Let

$$A = \begin{bmatrix} 1 & -1 & 0 \\ 0 & 2 & 1 \\ 2 & 0 & 1 \end{bmatrix}, \quad X = \begin{bmatrix} x_1 \\ x_2 \\ x_3 \end{bmatrix}, \quad \text{and} \quad P = \begin{bmatrix} 3 \\ 1 \\ 7 \end{bmatrix}.$$

Then

$$AX = P, \quad \text{and} \quad X = A^{-1}P.$$

In trying to find the inverse of A we discover that the determinant of A is zero. This means that the matrix is singular and therefore has no inverse. It also means that the left-hand sides of the equations (represented by the rows of A) are dependent on each other.

There are two possible situations that may arise. If the constants on the right-hand side of the equations are also dependent on each other in exactly the *same* way, the equations in their entireties are then dependent on each other, and there are an infinite number of solutions. The other possibility is that the constants are not dependent in exactly the same way. The equations then are not dependent on but *inconsistent* with each other, and there is no solution.

To determine whether the constants are related in the same way as the rows in A, we will construct a new square matrix, B, with one of the original columns (any column) replaced by the constant column. For instance

$$B = \begin{bmatrix} 1 & -1 & 3 \\ 0 & 2 & 1 \\ 2 & 0 & 7 \end{bmatrix}$$ where the 3rd column of A is replaced by P.

Now, if the same relationship exists among the constants, then the rows of B are again dependent and the determinant of B should accordingly be zero. Otherwise, it should not be zero. In the first case, an infinite number of solutions are possible whereas in the second, no solution can be found. Let us evaluate $|B|$ by using Laplace's expansion on the third row. We have:

$$|B| = 2(-7) + 7\,(2) = 0.$$

The singularity of B establishes the fact that an infinite number of solutions exist. To obtain these solutions, we search matrix A for a non-singular 2×2 submatrix. Such a submatrix is

$$A' = \begin{bmatrix} 1 & -1 \\ 0 & 2 \end{bmatrix}.$$

Since A' is made up of coefficients of x_1 and x_2, the third variable x_3 can be treated as a constant and shifted to the right-hand side of the first and second equations. Thus we have

$$A'X' = P' \quad \text{where } A' \text{ is the submatrix defined above;}$$

$$X' = \begin{bmatrix} x_1 \\ x_2 \end{bmatrix}, \quad \text{and} \quad P' = \begin{bmatrix} 3 \\ 1 - x_3 \end{bmatrix}.$$

It can easily be verified that the inverse of A' is $(A')^{-1} = \begin{bmatrix} 1 & \frac{1}{2} \\ 0 & \frac{1}{2} \end{bmatrix}$.

Therefore

$$X' = (A')^{-1}P' = \begin{bmatrix} 1 & \frac{1}{2} \\ 0 & \frac{1}{2} \end{bmatrix} \begin{bmatrix} 3 \\ 1 - x_3 \end{bmatrix} = \begin{bmatrix} \frac{7}{2} - \frac{1}{2}x_3 \\ \frac{1}{2} - \frac{1}{2}x_3 \end{bmatrix}.$$

Let x_3 be equal to any number k, then the infinite set of solutions is represented by:

$$x_1 = \tfrac{7}{2} - \tfrac{1}{2}k,$$
$$x_2 = \tfrac{1}{2} - \tfrac{1}{2}k,$$
$$x_3 = k.$$

The rationale of selecting a 2×2 submatrix as shown above is that at most two of the three equations are independent—a proposition established by the fact that $|A| = 0$. The third equation can be overlooked in solving the system if two of the three can be established as being independent, which is the purpose of evaluating the determinant of A'.

Example 3. If the column vector P is $\begin{bmatrix} 1 \\ 5 \\ -1 \end{bmatrix}$ instead of $\begin{bmatrix} 3 \\ 1 \\ 7 \end{bmatrix}$ as given in Example 2, then the B matrix defined above will be non-singular, i.e.

$$|B| = \begin{vmatrix} 1 & -1 & 1 \\ 0 & 2 & 5 \\ 2 & 0 & -1 \end{vmatrix} = 0 + 2(-1-2) - 5(2) = -16.$$

This means that the constants of the equations are not related in the same manner as the coefficients are (since $|A| = 0$). Therefore the equations are contradictory or inconsistent and no solution is possible.

To show this inconsistency we solve, (1) the first and second equations for x_1 and x_2 in terms of x_3, and (2) the second and third equations for x_1 and x_2 in terms of x_3.* If the solutions of (1) and (2) do not agree, then obviously there is inconsistency and no solution will satisfy all three equations simultaneously.

The submatrix we will use for the first solution is $A' = \begin{bmatrix} 1 & -1 \\ 0 & 2 \end{bmatrix}$ and the constant vector is $P' = \begin{bmatrix} 1 \\ 5 - x_3 \end{bmatrix}$.

From Example 2, $(A')^{-1} = \begin{bmatrix} 1 & \frac{1}{2} \\ 0 & \frac{1}{2} \end{bmatrix}$.

Hence,

$$X' = \begin{bmatrix} x_1 \\ x_2 \end{bmatrix} = (A')^{-1}P' = \begin{bmatrix} 1 & \frac{1}{2} \\ 0 & \frac{1}{2} \end{bmatrix}\begin{bmatrix} 1 \\ 5 - x_3 \end{bmatrix} = \begin{bmatrix} \frac{7}{2} - \frac{1}{2}x_3 \\ \frac{5}{2} - \frac{1}{2}x_3 \end{bmatrix}.$$

Or, $x_1 = \frac{7}{2} - \frac{1}{2}x_3$ and $x_2 = \frac{5}{2} - \frac{1}{2}x_3$.

To obtain the second solution, we use the submatrix $A'' = \begin{bmatrix} 0 & 2 \\ 2 & 0 \end{bmatrix}$ whose inverse is $(A'')^{-1} = \begin{bmatrix} 0 & \frac{1}{2} \\ \frac{1}{2} & 0 \end{bmatrix}$. The new constant vector is $P'' = \begin{bmatrix} 5 - x_3 \\ -1 - x_3 \end{bmatrix}$.

Therefore,

$$X'' = \begin{bmatrix} x_1 \\ x_2 \end{bmatrix} = (A'')^{-1}P'' = \begin{bmatrix} 0 & \frac{1}{2} \\ \frac{1}{2} & 0 \end{bmatrix}\begin{bmatrix} 5 - x_3 \\ -1 - x_3 \end{bmatrix} = \begin{bmatrix} -\frac{1}{2} - \frac{1}{2}x_3 \\ \frac{5}{2} - \frac{1}{2}x_3 \end{bmatrix}.$$

Or, $x_1 = -\frac{1}{2} - \frac{1}{2}x_3$ and $x_2 = \frac{5}{2} - \frac{1}{2}x_3$. These do not agree with the first solution. Thus, the contradiction is verified.

Example 4. We are given the vectors, $V_1 = \begin{bmatrix} 3 \\ -1 \\ 1 \end{bmatrix}$, $V_2 = \begin{bmatrix} 0 \\ 2 \\ 1 \end{bmatrix}$ and $V_3 = \begin{bmatrix} 0 \\ 0 \\ -2 \end{bmatrix}$. Determine if these three vectors constitute a basis of the 3-dimensional vector space. If so, determine the coefficients for expressing the vector, $V_4 = \begin{bmatrix} 6 \\ 0 \\ 5 \end{bmatrix}$, in terms of the basis.

---

* It is possible that either (1) or (2) or both cannot be solved. In such cases, inconsistency exists within part of the linear system, and hence, *a fortiori* in the whole system. The system, therefore, has no solution.

The meaning and significance of a basis have already been discussed in Section 2.13. Whether V_1, V_2 and V_3 constitute a basis or not depends on (1) whether they are independent of each other and (2) whether they cover the entire 3-dimensional vector space.

To answer (1), we evaluate the determinant of the matrix composed of V_1, V_2 and V_3. Let

$$B = [V_1\,V_2\,V_3] = \begin{bmatrix} 3 & 0 & 0 \\ -1 & 2 & 0 \\ 1 & 1 & -2 \end{bmatrix},$$

$$|B| = 3\begin{vmatrix} 2 & 0 \\ 1 & -2 \end{vmatrix} + 0 + 0 = -12.$$

Since $|B| \neq 0$, the vectors V_1, V_2 and V_3 represent an independent set. To answer (2), we have to show that any 3-dimensional vector, V_k, can be expressed as a linear combination of V_1, V_2 and V_3, as stated in (2.13.1). In other words, we must show that there is a set of coefficients, c_1, c_2 and c_3, such that

$$c_1V_1 + c_2V_2 + c_3V_3 = V_k.$$

If we define a column vector, $C = \begin{bmatrix} c_1 \\ c_2 \\ c_3 \end{bmatrix}$, then the above equation can be written in the matrix form, $BC = V_k$, where B is the matrix consisting of V_1, V_2 and V_3. Since B has already been found to be non-singular, B^{-1} exists. The solution is, therefore, $C = B^{-1}V_k$. Hence the unknown vector C can be uniquely determined and V_k can be expressed as a linear combination of V_1, V_2 and V_3. Since V_k is any 3-dimensional vector, all 3-dimensional vectors can be expressed as linear combinations of V_1, V_2 and V_3, and the 3-dimensional vector space is entirely covered by them. Therefore, we conclude that V_1, V_2, V_3 do constitute a basis. We now proceed to determine the C vector for V_4, as given in this example, by the formula $C = B^{-1}V_4$.

It will be noticed that the basis B has a peculiar structure, viz., the elements above the main diagonal are all zeroes. Such a matrix is called a *triangular* matrix and can be inverted quite simply without resorting to the use of the adjoint. Let

$$B^{-1} = \begin{bmatrix} b_{11} & b_{12} & b_{13} \\ b_{21} & b_{22} & b_{23} \\ b_{31} & b_{32} & b_{33} \end{bmatrix}.$$

Then $BB^{-1} = I$, or

$$\begin{bmatrix} 3 & 0 & 0 \\ -1 & 2 & 0 \\ 1 & 1 & -2 \end{bmatrix} \begin{bmatrix} b_{11} & b_{12} & b_{13} \\ b_{21} & b_{22} & b_{23} \\ b_{31} & b_{32} & b_{33} \end{bmatrix} = \begin{bmatrix} 1 & 0 & 0 \\ 0 & 1 & 0 \\ 0 & 0 & 1 \end{bmatrix}$$

By the rule of matrix multiplication, the first row of B multiplied by the first column of B^{-1} must be equal to the element in the first row and first column of the product matrix, which is 1 in this case. Hence, $3b_{11} + 0 + 0 = 1$, which yields $b_{11} = \frac{1}{3}$. By the same rule, the product of the second row of B and the first column of B^{-1} must be equal to 0. Hence, $-b_{11} + 2b_{21} + 0 = 0$. Since $b_{11} = \frac{1}{3}$, $b_{21} = \frac{1}{6}$. Similarly, $b_{11} + b_{21} - 2b_{31} = 0$, which yields $b_{31} = \frac{1}{4}$. Repeating the process for the second and third columns of B^{-1}, we finally obtain the inverse, $B^{-1} = \begin{bmatrix} \frac{1}{3} & 0 & 0 \\ \frac{1}{6} & \frac{1}{2} & 0 \\ \frac{1}{4} & \frac{1}{4} & -\frac{1}{2} \end{bmatrix}$.*

Hence

$$C = B^{-1}V_4 = \begin{bmatrix} \frac{1}{3} & 0 & 0 \\ \frac{1}{6} & \frac{1}{2} & 0 \\ \frac{1}{4} & \frac{1}{4} & -\frac{1}{2} \end{bmatrix} \begin{bmatrix} 6 \\ 0 \\ 5 \end{bmatrix} = \begin{bmatrix} 2 \\ 1 \\ -1 \end{bmatrix},$$

which means that $V_4 = 2V_1 + V_2 - V_3$.

Perhaps the more perceptive can already see the advantage of using the matrix method to solve simultaneous linear equations. It lies not in the solution of one set of equations (the process of inverting a matrix is just as lengthy and tedious as the method of elimination used in elementary algebra); rather, it lies in the repeated use of the same inverse to obtain solutions to systems of equations which have the same coefficient matrices but different constant vectors. In fact, it has been estimated that if we are to solve m simultaneous equations in m unknowns n times for n sets of different constants, it will call for roughly $n\left(\frac{m^3}{2}\right)$ multiplications and divisions by the elimination method as compared with $m^3 + nm$ multiplications and divisions by the matrix method. A quick calculation will show that the number of such operations is the same for both methods when $n = \frac{2m^2}{m^2 - 2}$. For

* Readers will realize that this special inversion method is essentially the classical method of solving simultaneous equations.

a 2-equation system, therefore, the matrix method would be more efficient when $n \geq 4$. For a 3-equation system, it would be more efficient when $n \geq 3$. As m increases further, n will approach 2 as a limit, which means that it takes no more than three repeated uses of the inverse to justify the solving of simultaneous equations by the matrix method. This is, of course, not a very difficult condition to satisfy. Therefore in using electronic digital computers to solve simultaneous equations, where multiplication and division take anywhere from five to twenty times as much time as addition and subtraction, the matrix method is almost invariably used.

EXERCISES

1. Given the following matrices:

$$A = \begin{bmatrix} 2 & 0 & 1 \\ -1 & 2 & 1 \end{bmatrix}, \qquad B = \begin{bmatrix} 1 & 2 \\ 0 & -1 \end{bmatrix},$$

$$C = \begin{bmatrix} 5 & 1 \\ 0 & -2 \\ 3 & 1 \end{bmatrix}, \qquad D = \begin{bmatrix} 0 & 1 & 1 \\ -2 & 1 & 0 \\ 3 & 1 & -1 \end{bmatrix},$$

$$U = [4 \quad -3 \quad 1], \qquad V = \begin{bmatrix} 0 \\ 1 \\ -1 \end{bmatrix},$$

$$X = \begin{bmatrix} x_1 \\ x_2 \\ x_3 \end{bmatrix}, \qquad W = [w_1 \quad w_2 \quad w_3],$$

perform the matrix operations specified below and comment on the significance of the results.

(a) AB, AC, AD, BA, BC, BD. Which multiplications are impossible?
(b) CA, CB, CD, DA, DB, DC. Which of these are impossible?
(c) UA, UC, UV, UX, XU, XW. Which of these are impossible?
(d) WA, WC, WD, DW, XD, DX. Which of these are impossible?
(e) AA^T, A^2 $(= AA)$, D^2 $(= DD)$.
(f) X^TDX, VUV.
(g) $AC - B$.
(h) $(A - C^T)D$.
(i) $(VU - D)X$.
(j) $CBA - CAD$.

2. Given:

$$I = \begin{bmatrix} 1 & 0 & 0 \\ 0 & 1 & 0 \\ 0 & 0 & 1 \end{bmatrix}, \quad O = \begin{bmatrix} 0 & 0 & 0 \\ 0 & 0 & 0 \\ 0 & 0 & 0 \end{bmatrix}, \quad A = \begin{bmatrix} 2 & -1 & 3 \\ 1 & 1 & 1 \\ 0 & 5 & -3 \end{bmatrix},$$

find the results of the following matrix operations. Comment on their significance.

(a) IA, AI.
(b) OI, OA.
(c) $I + O$, $A + O$.
(d) I^2, I^3, O^2.

3. Write in full the following 4×4 matrices:

(a) $A = [a_{ij}]$ where $a_{ij} = \begin{cases} 1 & \text{for } i + j = 5, \\ 0 & \text{for } i + j \neq 5. \end{cases}$

(b) $A = [a_{ij}]$ where $a_{ij} = (-1)^{i+j}$.

4. Let $X_1 = [-2 \quad 3]$, $X_2 = [0 \quad 2]$. Draw on graphs the results of the following operations and observe their significance.

(a) $aX_1 + (1 - a)X_2$ for $a = 1, \frac{3}{4}, \frac{1}{2}, \frac{1}{4}, 0$.
(b) $aX_1 + (1 - a)X_2$ for $a = 3, 2, 1, 0$.
(c) $aX_1 + (1 - a)X_2$ for $a = 1, 0, -1, -2$.
(d) $a_1X_1 + a_2X_2$ for any arbitrary values of a_1, a_2 provided that $a_1 + a_2 \neq 1$.

5. Determine whether the following matrices are composed of rows and columns that are linearly dependent on each other. If they are, show how the rows, or columns are related to each other.

(a) $\begin{bmatrix} 4 & 0 & 1 \\ 2 & -1 & 0 \end{bmatrix}$. (b) $\begin{bmatrix} -1 & 1 & 0 \\ 1 & 1 & 1 \end{bmatrix}$. (c) $\begin{bmatrix} 0 & 4 \\ 3 & -1 \\ -2 & 2 \end{bmatrix}$.

What inference can you draw from the above regarding row or column dependence as related to the row-column dimensions of a matrix?

6. A square matrix which has only one 1 in each row and each column with 0 everywhere else is known as a *permutation* matrix. For instance, P_1, P_2 shown below are examples of such matrices.

$$P_1 = \begin{bmatrix} 0 & 1 & 0 \\ 0 & 0 & 1 \\ 1 & 0 & 0 \end{bmatrix}, \quad P_2 = \begin{bmatrix} 0 & 0 & 1 \\ 0 & 1 & 0 \\ 1 & 0 & 0 \end{bmatrix}, \quad B = \begin{bmatrix} 2 & 1 & 0 \\ 1 & 0 & 1 \\ 1 & -1 & 1 \end{bmatrix}.$$

Perform the following operations and draw appropriate conclusions as to the role of permutation matrix in matrix multiplication.

(a) BP_1, BP_2. (b) P_1B, P_2B. (c) $P_1^TBP_1$.
(d) $P_1P_1^T$, $P_2P_2^T$. (e) P_1^2, P_2^2.

7. Derive the general expressions of the elements of the inverse of $A = \begin{bmatrix} a_{11} & a_{12} \\ a_{21} & a_{22} \end{bmatrix}$. Using the formulas obtained above, determine the inverses of: (a) $\begin{bmatrix} 0 & 1 \\ 1 & 0 \end{bmatrix}$. (b) $\begin{bmatrix} 2 & -1 \\ 0 & 3 \end{bmatrix}$. (c) $\begin{bmatrix} 1 & -1 \\ -1 & 1 \end{bmatrix}$.

8. Determine whether the following square matrices are singular. If so, determine further the relationship among the rows.

(a) $\begin{bmatrix} -4 & -1 & -2 \\ 2 & 1 & 4 \\ 1 & 0 & 3 \end{bmatrix}$. (b) $\begin{bmatrix} 4 & 0 & -1 \\ 0 & 0 & 0 \\ 2 & 1 & 1 \end{bmatrix}$.

(c) $\begin{bmatrix} 3 & 0 & -2 \\ -6 & 3 & 0 \\ 2 & 2 & -4 \end{bmatrix}$. (d) $\begin{bmatrix} 2 & 1 & 0 & -1 \\ 1 & 0 & -2 & 1 \\ 0 & 0 & 1 & 1 \\ 0 & 0 & 0 & -1 \end{bmatrix}$.

(e) $\begin{bmatrix} 1 & -2 & 0 & -2 \\ 0 & -5 & 0 & -2 \\ 1 & 1 & -1 & 2 \\ -2 & 3 & 2 & -1 \end{bmatrix}$.

9. (a) Write the full expression of the determinant of the 4×4 matrix $A = [a_{ij}]$, $i, j = 1, \ldots, 4$.
(b) Write all terms containing the elements a_{31} and a_{22} in the determinant of a 5×5 matrix $A = [a_{ij}]$.

10. Show by Laplace expansion that any square matrix with a row (or column) of 0, is necessarily singular.

11. (a) Given the $n \times n$ matrices A and aA, where a is a scalar, find $|aA|$ in terms of $|A|$.
(b) Specifically, what is $|-A|$ in terms of $|A|$?

12. Evaluate the determinants of the following matrices:

(a) $\begin{bmatrix} 1 & 5 & 4 \\ 2 & -2 & 6 \\ 1 & -2 & 1 \end{bmatrix}$. (b) $\begin{bmatrix} 1 & 2 & -1 & 3 \\ 0 & 0 & 1 & 0 \\ 1 & -3 & 0 & -2 \\ 3 & -1 & 0 & 4 \end{bmatrix}$.

(c) $\begin{bmatrix} 2 & 0 & 0 & 5 \\ 1 & 2 & 0 & 1 \\ 0 & 3 & -1 & 1 \\ -2 & 1 & 4 & 1 \end{bmatrix}$.

13. Find the inverses of the matrices in Exercise 12 by the use of their adjoints and determinants.

14. Suppose that a non-singular square matrix M is partitioned in the following manner:

$$M = \begin{bmatrix} A & B \\ C & D \end{bmatrix},$$

where A and D are *square* and *non-singular*. If we partition the inverse in the same manner and denote it by

$$M^{-1} = \begin{bmatrix} E & F \\ G & H \end{bmatrix},$$

then, by the definition of an inverse

$$MM^{-1} = \begin{bmatrix} A & B \\ C & D \end{bmatrix} \begin{bmatrix} E & F \\ G & H \end{bmatrix} = \begin{bmatrix} I & O \\ O & I \end{bmatrix}.$$

(a) Determine the submatrices E, F, G, and H in terms of the submatrices A, B, C and D.

(b) Apply the results in (a) to the inversion of the following matrices:

$$\begin{bmatrix} 1 & 0 & 0 & 0 \\ 0 & 1 & 0 & 0 \\ 0 & 0 & 1 & 0 \\ -1 & -1 & -1 & 1 \end{bmatrix}, \qquad \begin{bmatrix} 1 & 0 & 0 & 1 \\ 0 & 1 & 0 & 1 \\ 0 & 0 & 1 & 1 \\ -1 & -1 & -1 & 0 \end{bmatrix},$$

$$\begin{bmatrix} 1 & 0 & 2 & -1 \\ -1 & -2 & 0 & 2 \\ 3 & 3 & -2 & 4 \\ 2 & -1 & -1 & 3 \end{bmatrix}.$$

15. Given three non-singular matrices A, B and C, all of the same dimensions, prove that (a) $(AB)^{-1} = B^{-1}A^{-1}$, and (b) $(ABC)^{-1} = C^{-1}B^{-1}A^{-1}$.

16. Let A and B be two $n \times n$ non-zero matrices such that $AB = O$. Prove that both A and B must be singular. That is, neither A nor B have an inverse.

17. Show that the inverse of a permutation matrix is its own transpose.

18. Utilizing the results of Exercises 6, 14, 15 and 17, invert the matrix

$$M = \begin{bmatrix} -1 & 1 & 1 & 2 \\ 0 & 1 & 1 & 0 \\ 0 & 2 & 0 & 1 \\ 1 & 0 & 0 & 0 \end{bmatrix}.$$

(Hint: by defining suitable permutation matrices P_1 and P_2, the matrix M may be written as

$$M = P_1 M' P_2 = P_1 \begin{bmatrix} 1 & 0 & 0 & 1 \\ 0 & 1 & 0 & 2 \\ 0 & 0 & 1 & 0 \\ 1 & 2 & -1 & 1 \end{bmatrix} P_2.$$

It is then a simple matter to invert P_1, M', P_2 separately, and to obtain M^{-1} from P_1^{-1}, M'^{-1} and P_2^{-1}.

19. Solve the following simultaneous equations by the matrix-inversion method:

(a)
$$\begin{aligned} x_1 + x_2 &= 2, \\ x_2 + x_4 &= 3, \\ x_3 + x_4 &= 1, \\ -x_1 - x_2 - x_3 &= 0. \end{aligned}$$

(b)
$$\begin{aligned} x_1 - x_2 + x_3 &= 0, \\ 5x_1 + x_3 - x_4 &= 0, \\ -2x_1 + x_2 + x_4 &= 0, \\ x_1 + 3x_2 - x_3 - x_4 &= 1. \end{aligned}$$

(c)
$$\begin{aligned} x_1 - x_2 + 2x_3 - 4x_4 &= 0, \\ x_1 + x_2 + x_3 + x_4 &= -4, \\ x_1 + x_2 + 4x_4 &= -1, \\ 2x_1 + x_3 + 3x_4 &= 2. \end{aligned}$$

20. Determine the values of a and b which will make the following system of linear equations (a) inconsistent, and (b) indeterminate.

$$\begin{aligned} ax_1 + 2x_2 - x_3 &= 1, \\ -x_1 + x_2 &= -2, \\ -4x_1 + bx_2 + x_3 &= -3. \end{aligned}$$

21. Given the following matrices:

$$\begin{bmatrix} 1 & -1 & 1 & 0 \\ 5 & 0 & 1 & -1 \\ -2 & 1 & 0 & 2 \\ 1 & 3 & -1 & 1 \end{bmatrix}, \qquad \begin{bmatrix} 1 & 2 & -1 & 3 \\ 0 & 0 & 1 & 0 \\ 1 & -3 & 0 & -2 \\ 3 & -1 & 0 & 4 \end{bmatrix},$$

(a) Determine which one constitutes a basis.

(b) Express the following vectors as a linear combination of the column vectors in the basis:

$$U = \begin{bmatrix} 3 \\ 2 \\ 1 \\ 0 \end{bmatrix}, \qquad V = \begin{bmatrix} 0 \\ 0 \\ 0 \\ -2 \end{bmatrix}, \qquad W = \begin{bmatrix} 1 \\ -1 \\ 0 \\ 2 \end{bmatrix}.$$

(c) Try the same in terms of the column vectors in the other matrix.

PROPERTIES OF LINEAR PROGRAMMING SOLUTIONS

3.1. The Linear Programming Problem—A Restatement in Matrix Notation.

In consequence of the discussions in Chapter II, we are now in a position to define a linear programming problem in its alternative matrix and vector forms. We recall from Chapter I that such a problem is one of finding a set of values for the unknowns $x_1, x_2, \ldots, x_n$, which

$$\text{maximizes } f(X) = c_1x_1 + c_2x_2 + \ldots + c_nx_n, \tag{3.1.1}$$

subject to the linear structural constraints,

$$\begin{matrix} a_{11}x_1 + a_{12}x_2 + \ldots + a_{1n}x_n \leq b_1, \\ a_{21}x_1 + a_{22}x_2 + \ldots + a_{2n}x_n \leq b_2, \\ \cdot \quad \cdot \quad \cdot \quad \cdot \\ \cdot \quad \cdot \quad \cdot \quad \cdot \\ \cdot \quad \cdot \quad \cdot \quad \cdot \\ a_{m1}x_1 + a_{m2}x_2 + \ldots + a_{mn}x_n \leq b_m, \end{matrix} \tag{3.1.2}$$

as well as the non-negativity constraints,

$$x_1 \geq 0, x_2 \geq 0, \ldots, x_n \geq 0. \tag{3.1.3}$$

Upon inspection of (3.1.1) we realize that $c_1x_1 + c_2x_2 + \ldots + c_nx_n$

is actually the product of a row vector in c's and a column vector in x's. Similarly, the left-hand side of (3.1.2) is the product of a matrix in a's and the column vector in x's. The right-hand side of (3.1.2), in the meantime, can be expressed as an m-component column vector. Thus, we can write the general linear programming problem as follows:

(3.1.4) Maximize $f(X) = CX,$

(3.1.5) subject to constraints $AX \leq P_0,$

(3.1.6) and $X \geq O,$

where:

$$A = \begin{bmatrix} a_{11} & a_{12} & \cdots & a_{1n} \\ a_{21} & a_{22} & \cdots & a_{2n} \\ \cdot & \cdot & & \cdot \\ \cdot & \cdot & & \cdot \\ \cdot & \cdot & & \cdot \\ a_{m1} & a_{m2} & \cdots & a_{mn} \end{bmatrix}, \qquad X = \begin{bmatrix} x_1 \\ x_2 \\ \cdot \\ \cdot \\ \cdot \\ x_n \end{bmatrix},$$

$$P_0 = \begin{bmatrix} b_1 \\ b_2 \\ \cdot \\ \cdot \\ \cdot \\ b_m \end{bmatrix}, \qquad O = \begin{bmatrix} 0 \\ 0 \\ \cdot \\ \cdot \\ \cdot \\ 0 \end{bmatrix} \quad \text{and} \quad C = [c_1, c_2, \ldots, c_n].$$

A third way in which a general linear programming problem can be written is as follows:

(3.1.7) Maximize $f(X) = \sum_{j=1}^{n} c_j x_j,$

(3.1.8) subject to $\sum_{j=1}^{n} x_j P_j \leq P_0,$ where $P_j = \begin{bmatrix} a_{1j} \\ a_{2j} \\ \cdot \\ \cdot \\ \cdot \\ a_{mj} \end{bmatrix}, \quad j = 1, \ldots, n,$

$$P_0 = \begin{bmatrix} b_1 \\ b_2 \\ \cdot \\ \cdot \\ \cdot \\ b_m \end{bmatrix},$$

(3.1.9) and $x_j \geq 0, \qquad j = 1, 2, \ldots, n.$

The equivalence of these three forms of expression is complete and should be firmly established in the mind of the reader. From here on we will use whichever form happens to be most convenient for the purpose at hand. In most cases this will mean the matrix form as stated in (3.1.4), (3.1.5) and (3.1.6).

To solve a linear programming problem such as defined above, it is generally agreed that the simplex method as developed by George B. Dantzig is the best tool available. Its chief advantages are: (1) in application it can be reduced to a few steps of mechanical arithmetic operations governed by a few simple rules—so that even a relatively uninformed person can, by following instructions, work out the correct solution; (2) with suitable modifications, it can be applied to a wide range of other problems which, strictly speaking, are not linear programs.

We will explain in this chapter some significant properties of a general linear programming problem. We will then develop the theoretical and computational aspects of the simplex method in Chapters IV and V respectively.

3.2. Graphical Solution of a Simple Linear Programming Problem.

Before discussing the properties of linear programming solutions, however, let us first take up a simple linear programming problem involving only two variables. Due to its simplicity, we may solve the problem by a graphical method which, though limited in applicability, is very helpful in providing insight into some of the abstract theoretical concepts developed later in connection with the general simplex method.

Consider the following production planning problem.

> The ABC Corporation is a small manufacturing firm faced with the problem of idle capacity in its foundry and machine departments during slack seasons. The company is currently contemplating using these idle capacities to produce two additional products, called A and B, which can be marketed at prevailing prices in any reasonable quantities without difficulty. It is known that product A requires 10 and 5 man-hours per unit in the foundry and machine departments re-

spectively and that product B requires only 6 and 4. However, the profit margin of A is \$30 per unit as compared with \$20 per unit for B. In the week immediately ahead, it is estimated that there will be 1,000 and 600 man-hours available in the foundry and machine departments respectively. How much of A and B should be produced in order to most profitably utilize the excess capacities?

To answer the question, let us first try to formulate the problem in precise mathematical terms. The first task is usually the most difficult step, viz., the selection of decision variables. Fortunately, the present case is an exception. We may let the amounts of A and B to be produced be our decision variables, and call them x_1 and x_2 respectively. Since A yields \$30 profit per unit and B yields \$20 per unit, the objective of maximizing total profits can be stated as:

$$\text{Maximize } f(X) = 30x_1 + 20x_2. \tag{3.2.1}$$

It is realized, however, that the amounts of A and B that can be produced are restricted by the excess capacities available in the foundry and machine departments. Therefore we can write,

$$10x_1 + 6x_2 \leq 1{,}000, \quad \text{and} \tag{3.2.2}$$

$$5x_1 + 4x_2 \leq 600. \tag{3.2.3}$$

In words, the first inequality means that the total number of foundry man-hours used in producing A and B should be less than or equal to the 1,000 man-hours available in the foundry department. Similarly, the second inequality says that the total number of machine man-hours used in producing A and B should be no greater than the 600 man-hours available in the machine department. These two inequalities constitute the so-called structural constraints of the problem. In addition, since production cannot be negative, we require the usual non-negativity constraints

$$x_1, x_2 \geq 0. \tag{3.2.4}$$

With this we have completed the mathetical formulation of the production planning problem.

Since the problem has only two variables, we may show the objective function and the constraints on a 2-dimensional diagram such as Figure 3.1. Line 1 represents the equation $10x_1 + 6x_2 = 1{,}000$. It is a simple matter to verify that all the points to the left of Line 1 satisfy the inequality $10x_1 + 6x_2 < 1{,}000$. For example, $x_1 = 0$ and $x_2 = 0$ is a point (the origin) to the left of Line 1, and satisfies the inequality. Therefore, the entire area to the left of Line 1 including Line 1 itself contains all the values of x_1 and x_2 that satisfy the in-

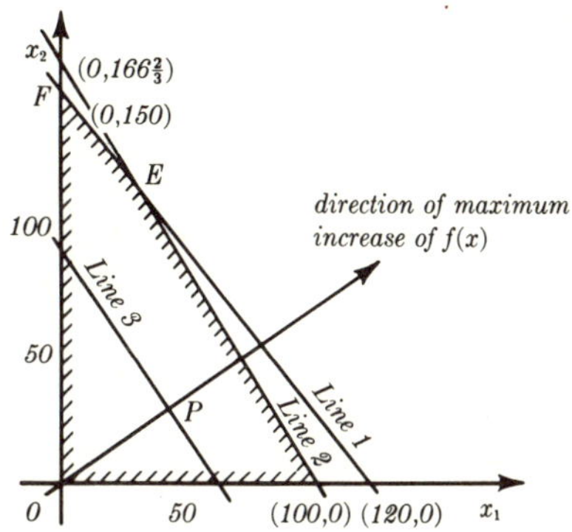

FIGURE 3.1

equality (3.2.2). Similarly, the area to the left of Line 2 ($5x_1 + 4x_2 = 600$), including the line itself, contains all the values of x_1 and x_2 that satisfy the inequality (3.2.3). Further, the non-negativity constraints (3.2.4) stipulate that x_1 and x_2 must lie in the first quadrant including its axis-boundaries. Consequently every pair of x_1 and x_2 that satisfies all four constraints is contained in the polygon defined by the shaded region in Figure 3.1. It should be emphasized that the points on the boundaries of the polygon also satisfy the four constraints stated in (3.2.2), (3.2.3) and (3.2.4). From this graphical analysis we realize that there is an infinite number of solutions that will satisfy all the constraints. The question is which particular solution(s) is the most profitable.

This question, of course, cannot be answered without reference to the objective function, (3.2.1). It is clear that the objective function, not being an equation, cannot be drawn as such on the graph. But, fortunately, what is crucial is not the graphical representation of the function but the direction of change in the value of the function. The latter as we will see presently can be established quickly on the graph and related directly to the polygon of solutions already determined.

To do this, let us arbitrarily assume a specific value, say 1,800, for the objective function, i.e. let $f(X) = 30x_1 + 20x_2 = 1{,}800$. We now have a linear equation which can be graphed. This is Line 3 in Figure 3.1. We want to show that the perpendicular distance between Line 3 and the origin is proportional to the assumed value of 1,800 by a proportionality factor which is a constant for the function under consideration. Once this relationship is established, we can assert that the perpendicular distance from the origin is proportional to the value of that function. Hence the direction of the perpendicular is the direction of the maximum rate of change in the value of that function.

Following the above line of reasoning, let us construct a perpendicular from the origin to Line 3, intersecting it at P. Let the coordinates of P be x_1 and x_2. Since P is on Line 3, $30x_1 + 20x_2 = 1{,}800$. Also, since P is on the perpendicular, OP, $20x_1 - 30x_2 = 0$. Solving these two equations, we obtain

$$x_1 = \frac{30}{30^2 + 20^2}(1{,}800), \qquad x_2 = \frac{20}{30^2 + 20^2}(1{,}800),$$

and the distance from the origin to Line 3 is

$$\overline{OP} = \sqrt{x_1^2 + x_2^2} = \frac{\sqrt{30^2 + 20^2}}{(30^2 + 20^2)}(1{,}800)$$

$$= k(1{,}800) \quad \text{where } k = \frac{\sqrt{30^2 + 20^2}}{(30^2 + 20^2)}.$$

This shows that the perpendicular distance from the origin to Line 3 is proportional to the value of the objective function by a factor k whose value is dependent on the coefficients of x_1 and x_2 in the function and is therefore invariant for a given function. This means that as Line 3 is shifted to the upper right or to the lower left parallel to itself, the value of the function changes in accordance with the lengthening or shortening of its perpendicular distance from the origin. Since both coefficients are positive in this case, moving the line to the upper right increases the value of the objective function, while moving it to the lower left towards the origin decreases the value.* Clearly

* A more expedient method of determining the direction of maximum increase in the value of a *linear* function such as, $f(X) = c_1x_1 + c_2x_2$, is to define a

then, if we want to find a pair of values for x_1 and x_2 which satisfies the constraints, i.e., located within or on the boundaries of the polygon in Figure 3.1, while at the same time maximizing the objective function, we can do so by shifting Line 3 to its upper right (indicated by arrow in Figure 3.1) until it breaks away from the polygon. The last point of contact between the line and the polygon represents the desired optimal solution. In Figure 3.1, this optimal solution is represented by point E which is a corner point of the polygon. Since it is also the intersection of Lines 1 and 2, the coordinates of E can be obtained by solving the two constraints, (3.2.2) and (3.2.3), regarded as equations. Accordingly, by solving $10x_1 + 6x_2 = 1{,}000$ and $5x_1 + 4x_2 = 600$, we obtain $x_1 = 40$ and $x_2 = 100$. Substituting these values into the objective function (3.2.1), we find that the maximum profit is $f(X) = 30x_1 + 20x_2 = 30(40) + 20(100) = 3{,}200$ dollars.

The reader may test his understanding of the graphical solution discussed above by verifying that if the unit profits of A and B are

---

gradient vector, $G = \left[\dfrac{\partial f(X)}{\partial x_1} \quad \dfrac{\partial f(X)}{\partial x_2}\right] = [c_1 \quad c_2]$. If the magnitudes of c_1 and c_2 are great, scale them down for easy graphical construction. For instance, let $f(X) = -15x_1 + 9x_2$. The gradient vector is $G = [-15 \quad 9]$, which is shown graphically below with the arrow pointing at the direction of maximum increase of the value of $f(X)$.

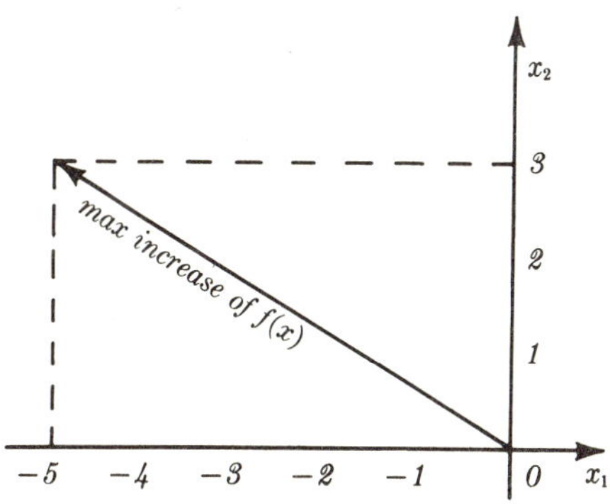

\$20 and \$30 respectively, i.e., $f(X) = 20x_1 + 30x_2$, then the optimal solution is the corner point F in Figure 3.1. This means that the best production plan should consist of no A and 150 units of B for a total profit of \$4,500. On the other hand, if the unit profits are \$30 and \$10 for products A and B respectively, then the optimal production plan would consist of 100 units of A and no B for a total profit of \$3,000. Finally, if the unit profits of A and B are −\$10 and −\$10 (losses), the optimal solution will be the point of the origin which is also a corner point of the polygon. It represents no production of either A or B, hence no profit whatsoever.

Most linear programming problems probably have a unique optimal solution such as the one illustrated in this section. It is not unusual however for more than one solution to be equally optimal from the standpoint of the objective function. In our present example, if the unit profits are \$30 and \$18 for A and B respectively, i.e., if the objective function has the same slope as Line 1, then the boundary line EG becomes optimal. This means any production plan that satisfies Line 1 between E and G is optimal and yields the same maximal profit of \$3,000.

3.3. Convex Sets and Extreme Points.

Although the graphical method explained in the preceding section is ill-adapted for the solution of complex linear programming problems involving more than two variables, it is highly suggestive of the two important properties of linear programming problems which will be discussed in this section. One is that the solutions to the set of constraints of a linear programming problem form a convex set, and the other is that the optimal solution which also maximizes or minimizes the objective function is an extreme point of the set. In terms of the example used in Section 3.2 (Figure 3.1), the multiplicity and convexity of solutions is reflected by the fact that the solution is a polygonal area ($OFEG$) rather than a point, and secondly, the optimal solution is point E which is an extreme point of the convex polygonal area. We will now proceed to a careful explanation of the concepts of convex set and extreme point.

A *set* in mathematics is simply a collection of well-defined objects. It can be a finite set or an infinite set depending on whether the collection is finite or not. For instance, the collective total of all teenagers in the United States at any instant of time is a finite set. On the other hand, the collection of all integers is an infinite set. Furthermore, we refer to sets containing only one element each as *unit sets*, and sets containing no elements at all as *empty sets*.

In linear programming, we use the word "set" to refer primarily to a collection of points (or vectors) in an n-dimensional space (or vector space). Suppose we have two points in a given set,

$$X_1 = \begin{bmatrix} x_{11} \\ x_{21} \\ \cdot \\ \cdot \\ \cdot \\ x_{n-1,1} \\ x_{n1} \end{bmatrix}, \qquad \text{and} \quad X_2 = \begin{bmatrix} x_{12} \\ x_{22} \\ \cdot \\ \cdot \\ \cdot \\ x_{n-1,2} \\ x_{n2} \end{bmatrix}.$$

We define as a *segment* joining X_1 and X_2, the locus of all points X obtained by combining X_1 and X_2 in the following way (cf. Chapter II Exercise 4):

$$X = aX_1 + (1 - a)X_2 \quad \text{where,} \quad 0 \leq a \leq 1. \tag{3.3.1}$$

Using this definition of a segment, we may now define a *convex set* as a set of points which has the property that the segment joining any two points in the set is also entirely in the set. A few specific examples will make these definitions clear.

Example 1. Let S be the set of all real numbers that are greater than or equal to k and smaller than or equal to l. Symbolically, let $S = \{x_i\}$, where $k \leq x_i \leq l$. Then, S is a convex set. To prove this, let us choose any two points in the set, say x_1 and x_2. By definition of

the set S $\quad x_1 \geq k$. $\quad$ Hence, $\quad ax_1 \geq ak$, where $0 \leq a \leq 1$.
Also $\quad x_2 \geq k$. $\quad$ Hence, $\quad (1 - a)x_2 \geq (1 - a)k$.
Therefore, $\quad ax_1 + (1 - a)x_2 \geq ak + (1 - a)k$.

We see that the left-hand side of the above inequality is any point on the segment defined in (3.3.1), and the right-hand side is simply k. Therefore, any point on the segment joining x_1 and x_2 is greater than or equal to k. Similarly, we can show that it is smaller than or equal to l. From these two results we conclude that any point on the segment joining x_1 and x_2 belongs to the set S. Hence the whole segment belongs to S, and we have proved that S is a convex set. Such a set may be represented graphically as in Figure 3.2 (a). It is a continuous line segment terminating at k on the left and at l on the right. In terms of the diagram, the segment defined by $ax_1 + (1 - a)x_2$ is simply the line segment between x_1 and x_2, which is clearly contained in the line segment between k and l. Therefore S is a convex set.

Example 2.

Let us now define a slightly more complicated set S which is a collection of points in the 2-dimensional space. These points, designated X_1, X_2, etc., are all two-component column vectors that satisfy the following matrix inequality:

$$(3.3.2) \qquad AX_i \leq P_0, \quad \text{where} \quad A = \begin{bmatrix} 1 & 1 \\ -1 & 1 \\ 0 & -1 \end{bmatrix}, \qquad X_i = \begin{bmatrix} x_{1i} \\ x_{2i} \end{bmatrix},$$

$$P_0 = \begin{bmatrix} 8 \\ 2 \\ -2 \end{bmatrix}, \qquad \text{and } i = 1, 2, \text{etc.}$$

Then, this set S is again a convex set.

To prove this, let us observe first of all that (3.3.2) is nothing more than a simpler way of writing three linear inequalities:

$$\begin{aligned} x_{1i} + \quad x_{2i} &\leq 8, \\ -x_{1i} + \quad x_{2i} &\leq 2, \\ -x_{2i} &\leq -2. \end{aligned}$$

When we interpret these inequalities graphically as shown in Figure 3.2 (b) we find that they represent simply the triangle $P_1P_2P_3$. The set S defined by (3.3.2) is thus simply the triangular area $P_1P_2P_3$. Since an area contains an infinite number of points, S is an infinite set. Once S is identified geometrically, it becomes a simple matter to show that S is convex. Select any two points in the triangle, such as X_1 and X_2. The segment joining X_1 and X_2 as defined by (3.3.1) is simply the line segment between them, which obviously is contained entirely in the triangle. Hence, S is a convex set. The algebraic proof is more involved but parallels exactly the proof offered in Example 1 and so will not be repeated. However, the reader is encouraged to work it out in order to test his grasp of the concept of convex set.

---

* Since x_{1i} and x_{2i} are the elements of any vector X_i, the second subscript i may be dropped in graphing the inequalities.

Figure 3.2 (c) offers an example of a non-convex set in 2-dimensional space. The star-shaped area is not a convex set for the simple reason that if we choose any pair of points in the set, the line segment joining them does not necessarily fall entirely within the area, e.g., X_1 and X_2 in Figure 3.2(c). As long as such an exception can be found, the definition of a convex set remains unsatisfied and the set is a non-convex set.

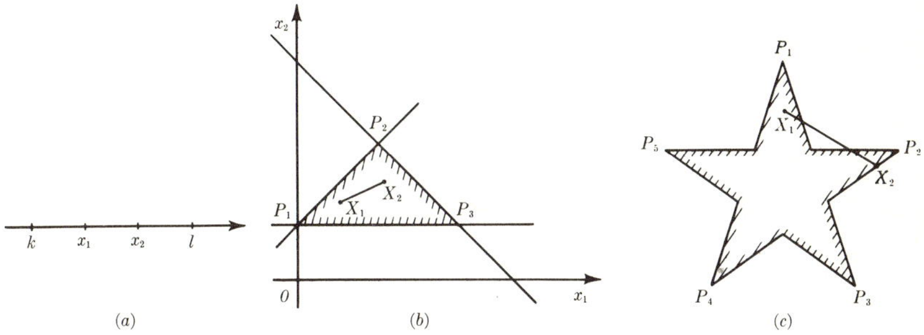

FIGURE 3.2

By relating the abstract concept of convex set to geometrical figures as we have done in Examples 1 and 2 we hope we have smoothed the path for the reader to acquire at least an intuitive understanding of this important concept. But we must hasten to point out possible pitfalls. For instance, from the above examples, the reader may have acquired the impression that the segment as defined in (3.3.1) is always a section of a straight line. This is obviously inappropriate since a line has physical meaning only in 2- or 3-dimensional space. But the concept of convex set is perfectly general for n-dimensional space. It is, therefore, important always to orient one's thinking on convex set toward its general algebraic definition stated in (3.3.1).

We may at this point hint at the connection between convex set and linear programming problems. Referring to (3.3.2), we notice that the matrix inequality represents the same thing as the system of linear constraints found in all linear programming problems. Since any point X_i in the set discussed in Example 2 must satisfy $AX_i \leq P_0$, it is, in fact, a solution to the system of constraints, and therefore a solution to the linear programming problem. Since we have already indicated that the set consisting of these points is convex, it follows that the set consisting of all solutions to the constraints of a linear programming problem is convex. This is the first of the important properties of linear programming problems we mentioned in the opening paragraph of this section. On the other hand, a moment's reflection will reveal that the star in Figure 3.2 (c), which is not a convex set, cannot be a solution set of a linear programming problem. The reason is simply that such a solution set cannot be defined by a system of linear inequalities such as (3.1.2). For instance, the points X_1 and X_2 in the set are on opposite sides of the line passing through P_5, P_2; hence, cannot be covered simultaneously by one inequality.

Finally, it remains for us to define *extreme points*. In the light of the earlier definition of a segment, we may define an extreme point of a convex set as a point in the set which is not on any segment joining two other points in the set. Referring to (a) and (b) in Fig. 3.2, k, l, P_1, P_2, and P_3 are extreme points of their respective convex sets. Algebraically, the definition means that if X is an extreme point, and X_2 is any other point in the set, then, the only point X_1 we can find to satisfy $X = aX_1 + (1 - a)X_2$ (3.3.1) is $X_1 = X$ with $a = 1$. Geometrically speaking, in a convex set defined by linear inequalities, an extreme point in a 2-dimensional space is the intersection of two lines; in a 3-dimensional space it is the intersection of three planes. In general, in an n-dimensional space with $n > 3$, the extreme points of such a convex set would be intersections of n *hyperplanes*—a hyperplane being a linear equation of the type, $a_1x_1 + a_2x_2 + \ldots + a_nx_n = b$.

3.4. Convex Combination.

Convex combination is simply a special form of linear combination. We have already seen that a vector X in an n-dimensional space can always be expressed in terms of a set of n independent vectors (in the same space) as follows: (cf. Chapter II, Section 2.13)

$$X = a_1X_1 + a_2X_2 + \ldots + a_nX_n, \tag{3.4.1}$$

where $X_1, X_2, \ldots, X_n$ represent the set of n independent vectors and

$a_1, a_2, \ldots, a_n$ are scalars or coefficients. When (3.4.1) holds, we say the vector X is a linear combination of the set of vectors, $X_1, X_2, \ldots, X_n$ in the n-dimensional space. To go one step further, if the coefficients $a_1, a_2, \ldots, a_n$ are all non-negative and add up to one, then we say X is a convex combination of the set of independent vectors, $X_1, X_2, \ldots, X_n$. In other words, *convex combination* holds if

$$(3.4.2) \qquad X = a_1X_1 + a_2X_2 + \ldots + a_nX_n,$$

$$\text{where} \quad a_1, a_2, \ldots, a_n \geq 0,$$

$$\text{and} \quad \sum_{i=1}^{n} a_i = 1.$$

The significance of convex combination in relation to convex set is that given a convex set S with extreme points $E_1, E_2, \ldots, E_n$, all the points in the set S can be expressed as a convex combination of its extreme points. Let us illustrate with the following simple example. Consider the quadrangle with extreme points E_1, E_2, E_3, and E_4 shown in Figure 3.3. It is clearly a convex set because a segment linking any

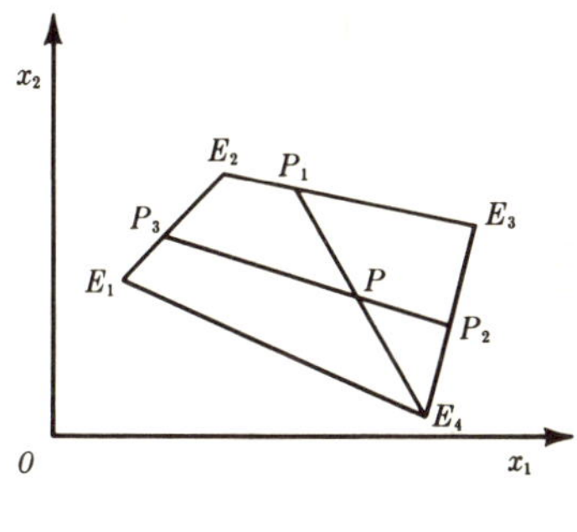

FIGURE 3.3

two points in the quadrangle will lie wholly in it. Now, select any point P in the set. We will show that such a point can be expressed as a convex combination of the extreme points of the set. First, pass a line through P intersecting two sides of the quadrangle at P_2 and P_3 as shown. Then, P is on the segment linking P_2 and P_3; hence is a convex combination of P_2 and P_3 according to (3.3.1). Thus,

$$(3.4.3) \qquad P = a_2P_2 + a_3P_3, \quad \text{where} \quad a_2, a_3 \geq 0, \quad \text{and} \quad a_2 + a_3 = 1.$$

In turn, we observe that P_2 is on the segment linking E_3 and E_4, and P_3 is on the segment linking E_1 and E_2. Therefore we can write

$$(3.4.4) \qquad P_2 = e_3E_3 + e_4E_4, \quad \text{where} \quad e_3, e_4 \geq 0, \quad \text{and} \quad e_3 + e_4 = 1;$$

$$(3.4.5) \qquad P_3 = e_1E_1 + e_2E_2, \quad \text{where} \quad e_1, e_2 \geq 0, \quad \text{and} \quad e_1 + e_2 = 1.$$

Substituting (3.4.4) and (3.4.5) into (3.4.3) we have

$$(3.4.6) \qquad P = a_2e_3E_3 + a_2e_4E_4 + a_3e_1E_1 + a_3e_2E_2,$$

which shows P as a linear combination of the four extreme points with non-negative coefficients. If we can prove that these coefficients add up to one, we can further assert that P is a convex combination of the four extreme points. To do this, we note that

$$\begin{aligned} a_2e_3 + a_2e_4 + a_3e_1 + a_3e_2 &= a_2(e_3 + e_4) + a_3(e_1 + e_2) \\ &= a_2(1) + a_3(1) = 1. \end{aligned}$$

Therefore, P is a convex combination of the extreme points E_1, E_2, E_3, and E_4.

It will be noted that (3.4.6) is not the only convex combination for P. Referring to Figure 3.3, we see that if the segment E_4P_1 is drawn through P, we can express P as a convex combination of E_4 and P_1. But P_1 is a convex combination of E_2 and E_3. Therefore P can be expressed as a convex combination of the three extreme points E_2, E_3, and E_4.

3.5. Classification and Properties of Linear Programming Solutions.

We have seen that a linear programming problem actually consists of two parts: (1) finding the solutions that satisfy the constraints, and (2) selecting from these solutions the one that also maximizes (or minimizes) the objective function. Thus, there is a need for distinguishing between the several kinds of solutions we encounter in a linear programming problem.

Referring to the matrix formulation of linear programming problems as shown in (3.1.4), (3.1.5) and (3.1.6) in Section 3.1, we define the following three types of solutions:

a. Feasible solution—an n-component column vector X that satisfies (3.1.5) and (3.1.6), or a set of specific values of the n variables, none negative, which satisfies the inequality constraints contained in (3.1.5).

b. Basic feasible solution—a feasible solution that has no more than m positive components.
c. Optimal solution—either a feasible solution or a basic feasible solution that also optimizes the value of the linear objective function.

Of the above, (b) and (c) will be discussed in the next section. In the remainder of this section, we will try to develop some generalizations on the properties of (a)—the feasible solutions.

The most significant property of the set of feasible solutions to a linear programming problem is that they form a convex set. We have already indicated by a 2-dimensional example in Section 3.3 that the set of points X_i satisfying the matrix inequation $AX_i \leq P_0$ (3.3.2), forms a convex set. The constraints (3.1.5) of a linear programming problem are of exactly the same type. The non-negativity constraints (3.1.6) can be rewritten as $-X \leq 0$, and therefore again are of the same type. Thus, the two sets of constraints (3.1.5) and (3.1.6) can be combined into one matrix inequality:

$$A^*X \leq P_0^*, \tag{3.5.1}$$

$$\text{where} \quad A^* = \begin{bmatrix} A \\ -I \end{bmatrix}, \quad X = \begin{bmatrix} x_1 \\ \cdot \\ \cdot \\ \cdot \\ x_n \end{bmatrix}, \quad \text{and} \quad P_0^* = \begin{bmatrix} P_0 \\ 0 \end{bmatrix}.$$

Thus any X that satisfies (3.5.1) is a feasible solution of the original linear programming problem and all such solutions together constitute a convex set.

It is a relatively simple matter to prove the convexity of the linear programming feasible solution set. Assume that X_1 and X_2 are two arbitrary feasible solutions of (3.5.1) in the sense that

$$A^*X_1 \leq P_0^*, \quad \text{and} \quad A^*X_2 \leq P_0^*.$$

Let X by any convex combination of X_1 and X_2, i.e.

$$X = a_1X_1 + a_2X_2, \quad \text{where} \quad a_1, a_2 \geq 0, \quad \text{and} \quad a_1 + a_2 = 1.$$

Then, $\quad A^*X = A^*(a_1X_1 + a_2X_2) = a_1A^*X_1 + a_2A^*X_2.$

Substituting $a_2 = 1 - a_1$ into the above equation, we have

$$A^*X = a_1A^*X_1 + (1 - a_1)A^*X_2 \leq a_1P_0^* + (1 - a_1)P_0^*.$$

Or, $\quad A^*X \leq P_0^*.$

The above inequality simply means that X is also a point in the set of feasible solutions. Since X_1 and X_2 are any two points in the set of feasible solutions, the set is a convex set† (cf. Sections 3.3 and 3.4). This is a very important characteristic of the set of feasible solutions to a linear programming problem and must be thoroughly grasped.

3.6. Extreme-point Solution and Basic Feasible Solution of a Linear Programming Problem.

Since the set of feasible solutions of a linear programming problem is a convex set, according to our discussion in Section 3.4, all feasible solutions must be capable of being expressed as a convex combination of the extreme points of the set. Let us, therefore, turn our attention to the problem of identifying these extreme-point solutions.

It has been previously suggested that the extreme points of the convex set defined by linear inequalities are intersections of pairs of lines in a 2-dimensional space and intersections of sets of three planes in a 3-dimensional space. Lines and planes of course are linear equations in two and three variables respectively. For example, in the production planning problem discussed in Section 3.2, a feasible solution is a point in the set of (x_1, x_2) defined by inequalities (3.2.2), (3.2.3) and (3.2.4). Such a set, we have already seen in Section 3.5, must be a convex set. Graphically it is the polygon $OFEG$ in Figure 3.1. The corner points of the polygon, O, F, E, G, are clearly the extreme points of the convex set since none of these points lie on segments linking two other points in the set. Furthermore, these points represent intersections of two adjacent boundary lines of the polygon, e.g., O is the intersection of the two lines, $x_1 = 0$ and $x_2 = 0$, which are respectively the equality parts of the inequalities $x_1 \geq 0$ and $x_2 \geq 0$ (3.2.4). Similarly, E is the intersection of Line 1 and Line 2 which are, respectively, the equality parts of the inequalities (3.2.2) and (3.2.3); F is the intersection of Line 2 and $x_1 = 0$; and finally G is the intersection of Line 1 and $x_2 = 0$. Thus, our earlier statement concerning the extreme points of a convex set defined by linear inequalities in 2-dimensional space is verified. Consequently, we can

† Such a set is called a convex polyhedral set or, equivalently, *convex polyhedron* in mathematics. A polyhedron is simply a many-sided solid in n-dimensional space bounded by surfaces called hyperplanes which are the geometrical counterparts of linear equations in n variables.

set forth the following working procedure for the determination of the extreme-point solutions to a *2-variable* linear programming problem with m linear inequality constraints.

First, write down all the constraints including the non-negativity constraints in their equality forms. There will be $m + 2$ equations which barring redundancy represent the boundary lines of the polygonal convex set of feasible solutions.

Second, obtain all possible points of intersection by solving simultaneously two of the $m + 2$ equations at a time in all possible combinations.

Third, check each solution against the original set of constraints (two of these will be automatically satisfied since they are responsible for the particular solution). If it satisfies all the constraints including the non-negativity ones, it is, then, an extreme point of the convex polygonal solution set and, hence, an extreme-point solution. This step is essential because although all extreme points are intersections of two boundary lines, the converse is not true.

The significance of the above procedure does not, however, lie in its use in the determining of all extreme-point solutions to a linear programming problem. It is significant rather in its contribution to our better understanding of the nature of extreme-point solutions and through that the development of a powerful and efficient method of attacking linear programming problems.

The above discussion of a 2-variable m-constraint problem can be readily generalized and applied to an n-variable m-constraint linear programming problem in the following manner. We first realize that the total number of constraints in the generalized problem including the non-negativity constraints is $m + n$. An extreme-point solution of this problem, therefore, must be the simultaneous solution of n of the $m + n$ constraints taken as equations.* However, this procedure is not satisfactory because the feasibility of the solution remains to be confirmed by a check against the other m constraints not employed in reaching the solution. It would be highly desirable if we could

---

* The proof, which is omitted here, in essence is as follows: if the solution were not an extreme point, it could be expressed as a convex combination of two other extreme points. But one of the latter cannot be a solution because some of the constraints are bound to be violated. Hence, the contradiction and the proof.

devise a procedure which would automatically reveal whether the m remaining constraints are satisfied. In the following paragraphs we will trace the development of such a procedure leading to the formation of the concept of basic feasible solution as defined in Section 3.5 above.

First, we note that an inequality of the type,

$$a_{11}x_1 + a_{12}x_2 + \ldots + a_{1n}x_n \leq b_1, \tag{3.6.1}$$

can be converted into an equation,

$$a_{11}x_1 + a_{12}x_2 + \ldots + a_{1n}x_n + x_{n+1} = b_1, \tag{3.6.2}$$

by means of adding the *non-negative* variable x_{n+1} (called *slack* variable). The effect of this change (and its advantage) is that the somewhat indefinite relationship represented by the sign $\leq$, is made explicit and precise by the value which the slack variable x_{n+1} assumes. A positive x_{n+1} in (3.6.2) implies that the left-hand side of (3.6.1) is smaller than the right-hand side; a zero x_{n+1} means that the two sides are equal; and finally a negative x_{n+1} indicates violation of (3.6.1). Hence we convert all inequalities representing structural constraints into equations as follows:

$$\text{Maximize } f(X) = c_1x_1 + c_2x_2 + \ldots + c_nx_n, \tag{3.6.3}$$

subject to:

$$\begin{array}{lllll} a_{11}x_1 + a_{12}x_2 + \ldots + a_{1n}x_n & + x_{n+1} & & & = b_1, \\ a_{21}x_1 + a_{22}x_2 + \ldots + a_{2n}x_n & & + x_{n+2} & & = b_2, \\ \vdots & & & & \vdots \\ a_{m1}x_1 + a_{m2}x_2 + \ldots + a_{mn}x_n & & & + x_{n+m} & = b_m, \end{array} \tag{3.6.4}$$

(3.6.5) and

$$x_j \geq 0, \qquad j = 1, 2, \ldots, n, n+1, \ldots, n+m.$$

Two important consequences follow: (1) any violation of a structural constraint will automatically be revealed by a negative solution-value of its associated slack variable, and (2) the use of any structural constraint to determine an extreme-point solution must be associated with the fact that its corresponding slack variable is zero. Recalling that an extreme-point solution is the result of solving simultaneously n constraints—in general, say, $k(\leq m)$ constraints from the structural group and $n - k$ from the non-negativity group—in their equality form, we realize now that such a solution must have k slack variables and $n - k$ structural variables equal to zero. Since there are altogether $n + m$ variables in (3.6.4), and $k + (n - k) = n$ of them are zero, we arrive at the important conclusion that an extreme-point solution is simply a simultaneous solution of the m equations in (3.6.4) in m variables. Such a solution is, of course, what we have defined as a basic feasible solution.

Mathematically, the requirement that n variables, say $x_{m+1}, \ldots, x_{m+n}$, have to be assumed equal to zero reduces the constraint system in (3.6.4) to the following:

(3.6.6) $$P_1x_1 + P_2x_2 + \ldots + P_mx_m = P_0,$$

$$\text{where} \quad P_0 = \begin{bmatrix} b_1 \\ b_2 \\ \cdot \\ \cdot \\ \cdot \\ b_m \end{bmatrix} \quad \text{and} \quad P_j = \begin{bmatrix} a_{1j} \\ a_{2j} \\ \cdot \\ \cdot \\ \cdot \\ a_{mj} \end{bmatrix}, \quad j = 1, \ldots, m.$$

Or, in matrix form

(3.6.7) $$B\tilde{X} = P_0 \quad \text{where} \quad B = [P_1\, P_2 \ldots P_m] \quad \text{and} \quad \tilde{X} = \begin{bmatrix} x_1 \\ x_2 \\ \cdot \\ \cdot \\ \cdot \\ x_m \end{bmatrix}.$$

Now, the question is: is B a basis? In other words, are the vectors P_1, $P_2, \ldots, P_m$ linearly independent? To answer this, let us recall that an extreme point by definition (Section 3.31) is not on any segment joining two other points in the set. An extreme-point solution, therefore, cannot be expressed as a convex combination of two other solutions; consequently it must be a unique solution of (3.6.7). From this we infer that B must be a set of m linearly independent vectors. For if it were not, (3.6.7) would have either no solution or an infinite number of solutions. Neither case agrees with our original assumption of an extreme-point solution. Thus we arrive at a second important conclusion, i.e., all extreme-point solutions of a linear programming problem are associated with a set of m independent P vectors.

Based on the two conclusions obtained above, we may confidently engage in the search for extreme-point solutions by selecting m independent P vectors from the m structural constraints and solving the resulting equation system as indicated by (3.6.7). These independent vectors which comprise a basis are known as *basis vectors*, and the variables with which they are associated, *basis variables*. Of course, there are the questions of how to select the m vectors which are independent of each other, and how to predict that they will yield a solution that is feasible even if they are independent. These questions will be answered when we discuss the simplex method in the next chapter.

Although the importance of extreme-point solutions is clearly visible in Figure 3.1, in the next section, we shall further justify the close attention given them here by proving their relationship with the optimal solution. In the meantime, let us consider the following numerical example so as to strengthen our grasp of the concepts developed in this section.

Example. Determine the extreme-point solutions of the following linear programming problem.

$$\text{Maximize } f(X) = 2x_1 - x_2 + 3x_3,$$

$$\text{subject to:} \quad x_1 + 4x_2 + x_3 \leq 6, \tag{1}$$

$$x_1 + x_2 + x_3 \leq 3, \tag{2}$$

$$-2x_1 + 6x_2 - x_3 \leq 4, \tag{3}$$

$$\text{and} \quad x_1 \geq 0, \tag{4}$$

$$x_2 \geq 0, \tag{5}$$

$$x_3 \geq 0. \tag{6}$$

Since we are not concerned with the optimal solution in this example, we will overlook the objective function and write the constraints (1) through (3) in the following modified form:

$$x_1 + 4x_2 + x_3 + x_4 = 6, \tag{7}$$

$$x_1 + x_2 + x_3 + x_5 = 3, \tag{8}$$

$$-2x_1 + 6x_2 - x_3 + x_6 = 4; \tag{9}$$

or equivalently

$$P_1x_1 + P_2x_2 + P_3x_3 + P_4x_4 + P_5x_5 + P_6x_6 = P_0, \tag{10}$$

where

$$P_1 = \begin{bmatrix} 1 \\ 1 \\ -2 \end{bmatrix}, \quad P_2 = \begin{bmatrix} 4 \\ 1 \\ 6 \end{bmatrix}, \quad P_3 = \begin{bmatrix} 1 \\ 1 \\ -1 \end{bmatrix},$$

$$P_4 = \begin{bmatrix} 1 \\ 0 \\ 0 \end{bmatrix}, \qquad P_5 = \begin{bmatrix} 0 \\ 1 \\ 0 \end{bmatrix}, \qquad P_6 = \begin{bmatrix} 0 \\ 0 \\ 1 \end{bmatrix}, \qquad P_0 = \begin{bmatrix} 6 \\ 3 \\ 4 \end{bmatrix}.$$

According to our discussion, since there are three variables in the original problem, an extreme-point solution can be obtained by solving three of the six constraints [from (1) through (6)] as equalities.

(a) Let us first try constraints (1), (2) and (6). We have, then, the following equations:

$$\begin{aligned} x_1 + 4x_2 + x_3 &= 6, \\ x_1 + x_2 + x_3 &= 3, \\ x_3 &= 0, \end{aligned}$$

which has the solution: $x_1 = 2$, $x_2 = 1$, $x_3 = 0$. Since these answers are non-negative and satisfy (3) also, they are indeed an extreme-point solution. This is point E in Figure 3.4.

Using the alternative approach, we realize that the assumption that (1), (2) and (6) are the limiting constraints implies that x_4, x_5 and x_3 respectively are zero. Hence, from (7), (8) and (9), we obtain:

$$\begin{aligned} x_1 + 4x_2 \qquad &= 6, \\ x_1 + x_2 \qquad &= 3, \\ -2x_1 + 6x_2 + x_6 &= 4. \end{aligned}$$

The solution to this set of equations is $x_1 = 2$, $x_2 = 1$ and $x_6 = 2$. The basic feasible solution is therefore $x_1 = 2$, $x_2 = 1$, $x_3 = x_4 = x_5 = 0$, $x_6 = 2$. This agrees with the extreme-point solution obtained earlier. To save space, in the ensuing illustrations we shall write a basic feasible solution as the transpose of the column vector X. For this particular illustration therefore, $X^T = [2\ 1\ 0\ 0\ 0\ 2]$.

(b) Assume that inequalities (1), (3) and (6) are the limiting constraints. We have

$$\begin{aligned} x_1 + 4x_2 + x_3 &= 6, \\ -2x_1 + 6x_2 - x_3 &= 4, \\ x_3 &= 0, \end{aligned}$$

which yield the solution $x_1 = \frac{10}{7}$, $x_2 = \frac{8}{7}$, and $x_3 = 0$. These answers also satisfy the remaining constraints, i.e., (2), (4) and (5), and hence represent an extreme-point solution (point D in Fig. 3.4).

Using the basic feasible solution approach, we recognize that the above specific assumption regarding limiting constraints implies that $x_4 = x_6 = x_3 = 0$ in equations (7), (8) and (9). Hence we have:

$$\begin{aligned} x_1 + 4x_2 \qquad &= 6, \\ x_1 + \ x_2 + x_5 &= 3, \\ -2x_1 + 6x_2 \qquad &= 4, \end{aligned}$$

which yield the solution $x_1 = \frac{10}{7}$, $x_2 = \frac{8}{7}$, and $x_5 = \frac{3}{7}$. Since they are all non-negative, the basic feasible solution is $X^T = [\frac{10}{7}\ \frac{8}{7}\ 0\ 0\ \frac{3}{7}\ 0]$, which agrees with the extreme-point solution obtained earlier.

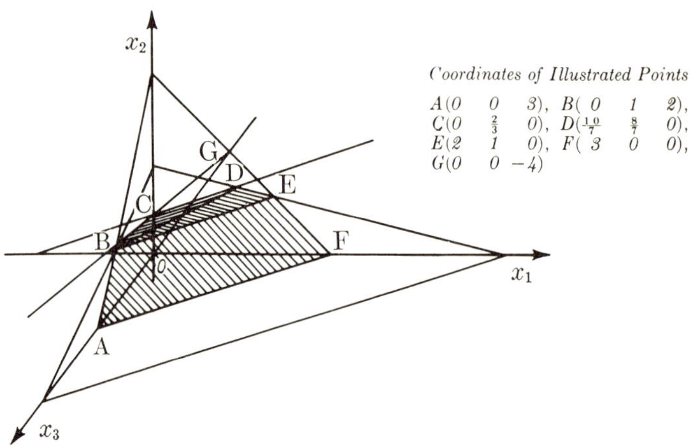

FIGURE 3.4

(c) Assume that (1), (2) and (3) are the limiting constraints. We have:

$$\begin{aligned} x_1 + 4x_2 + x_3 &= 6, \\ x_1 + \ x_2 + x_3 &= 3, \\ -2x_1 + 6x_2 - x_3 &= 4, \end{aligned}$$

which give the solution $x_1 = 0$, $x_2 = 1$ and $x_3 = 2$. Since they are all non-negative, (4), (5) and (6) are also satisfied, and the solution is an extreme-point solution (point B in Figure 3.4).

Following the basic feasible solution approach, we realize that the assumption made above implies that $x_4 = x_5 = x_6 = 0$ in (7), (8) and (9). Thus we end up with the same equations and the same solution as above. The basic feasible solution in this instance is $X^T = [0\ 1\ 2\ 0\ 0\ 0]$.

Notice that this particular solution has only two positive components instead of three as in the case of the two preceding solutions. Such a solution, i.e., a basic feasible solution with less positive components than the number of structural constraints, is called a *degenerate basic feasible solution.* Upon investigation we find that P_0 can be expressed as a linear combination of only two of the three basis vectors implicit in the above equations, viz., $P_0 = P_2 + 2P_3$, which is the cause of the degeneracy.

(d) If we assume now that (2), (3) and (4) are the limiting constraints, the following equation system is obtained:

$$\begin{aligned} x_1 + x_2 + x_3 &= 3, \\ -2x_1 + 6x_2 - x_3 &= 4, \\ x_1 &= 0, \end{aligned}$$

yielding the solution $x_1 = 0$, $x_2 = 1$ and $x_3 = 2$. Since they are all non-negative, constraints (5) and (6) are satisfied. Substituting these values into (1), we see that (1) is also satisfied. Hence, the solution is an extreme-point solution which is the same point B in Fig. 3.4 as obtained in (c).

Following the basic feasible solution approach, we realize that the specific assumption made here implies that $x_5 = x_6 = x_1 = 0$ in (7), (8) and (9). We have, therefore,

$$\begin{aligned} 4x_2 + x_3 + x_4 &= 6, \\ x_2 + x_3 &= 3, \\ 6x_2 - x_3 &= 4, \end{aligned}$$

and the solution is $x_2 = 1$, $x_3 = 2$ and $x_4 = 0$. Since all of them are non-negative, the basic feasible solution is $X^T = [0\ 1\ 2\ 0\ 0\ 0]$. Notice that (c) and (d) yield the same extreme-point solution. But the basis implicit in (c) is $[P_1\ P_2\ P_3]$, whereas in (d) it is $[P_2\ P_3\ P_4]$. Thus, a degenerate basic feasible solution is associated with more than one basis.

(e) To illustrate that the above procedure does not necessarily yield an extreme-point solution (or a basic feasible solution), let us assume that (3), (4) and (5) are the limiting constraints. We have:

$$\begin{aligned} -2x_1 + 6x_2 - x_3 &= 4, \\ x_1 \qquad\qquad &= 0, \\ x_2 \qquad &= 0. \end{aligned}$$

These equations yield the solution $x_1 = x_2 = 0$, $x_3 = -4$. The last answer violates constraint (6). Therefore, it is not an extreme-point solution. In Fig. 3.4, this solution is shown as point G.

Following the alternative approach, we realize that the above assumption means that $x_1 = x_2 = x_6 = 0$. Thus, we have, from (7), (8) and (9),

$$\begin{aligned} x_3 + x_4 \qquad &= 6, \\ x_3 \qquad + x_5 &= 3, \\ -x_3 \qquad\qquad &= 4, \end{aligned}$$

which yield the same infeasible solution $x_3 = -4$, $x_4 = 10$ and $x_5 = 7$.

(f) Finally, consider the assumption that (1), (2) and (5) are the limiting constraints. We have:

$$\begin{aligned} x_1 + 4x_2 + x_3 &= 6, \\ x_1 + x_2 + x_3 &= 3, \\ x_2 \qquad &= 0, \end{aligned}$$

which have no solution because of contradiction implied in the equations.

Following the basic feasible solution approach, we realize that the assumption means that $x_4 = x_5 = x_2 = 0$. Hence, (7), (8) and (9) become

$$\begin{aligned} x_1 + x_3 \qquad &= 6, \\ x_1 + x_3 \qquad &= 3, \\ -2x_1 - x_3 + x_6 &= 4, \end{aligned}$$

which again produce no solution as the first two equations contradict each other. This situation can only result from linear dependence among the vectors selected to form the basis. In the present case, the vectors selected are P_1, P_3 and P_6 [cf. (10)] which are obviously related in the following manner, $P_1 = P_3 - P_6$.

It is hoped that the foregoing illustrations have firmly convinced the reader of the one-to-one correspondence between the extreme-point solutions and the basic feasible solutions. By (c) and (d) we also hope to have illustrated the nature and cause of degenerate basic feasible solutions to linear programming problems. The haphazard nature of the procedures employed here, however, was made evident by (e) and (f) where no feasible solution was found. These failures serve to underscore the importance of developing a more efficient procedure which will automatically by-pass such dead-end situations. The simplex method is just such a procedure. It will, therefore, form the main body of our discussion in the next chapter.

3.7. Optimal Solution and Basic Feasible Solution.

Why are we so interested in basic feasible solutions? The answer, briefly, is that the optimal solution of a linear programming problem can be found among its basic feasible solutions. We shall prove this by showing that if an optimal solution is not a basic feasible solution, then there must be some basic feasible solution that is equally optimal.

According to Section 3.5, an optimal solution is a set of values of the unknowns, designated by the n-component column vector X which satisfies $AX \leq P_0$ (3.1.5) and $X \geq 0$ (3.1.6), and, at the same time optimizes the value of the objective function $f(X) = CX$ (3.1.4). For our proof, we will assume that optimization means maximization although the proof is equally applicable, *mutatis mutandis*, to a case of minimization.

First we recall from Section 3.4 that any point in a convex set can be expressed as a convex combination of the extreme points of the set. Since as we have shown in Section 3.5 that the solutions to a linear programming problem form a convex set, a non-extreme-point solution (or a non-basic feasible solution) can be expressed as a convex combination of the extreme-point solutions (or basic feasible solutions) to the problem.

Now, let us assume that X_0 is a maximal solution. It can be, of course, either a basic feasible solution or a non-basic feasible solution. If it is the former, the assertion in the preceding paragraph is obviously true and requires no proof. If it is the latter, then we can write*

$$X_0 = a_1X_1^e + a_2X_2^e + \ldots + a_tX_t^e, \tag{3.7.1}$$

$$\text{where} \quad 0 \leq a_i \leq 1,$$

$$\sum_{i=1}^{t} a_i = 1,$$

$$X_i^e = \text{extreme-point solutions,}$$

$$\text{and} \quad i = 1, 2, \ldots, t.$$

* See Gass, *op. cit.*, pp. 39–41.

Since $f(X)$ is a linear function

$$(3.7.2) \qquad \begin{aligned} f(X_0) &= f(a_1X_1^e + a_2X_2^e + \ldots + a_tX_t^e) \\ &= a_1f(X_1^e) + a_2f(X_2^e) + \ldots + a_tf(X_t^e). \end{aligned}$$

Of these $f(X_i^e)$ let us assume that $f(X_M^e)$ is the largest. Due to the condition that a_i are all no less than zero,

$$(3.7.3) \qquad \begin{aligned} & a_1f(X_1^e) + a_2f(X_2^e) + \ldots + a_tf(X_t^e) \\ & \qquad \leq a_1f(X_M^e) + a_2f(X_M^e) + \ldots + a_tf(X_M^e) \\ & \qquad \leq (a_1 + a_2 + \ldots + a_t)f(X_M^e). \end{aligned}$$

But as stated in (3.7.1), $(a_1 + a_2 + \ldots + a_t) = 1$. Consequently, by substituting (3.7.2) into (3.7.3) we obtain

$$(3.7.4) \qquad f(X_0) \leq f(X_M^e).$$

However, since $f(X_0)$ is assumed to be maximum, it cannot be less than $f(X_M^e)$. Hence, $f(X_0) = f(X_M^e)$. Thus, we have proved that if X_0 is optimal but not basic, there is at least one basic feasible solution (X_M^e) that is equally optimal. Hence, the assertion that the optimal solution can always be found among the basic feasible solutions is valid in all cases.

This conclusion has most important implications. It means that the search for an optimal solution (or solutions) to a linear programming problem can be confined to a search among its basic feasible solutions. The task would be impossible otherwise since there are in general an infinite number of solutions to this type of problem—a generalization based on the convex nature of its solution set. On the other hand, it has only a finite number of basic feasible solutions. Thus, by eliminating all non-basic feasible solutions from our concern, the task is reduced to manageable proportions. Even then, the number of basic feasible solutions may be so large that it would be impracticable to evaluate all of them. Fortunately, the simplex method makes it unnecessary to cover all of them because it always leads to a better basic feasible solution until the optimal is reached. The development of this method will be the concern of the next two chapters.

Example. Although it is possible for a non-basic feasible solution to be an optimal solution, in general, this is not true. We will illustrate this point by the following example. First we arbitrarily select a non-basic feasible solution. Then, we determine two other feasible solutions of which it is a convex combination and show that one of the two will yield a better value of the objective function then the initial solution (barring the exception that all three of them yield the same value). Finally, we show that there is a basic feasible solution which gives a higher value of the objective function than the initial solution.

We will use the numerical example in Section 3.6, which has the objective function $f(X) = 2x_1 - x_2 + 3x_3$. By solving the indeterminate linear system (7), (8) and (9), we obtain a non-basic feasible solution $X_0^T = [\frac{10}{9}\ \frac{7}{9}\ \frac{4}{9}\ \frac{12}{9}\ \frac{6}{9}\ \frac{18}{9}]$. Let X_0 be a convex combination of X_1 and X_2, two other feasible solutions. We have

$$\begin{aligned} X_0 &= a_1 X_1 + (1 - a_1)X_2, \qquad 0 < a_1 < 1, \\ AX_1 &= P_0, \\ AX_2 &= P_0, \end{aligned}$$

where A and P_0 are obtained from (7), (8) and (9) in Section 3.6,

$$A = \begin{bmatrix} 1 & 4 & 1 & 1 & 0 & 0 \\ 1 & 1 & 1 & 0 & 1 & 0 \\ -2 & 6 & -1 & 0 & 0 & 1 \end{bmatrix} \quad \text{and} \quad P_0 = \begin{bmatrix} 6 \\ 3 \\ 4 \end{bmatrix}.$$

The three matrix equations above contain 12 linear equations (6 in the first matrix equation, 3 each in the second and third) and 13 variables (a_1 and 6 components each of X_1 and X_2). By assuming $a_1 = \frac{1}{3}$, we obtain the solutions $X_1^T = [\frac{4}{3}\ \frac{1}{3}\ \frac{2}{3}\ \frac{8}{3}\ \frac{2}{3}\ \frac{16}{3}]$ and $X_2^T = [1\ 1\ \frac{1}{3}\ \frac{2}{3}\ \frac{2}{3}\ \frac{1}{3}]$ which are both feasible. Substituting these solutions into $f(X)$, we find $f(X_0) = 2(\frac{10}{9}) - (\frac{7}{9}) + 3(\frac{4}{9}) = \frac{25}{9}$, which is smaller than $f(X_1) = 2(\frac{4}{3}) - (\frac{1}{3}) + 3(\frac{2}{3}) = \frac{13}{3}$. In other words, we have found a feasible solution that is better in terms of the value of the objective function than the arbitrarily obtained initial solution.

To find a basic feasible solution which also yields a higher value than the initial non-basic feasible solution, let us try point B in Figure 3.4 whose coordinates have been determined in part (d) of the example in Section 3.5. They are $x_1 = 0$, $x_2 = 1$, $x_3 = 2$, $x_4 = x_5 = x_6 = 0$. The corresponding value of the objective function is, therefore, $0 - 1 + 3(2) = 5$ which is, of course, higher.

EXERCISES

1. Referring to the production planning problem described in Section 3.2; if in addition to the foundry and machine departments, the welding department is also involved as follows:

Product A requires $\frac{1}{4}$ man-hour of welding per unit
Product B requires $\frac{1}{10}$ man-hour of welding per unit

and if the available capacity of the welding department is 20 man-hours, what then is the most profitable combination of A and B that can be produced? Does this optimal product-mix make full use of the available welding capacity? If not, how much is still available for other uses?

2. Solve Exercise 2 of Chapter I graphically.

3. Draw on graph paper the polygon defined by the following inequalities:

$$\begin{aligned} 2x_1 + 3x_2 &\leq 12, \\ x_1 - 6x_2 &\leq 1, \\ x_1 + x_2 &\geq 2, \\ -2x_1 + x_2 &\leq 1. \end{aligned}$$

(a) Write the above system in matrix form.
(b) Ascertain algebraically whether the following points satisfy the matrix inequality in (a):

$$X_1 = \begin{bmatrix} x_1 \\ x_2 \end{bmatrix} = \begin{bmatrix} 1 \\ 2 \end{bmatrix}, \qquad X_2 = \begin{bmatrix} 2 \\ 4 \end{bmatrix}, \qquad X_3 = \begin{bmatrix} 4 \\ 1 \end{bmatrix}.$$

Note their graphical positions relative to the polygon.

(c) What does the polygon represent? Is it a convex set? Why?

4. Given the following inequalities:

$$\begin{aligned} 3x_1 - 2x_2 &\leq 12, \\ x_1 - 6x_2 &\leq 1, \\ -x_1 + 2x_2 &\leq 4, \\ x_1 \qquad &\geq 0, \\ x_2 &\geq 0; \end{aligned}$$

(a) determine their extreme-point solutions algebraically;
(b) verify these solutions graphically;
(c) verify both algebraically and graphically that the convex combination of any two extreme-point solutions is also a solution.
(d) find the extreme-point solutions whose convex combinations are the non-extreme-point solutions: (1) $X = \begin{bmatrix} 4 \\ 2 \end{bmatrix}$, (2) $X = \begin{bmatrix} 6 \\ 3 \end{bmatrix}$.

5. Work out the algebraic proof of Example 2 in Section 3.3.

6. Convert the inequalities in Exercise 4 into an equation system.

(a) Identify all bases that may be obtained with the coefficient vectors (P_j vectors) in this system.
(b) Which of these bases yield a basic feasible solution?
(c) Establish the correspondence between these basic feasible solutions and the extreme-point solutions obtained in Exercise 4.

7. Assume the following constraint system of a simple linear programming problem:

$$P_1x_1 + P_2x_2 + P_3x_3 + P_4x_4 = P_0,$$

where: $P_1 = \begin{bmatrix} 2 \\ 1 \end{bmatrix}, \quad P_2 = \begin{bmatrix} 1 \\ 3 \end{bmatrix}, \quad P_3 = \begin{bmatrix} 1 \\ 0 \end{bmatrix},$

$$P_4 = \begin{bmatrix} 0 \\ 1 \end{bmatrix}, \quad \text{and} \quad P_0 = \begin{bmatrix} 4 \\ 3 \end{bmatrix}.$$

(a) Find the solution associated with each of the following bases: $[P_1\, P_2]$, $[P_1\, P_3]$ and $[P_2\, P_4]$.
(b) Plot the basis vectors and P_0 on graph paper for each of the above bases.
(c) What significant relationship exists between P_0 on the one hand and the set of basis vectors on the other, when the solution is feasible?

8. Suppose we write the constraint system in (3.6.4) as follows:

$$P_1x_1 + P_2x_2 + \ldots + P_nx_n + P_{n+1}x_{n+1} + \ldots + P_{n+m}x_{n+m} = P_0$$

$$\text{where:} \quad P_1 = \begin{bmatrix} a_{11} \\ a_{21} \\ \cdot \\ \cdot \\ \cdot \\ a_{m1} \end{bmatrix}, \ldots, P_{n+1} = \begin{bmatrix} 1 \\ 0 \\ \cdot \\ \cdot \\ \cdot \\ 0 \end{bmatrix}, \ldots; P_0 = \begin{bmatrix} b_1 \\ b_2 \\ \cdot \\ \cdot \\ \cdot \\ b_m \end{bmatrix}.$$

Prove that if a feasible solution exists, then there is a certain constant k such that kP_0 may be expressed as a convex combination of $P_1, \ldots, P_{n+m}$.

9. Given four extreme-points of a convex polygon: $\begin{bmatrix} 2 \\ 3 \end{bmatrix}, \begin{bmatrix} 6 \\ 1 \end{bmatrix}, \begin{bmatrix} 1 \\ -4 \end{bmatrix},$ $\begin{bmatrix} -2 \\ -2 \end{bmatrix},$

(a) find the linear inequalities that define the polygon;

(b) find the convex combination of the four extreme points that yields (1) $\begin{bmatrix} 2 \\ 2 \end{bmatrix}$, (2) $\begin{bmatrix} 3 \\ 0 \end{bmatrix}$, (3) $\begin{bmatrix} 0 \\ 0 \end{bmatrix}$ and (4) $\begin{bmatrix} 4 \\ 2 \end{bmatrix}$.

10. Try to define the non-convex star-shaped area shown in Figure 3.2 by a set of linear inequalities. Can it be done? In what way does it reinforce the theorem that the solutions to a set of linear inequalities form a convex set?

11. Determine the basic feasible solutions of the following linear programming problem:

$$\begin{aligned} \text{Maximize } f(X) &= x_1 + x_2, \\ \text{subject to:} \quad & x_1 + 4x_2 - x_3 - x_4 = 4, \\ & 2x_1 - x_2 + x_3 + 2x_4 = 6, \\ & -x_1 + x_2 + 2x_3 = 0, \\ \text{and} \quad & x_j \geq 0, \qquad j = 1, 2, 3, 4. \end{aligned}$$

12. Show that a solution that maximizes the linear function $f(X)$ also maximizes the function $f(X) + c$, where c is a constant.

13. Take the five inequalities in Exercise 4 as constraints.
    (a) Find by graphical means the solutions that maximize $f(X) = x_1 + x_2$ subject to a progressively smaller set of the original five constraints, obtained by successively eliminating the last one, until finally only the first one is left.
    (b) What significant conclusion can you suggest regarding the relationship between the number of constraints and the maximal value of $f(X)$?

THE SIMPLEX METHOD—THEORETICAL

4.1. The Approach of the Simplex Method.

In the previous chapter, we have proved, explained and demonstrated by examples two important properties of solutions to linear programming problems. One is that these solutions, which must be non-negative, together form a convex set. The other is that the optimal solution can be found among the extreme points of the convex set (or a basic feasible solution). The simplex method is a scheme which systematically evaluates these extreme-point solutions in terms of the objective function until a terminal status is reached. This occurs when either (a) a finite optimal solution is found, (b) an infinite optimal solution is positively identified, or (c) conclusive evidence is obtained that the problem has no feasible solution. The whole process lends itself to such precise and complete description at all points that once a problem is soundly conceived and correctly structured, no further judgement will be required for its ultimate solution. This is an important advantage because it makes possible the use of the digital computer to carry out the lengthy and tedious computational details.

The approach of the simplex method may be briefly described as follows. Since each basic feasible solution is associated with a basis (cf. Chapter III Section 3.6), the simplex method starts with a basis that yields a basic feasible solution. It then moves on to a new basis associated with a better basic feasible solution by means of replacing one of the current basis vectors with a non-basis vector. This non-basis vector is selected because of its potential contribution to the improvement of the objective function, which can be determined beforehand. On the other hand, the current basis vector selected to be replaced is determined according to a rule which insures the continued feasibility of the new solution. Once the new basis (and the better basic feasible solution) is fully established, it forms the new starting point for repeating the same evaluative process to determine a still better basic feasible solution. This continues until one of the three terminal statuses previously mentioned is reached. Thus, the simplex method is an iterative method calling for the repeated application of the same set of rules in each stage of improvement.

It will be the main concern of this chapter to develop these rules —two decision rules and three transformation rules. The next chapter will be devoted to the problems of their systematic application.

4.2. Generating New Basic Feasible Solutions—Derivation of Decision Rule I.

As explained in Section 3.6 of the previous chapter, a linear programming problem can be written, in general, in the following modified form:

(4.2.1) $$\text{Maximize } f(X) = CX,$$

(4.2.2) $$\text{subject to: } P_1x_1 + P_2x_2 \ldots + P_nx_n + P_{n+1}x_{n+1} + \ldots + P_{n+m}x_{n+m} = P_0,$$

and $x_j \geq 0$, $j = 1, \ldots, n, n+1, \ldots, n+m$, where $x_{n+1}, \ldots, x_{n+m}$ and $P_{n+1}, \ldots, P_{n+m}$ are respectively the slack variables and their corresponding slack vectors.

Assume that a basis $B = [P_{(1)}\, P_{(2)} \ldots P_{(m)}]$ exists, where $P_{(1)}$, $P_{(2)}, \ldots, P_{(m)}$, represent an independent subset of the P vectors in (4.2.2), such that

(4.2.3) $$\theta_{(1)}P_{(1)} + \theta_{(2)}P_{(2)} + \ldots + \theta_{(m)}P_{(m)} = P_0 \quad \text{and} \quad \theta_{(i)} \geq 0.$$

Then, $x_{(1)} = \theta_{(1)}$, $x_{(2)} = \theta_{(2)}$, . . . , $x_{(m)} = \theta_m$ with the remaining n variables equal to zero represent a basic feasible solution.

Attention is called to the use of parentheses around the subscripts of vectors and variables in the basis. The number in the parentheses refers to the order of the vector in the basis and not to the order in the structural constraints. For instance, $P_{(1)}$ means the first vector in the basis and not P_1, the first vector in the constraints. $P_{(1)}$ may be P_3 in one basis and P_4 in another. In general, any $P_{(i)}$, $(i) = (1), (2), \ldots, (m)$ may be any P_j, $j = 1, 2, \ldots, n + m$. Whenever we need to refer to anything related to the basis in this chapter and in Chapter V and VI, we shall use the subscript with parentheses.

Now, let us return to our main topic. Since B is a basis, any of the non-basis vectors, say P_k, can be expressed as a linear combination of the basis vectors in B (cf. Section 2.13 of Chapter II). Thus

$$P_k = y_{(1)k}P_{(1)} + y_{(2)k}P_{(2)} + \ldots + y_{(m)k}P_{(m)} = BY_k, \qquad Y_k = \begin{bmatrix} y_{(1)k} \\ y_{(2)k} \\ \cdot \\ \cdot \\ \cdot \\ y_{(m)k} \end{bmatrix}, \tag{4.2.4}$$

or $Y_k = B^{-1}P_k$.

(cf. Example 4, Section 2.18, Chapter II). We will show in the following how a new basis and a new basic feasible solution can be obtained by the inclusion of this vector P_k in place of some existing basis vector $P_{(i)}$ selected to maintain feasibility of the solution.

Multiplying through **(4.2.4)** by an undetermined positive scalar θ'_k, and then adding the resulting equation to (4.2.3), we obtain

$$\theta_{(1)}P_{(1)} + \theta_{(2)}P_{(2)} + \ldots + \theta_{(m)}P_{(m)} + \theta'_kP_k = P_0 + \theta'_ky_{(1)k}P_{(1)} + \theta'_ky_{(2)k}P_{(2)} + \ldots + \theta'_ky_{(m)k}P_{(m)}$$

which after suitable transposing and regrouping yields

$$(\theta_{(1)} - \theta'_ky_{(1)k})P_{(1)} + (\theta_{(2)} - \theta'_ky_{(2)k})P_{(2)} + \ldots + (\theta_{(m)} - \theta'_ky_{(m)k})P_{(m)} + \theta'_kP_k = P_0 \tag{4.2.5}$$

Obviously, an infinite number of solutions could be obtained from the above by assigning different values to θ'_k.* But we must bear in mind

* θ'_k must be non-negative; otherwise it means that $x_k = \theta'_k$ will be negative, making the solution infeasible. On the other hand, θ'_k may be zero if the original basic feasible solution is already a degenerate one.

that the optimal solution is a basic feasible solution in linear programming. Therefore, we must reduce the $m + 1$ terms on the left-hand side of (4.2.5) to m terms since a basic feasible solution has no more than m positive values among the n variables. This means that one of the coefficients of the existing basis vectors, $P_{(1)}, P_{(2)}, \ldots, P_{(m)}$ has to be made zero. As to which coefficient will vanish, obviously the answer depends upon the non-negative value assigned to θ'_k. If $\theta'_k = 0$, all the coefficients will be positive and we have, in fact, the same solution as before. As θ'_k increases we find that $(\theta_{(i)} - \theta'_k y_{(i)k})$ increases or decreases, depending on whether $y_{(i)k}$ is negative or positive. Since only a decrease in the value of $\theta_{(i)}$ presents a threat to the condition of feasibility (non-negativity), we can concentrate on the positive $y_{(i)k}$'s and determine the minimal θ'_k which will make some $(\theta_{(i)} - \theta'_k y_{(i)k})$ the first to become zero. Let us assume that this occurs for $(i) = (l)$. Then

$$(4.2.6) \qquad \theta'_k = \frac{\theta_{(l)}}{y_{(l)k}} = \text{minimum}\left\{\frac{\theta_{(i)}}{y_{(i)k}}\right\}$$

for all $y_{(i)k} > 0$ Decision Rule I.

Once the value of θ'_k is determined, the new values of the basis variables (designated $\theta'_{(i)}$) can readily be computed and we have a new basic feasible solution, viz., $x_{(1)} = \theta'_{(1)}, \ldots, x_{(l)} = \theta'_{(l)}, \ldots, x_{(m)} = \theta'_{(m)}$, and the remaining n variables all equal to zero where

$$(4.2.7) \qquad \theta'_{(l)} = \theta'_k = \frac{\theta_{(l)}}{y_{(l)k}};$$

$$(4.2.8) \qquad \theta'_{(i)} = \theta_{(i)} - \frac{\theta_{(l)}}{y_{(l)k}} y_{(i)k} \quad \text{for } (i) = (1), \ldots, (m); (i) \neq (l).$$

Since the new solution is a basic feasible solution, it again satisfies (4.2.3) except for the replacement of $P_{(l)}$ by P_k. The following example should make the above discussion clear.

Example. We will use the familiar example in Section 2.6, Chapter II to illustrate Decision Rule I. Equations (7), (8) and (9) of the said example are reproduced below. The objective function is omitted since optimization is not our aim in this section.

$$\begin{aligned} x_1 + 4x_2 + x_3 + x_4 &= 6, \\ x_1 + x_2 + x_3 + x_5 &= 3, \\ -2x_1 + 6x_2 - x_3 + x_6 &= 4. \end{aligned}$$

These equations can be more conveniently presented in the vector form,

$$P_1x_1 + P_2x_2 + P_3x_3 + P_4x_4 + P_5x_5 + P_6x_6 = P_0$$

where $P_1, \ldots, P_6$ are column vectors composed of the coefficients of the variables $x_1, \ldots, x_6$ respectively and P_0, the vector of the constants.

It has already been shown in Figure 3.4 that point F is an extreme-point solution. The coordinates of F are $x_1 = 3$, $x_2 = x_3 = 0$. Substituting these values into the three equations above, we obtain $x_4 = 3$, $x_5 = 0$ and $x_6 = 10$. Hence, the basic feasible solution that corresponds to F is $X^T = [3\ 0\ 0\ 3\ 0\ 10]$ which is associated with the basis $B = [P_1\, P_4\, P_6]$.* If we desire a new basic feasible solution which includes x_2 as a variable in the basis, i.e. to have P_2 included in the basis, we must then use Decision Rule I (4.2.6) to determine which vector currently in the basis will have to be replaced.

To be able to apply Decision Rule I, the linear-combination coefficients of P_2 must be found. To determine the latter by (4.2.4), in turn, requires the inverse of the basis. This then will be our starting point.

$$B = [P_1\, P_4\, P_6] = \begin{bmatrix} 1 & 1 & 0 \\ 1 & 0 & 0 \\ -2 & 0 & 1 \end{bmatrix}.$$

This matrix can be easily inverted by (2.17.2) or other suitable methods and is found to be,

$$B^{-1} = \begin{bmatrix} 0 & 1 & 0 \\ 1 & -1 & 0 \\ 0 & 2 & 1 \end{bmatrix}.$$

By (4.2.4) then,

$$Y_2 = \begin{bmatrix} y_{12} \\ y_{42} \\ y_{62} \end{bmatrix} = B^{-1}P_2 = \begin{bmatrix} 0 & 1 & 0 \\ 1 & -1 & 0 \\ 0 & 2 & 1 \end{bmatrix} \begin{bmatrix} 4 \\ 1 \\ 6 \end{bmatrix} = \begin{bmatrix} 1 \\ 3 \\ 8 \end{bmatrix}.$$

Since all the coefficients are positive, according to Decision Rule I three ratios will have to be calculated. These are:

* In this and subsequent numerical examples, basis vectors and variables are directly identified by the subscript j. Bracketed subscripts (i) are not used since they merely serve to describe the order of the vectors in the basis, which is unnecessary in the examples.

$$\frac{\theta_1}{y_{12}} = \frac{3}{1} = 3, \quad \frac{\theta_4}{y_{42}} = \frac{3}{3} = 1, \quad \frac{\theta_6}{y_{62}} = \frac{10}{8} = \frac{5}{4}.$$

As the second ratio turns out to be the smallest, P_4 must be the basis vector to be replaced by P_2. Hence, according to (4.2.6), $x_2 = \theta_2' = 1$, and according to (4.2.8), $x_1 = \theta_1' = \theta_1 - \frac{\theta_4}{y_{42}} y_{12} = 3 - 1(1) = 2$, and similarly $x_6 = \theta_6' = 10 - 1(8) = 2$. Thus, the new basic feasible solution is $X^T = [2\ 1\ 0\ 0\ 0\ 2]$ which is associated with the new basis $B' = [P_1\ P_2\ P_6]$ and corresponds to extreme point E in Figure 3.4.

Let us try another example by including P_3 instead of P_2 in the original basis and observe how Decision Rule I determines the vector to be replaced in order that a different basic feasible solution with a positive x_3 may be found. Since the inverse of the old basis $B = [P_1\ P_4\ P_6]$ has already been obtained, we may directly solve for Y_3 as follows,

$$Y_3 = \begin{bmatrix} y_{13} \\ y_{43} \\ y_{63} \end{bmatrix} = B^{-1}P_3 = \begin{bmatrix} 0 & 1 & 0 \\ 1 & -1 & 0 \\ 0 & 2 & 1 \end{bmatrix} \begin{bmatrix} 1 \\ 1 \\ -1 \end{bmatrix} = \begin{bmatrix} 1 \\ 0 \\ 1 \end{bmatrix}.$$

Two ratios now have to be determined since there are two positive coefficients in Y_3. These are:

$$\frac{\theta_1}{y_{13}} = \frac{3}{1} = 3 \quad \text{and} \quad \frac{\theta_6}{y_{63}} = \frac{10}{1} = 10.$$

As the first ratio turns out to be the smallest, P_1 must be the vector to be replaced by P_3. Hence, according to (4.2.7), $x_3 = \theta_3' = 3$, and according to (4.2.8), $x_4 = \theta_4' = 3 - 3(0) = 3$, and $x_6 = \theta_6' = 10 - 3(1) = 7$. Thus, the new basic feasible solution is $X^T = [0\ 0\ 3\ 3\ 0\ 7]$ which is associated with the new basis $B' = [P_3\ P_4\ P_6]$ and corresponds to the extreme-point A in Figure 3.4.

What would have happened if we had acted contrary to Decision Rule I and picked P_6 (associated with the larger ratio) to be replaced by P_3? If we did that, the new solution would be, according to (4.2.7) and (4.2.8),

$$x_3 = \theta_3' = \frac{\theta_6}{y_{63}} = \frac{10}{1} = 10,$$

$$x_1 = \theta_1' = 3 - 10(1) = -7,$$
$$x_4 = \theta_4' = 3 - 10(0) = 3.$$

Whereas the above solution does satisfy the original equations, clearly it is non-feasible since x_1 is now negative. Hence, the smallest ratio must always be used.

The above examples show how we may obtain new basic feasible solutions from an old one through a transformation of the basis involving the replacement of one basis vector (the outgoing vector) by a non-basis vector (the incoming vector). The outgoing vector is selected according to Decision Rule I (4.2.6) which is based primarily on feasibility considerations. But as has been shown in the example, the outgoing vector varies depending on the vector being brought into the basis. Thus, a rule has to be developed to guide the selection of the incoming vector. Such a rule (Decision Rule II) will be derived in Section 4.5, based on the consideration of improving the value of the objective function.

Before we move on to the next section, let us pause momentarily to consider one question: is the set of vectors in the new basis necessarily independent after the substitution? If it is not, then the new solution may not be an extreme-point solution, and our attempt at a systematic evaluation of extreme-point solutions in order to arrive at the optimal solution may be side-tracked. Fortunately, the answer to the above question is in the affirmative, i.e., independence of the new set of vectors is always insured by the rule we use to select the outgoing vector. Suppose we have originally a set of independent vectors $\{P_{(1)}, \ldots, P_{(m)}\}$ forming a basis, and we introduce P_k, a non-basis vector, into the basis to replace $P_{(l)}$, a basis vector determined by Decision Rule I (4.2.6). The only way linear dependence can result from such a substitution is if P_k happens to be linearly dependent on some subset of the basis vectors that does not include $P_{(l)}$. This implies that $y_{(l)k} = 0$ in the expression of P_k as a linear combination of the m basis vectors, viz.,

$$(4.2.9) \qquad P_k = y_{(1)k}P_{(1)} + y_{(2)k}P_{(2)} + \ldots + 0 \cdot P_{(l)} + \ldots + y_{(m)k}P_{(m)}$$

If $y_{(l)k}$ is zero, however, according to Decsion Rule I which considers only positive $y_{(i)k}$, $P_{(l)}$ will never be replaced. Hence the new set of vectors will always be independent, always form a basis and yield a basic feasible solution that is an extreme-point solution of a linear programming problem, so long as some $y_{(i)k}$ is positive.

4.3. First Exception to Decision Rule I—The Case of Unbounded Solutions.

Decision Rule I (4.2.6) concerning the selection of basis vector to be ejected from the basis is clearly applicable only when there are some $y_{(i)k}$ in (4.2.4) that are positive. If they are all zero or negative, then we can assign any arbitrary positive value to θ'_k and obtain from (4.2.8) larger positive values for those basis variables whose $y_{(i)k}$ are negative, and no change in values for those whose $y_{(i)k}$ are zero. If we let θ'_k be positive infinity, then some or all of the basis variables will also be positive infinity. This means that feasibility no longer is a problem as in the case when some $y_{(i)k}$ are positive. It also means that a feasible solution (non-basic) with some variables assuming infinity can be obtained such that the value of the objective function will be infinity [either positive or negative depending on the c_j in the objective function (3.1.1)]. Usually in such cases there are $m + 1$ positive variables in an infinite solution, but it is possible that there may be only m or less if the solution is degenerate.* We will illustrate this special situation by the following simple example.

Example. Solve the problem

$$\begin{aligned} \text{Maximize } f(X) &= 2x_1 + 3x_2, \\ \text{subject to:} \quad x_1 - x_2 &\leq 2, \\ 3x_1 + x_2 &\geq 9, \\ \text{and} \quad x_1, x_2 &\geq 0. \end{aligned}$$

First, we convert the constraints into equations by adding a slack variable x_3 to the first structural constraint and subtracting a slack variable x_4 from the second structural constraint which is of the "$\geq$" type. Thus, we have,

$$\begin{aligned} x_1 - x_2 + x_3 \qquad &= 2, \\ 3x_1 + x_2 \qquad - x_4 &= 9, \end{aligned}$$

which can be written as,

$$P_1x_1 + P_2x_2 + P_3x_3 + P_4x_4 = P_0,$$

where,

* Referring to (4.2.8), if for $i = h$, $\theta_{(h)} = 0$ and $y_{(h)k} = 0$, then $\theta'_{(h)}$ will always be zero regardless of the value assigned to θ'_k. Hence it is possible that an infinite solution may contain m (or even less than m) positive variables.

$$P_1 = \begin{bmatrix} 1 \\ 3 \end{bmatrix}, \quad P_2 = \begin{bmatrix} -1 \\ 1 \end{bmatrix}, \quad P_3 = \begin{bmatrix} 1 \\ 0 \end{bmatrix}, \quad P_4 = \begin{bmatrix} 0 \\ -1 \end{bmatrix}, \quad P_0 = \begin{bmatrix} 2 \\ 9 \end{bmatrix}$$

As explained in Section 3.6, Chapter III, to obtain a basic feasible solution, let us assume that x_1 and x_2 are the basis variables and $x_3 = x_4 = 0$. We have then

$$P_1x_1 + P_2x_2 = P_0,$$

and the basis $B = [P_1\, P_2]$. To solve the above for x_1 and x_2, we first determine the inverse of the basis by (2.17.2), which gives

$$B^{-1} = \begin{bmatrix} 1 & -1 \\ 3 & 1 \end{bmatrix}^{-1} = \begin{bmatrix} \frac{1}{4} & \frac{1}{4} \\ -\frac{3}{4} & \frac{1}{4} \end{bmatrix}.$$

The solution is therefore $B^{-1}P_0$, or $x_1 = \frac{11}{4}$, $x_2 = \frac{3}{4}$.

If we choose P_4 to come into the basis, we then find

$$Y_4 = B^{-1}P_4 = \begin{bmatrix} \frac{1}{4} & \frac{1}{4} \\ -\frac{3}{4} & \frac{1}{4} \end{bmatrix} \begin{bmatrix} 0 \\ -1 \end{bmatrix} = \begin{bmatrix} -\frac{1}{4} \\ -\frac{1}{4} \end{bmatrix}, \text{ or } y_{14} = -\tfrac{1}{4},\ y_{24} = -\tfrac{1}{4}.$$

Since both are negative, neither the value of x_1 nor of x_2 will decrease as x_4 is introduced at a positive level. Consequently, we fail to obtain a basic feasible solution in either x_1 and x_4 or x_2 and x_4. But instead, we can obtain a non-basic feasible solution in x_1, x_2 and x_4 as follows: Let $x_4' = \theta_4' = 10$. Then according to (4.2.8), $x_1 = \theta_1' = \theta_1 - \theta_4' y_{14} = \frac{11}{4} - 10(-\frac{1}{4}) = \frac{21}{4}$. Similarly, $x_2 = \theta_2' = \theta_2 - \theta_4' y_{24} = \frac{3}{4} - 10(-\frac{1}{4}) = \frac{13}{4}$. On the other hand, if we let x_4 be positive infinity, then x_1 and x_2 will also be positive infinity. Therefore, $f(X)$ which is $2x_1 + 3x_2$ will likewise be infinity, and the problem has no finite maximal solution but an infinite maximal value of positive infinity.

The fact that a linear programming problem admits of solutions of this type, i.e. having some variables assume a value of positive infinity, suggests that its solution set, though convex, is not closed on all sides. This open convex solution set makes it possible for the linear objective function to be "pushed" indefinitely in the direction of the opening, and ultimately to assume a value of either positive or negative infinity depending on the coefficients in the function. Such a problem in linear programming is known as an unbounded problem. It is important to realize, however, that an open convex solution set does not necessarily mean that the objective function must ultimately assume an infinite value. It also depends on the c_j. For instance, maximizing an objective function with negative c_j over an open solution set will yield, nevertheless, a finite maximum.*

A graphic solution of the simple illustrative example above will make the concept of "unboundedness" clear. In Figure 4.1, the limit-

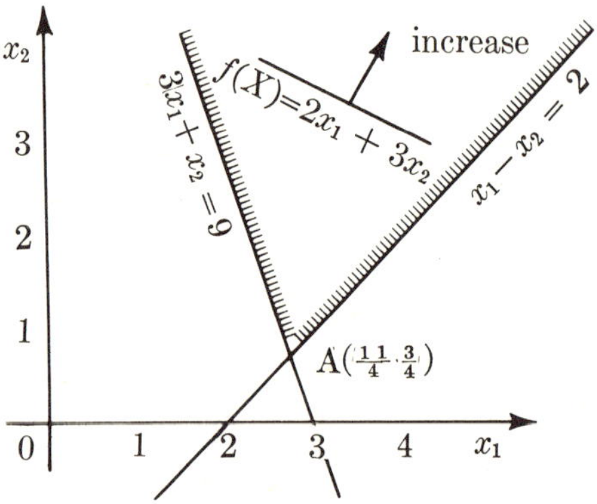

FIGURE 4.1

ing positions of the two structural constraints are represented by the two lines marked $x_1 - x_2 = 2$ and $3x_1 + x_2 = 9$. The convex polygonal solution set is indicated by the partially shaded area that opens at the top. The objective function is shown as the line $f(X) = 2x_1 + 3x_2$, with a gradient vector pointing in the direction of maximum increase. Clearly, then, $f(X)$ can be shifted indefinitely upward to the right and still remain in the solution set. Thus the maximal value of $f(X)$ is positive infinity. But the minimal value is finite as indicated by the extreme point A.

4.4. Second Exception to Decision Rule I—The Case of Degenerate Solutions.

The application of Decision Rule I calls for the search for the minimum $\left\{\frac{\theta_{(i)}}{y_{(i)k}}\right\}$ (4.2.6) in order to identify the basis vector $P_{(l)}$ to be

* For additional discussion on the question of "unboundedness," see Charnes, Cooper, and Henderson, *An Introduction to Linear Programming* (New York: John Wiley & Sons, Inc., 1953) pp. 61–62.

replaced. It is quite possible, however, that no unique minimum exists. In other words, there may be two or more basis vectors for which the ratios $\frac{\theta_{(i)}}{y_{(i)k}}$ are equally minimal. When this happens, if one of these basis vectors is arbitrarily chosen to be eliminated in the next basic feasible solution, the variables associated with the rest of these basis vectors (having the same ratios) will all assume the value of zero. For let us assume that $\frac{\theta_{(l)}}{y_{(l)k}} = \frac{\theta_{(h)}}{y_{(h)k}} = \min \left\{\frac{\theta_{(i)}}{y_{(i)k}}\right\}$ for $y_{(i)k} > 0$. Then, $\theta_{(h)} = \frac{\theta_{(l)}}{y_{(l)k}} y_{(h)k}$. If $P_{(l)}$ is arbitrarily chosen to be replaced, then according to (4.2.7),

$$\theta'_{(h)} = \theta_{(h)} - \frac{\theta_{(l)}}{y_{(l)k}} y_{(h)k} = 0.$$

The result will be a basic feasible solution with less than m (the number of basis vectors) positive components, and has been referred to in Chapter III Section 3.6 as a *degenerate basic feasible solution.* The cause of degeneracy lies in the relationship between the requirement vector P_0 and the structural and slack vectors P_j. From Equation (3.6.6) in Chapter III Section 3.6 we see that the nature of a basic feasible solution is that

$$P_0 = x_{(1)}P_{(1)} + x_{(2)}P_{(2)} + \ldots + x_{(m)}P_{(m)}.$$

If one of the basis variables, say $x_{(m)}$ is equal to zero as in the case of a degenerate basic feasible solution, then P_0 can be expressed as a linear combination of $m - 1$ basis vectors—a relationship which we previously noted in Chapter II, Section 2.13. Thus, whenever in a given linear programming problem the requirement vector P_0 can be expressed as a linear combination of an incomplete set of independent vectors, degenerate basic feasible solution exists although it may not be explicitly determined in the process of solving a problem by the simplex method.

Example. Consider the following linear programming problem with the objective function omitted,

$$\begin{aligned} 3x_1 + x_2 &\leq 7, \\ x_1 + 2x_2 &\leq 2, \\ x_1 - x_2 &\leq 2, \\ \text{and} \qquad x_1, x_2 &\geq 0. \end{aligned}$$

By adding slack variables we have

$$P_1x_1 + P_2x_2 + P_3x_3 + P_4x_4 + P_5x_5 = P_0,$$

where
$$P_1 = \begin{bmatrix} 3 \\ 1 \\ 1 \end{bmatrix}, \quad P_2 = \begin{bmatrix} 1 \\ 2 \\ -1 \end{bmatrix}, \quad P_3 = \begin{bmatrix} 1 \\ 0 \\ 0 \end{bmatrix},$$

$$P_4 = \begin{bmatrix} 0 \\ 1 \\ 0 \end{bmatrix}, \quad P_5 = \begin{bmatrix} 0 \\ 0 \\ 1 \end{bmatrix}, \quad \text{and} \quad P_0 = \begin{bmatrix} 7 \\ 2 \\ 2 \end{bmatrix}.$$

Since P_3, P_4, P_5 are clearly independent, we use them to form the initial basis and obtain

$$P_3x_3 + P_4x_4 + P_5x_5 = P_0,$$

$$\text{or} \quad B\tilde{X} = P_0, \quad \text{where} \quad B = [P_3\, P_4\, P_5] \quad \text{and} \quad \tilde{X} = \begin{bmatrix} x_3 \\ x_4 \\ x_5 \end{bmatrix}.$$

It is obvious that B is an identity matrix and is therefore its own inverse. The solution is simply $\tilde{X} = B^{-1}P_0 = IP_0 = P_0$, or $x_3 = 7$, $x_4 = 2$ and $x_5 = 2$. The full basic feasible solution (of which $\tilde{X}$ is the positive part) is thus $X^T = [0\ 0\ 7\ 2\ 2]$.

Suppose we want to obtain a new basic feasible solution by introducing P_1 into the basis. Since $B^{-1} = I$ (identity matrix), $Y_1 = B^{-1}P_1 = IP_1 = P_1$, which means that $y_{31} = 3$, $y_{41} = 1$, $y_{51} = 1$. Applying Decision Rule I (4.2.6), we find that $\dfrac{\theta_4}{y_{41}} = \dfrac{2}{1} = 2$ is the minimum. But so is $\dfrac{\theta_5}{y_{51}} = \dfrac{2}{1} = 2$. If we arbitrarily choose P_4 to be dropped, then according to (4.2.7), $x_3 = 7 - 2\cdot 3 = 1$ and $x_5 = 2 - 2\cdot 1 = 0$. The new basic feasible solution is thus $X^T = [2\ 0\ 1\ 0\ 0]$, which has only two positive components instead of the full complement of three (the number of structural constraints, m, in this problem is three). This is a degenerate basic feasible solution. On the other hand, if P_5 is chosen to be replaced, we will find that after similar calculations the identical solution is obtained. The only difference between these two degenerate solutions is that in the first case the basis consists of P_1, P_3, P_5, while in the second case it consists of P_1, P_3, P_4. From both of these solutions we note that

$$P_1x_1 + P_2x_2 + P_3x_3 + P_4x_4 + P_5x_5 = 2P_1 + P_3 = P_0,$$

which indicates that P_0 is a linear combination of only P_1 and P_3, hence is dependent on a partial set of any three independent vectors that includes these two.

Degeneracy *per se* actually does not present any problem in the solution of linear programming problems. It is only when more than one basis variable assumes a value of zero in a degenerate basic feasible solution that the possibility of serious trouble arises. In such cases there will be a tie in the ratio $\frac{\theta_{(i)}}{y_{(i)k}}$ among the zero-valued basis vectors whose $y_{(ik)}$ are positive. Moreover, the minimum of this ratio must be zero. Consequently, the value of the new basis variable must also be zero and the value of the objective function will remain the same for the new basic feasible solution. [See Section 4.5 for the effect of change of basis on the value of $f(X)$.] The possibility exists therefore that the old basis (and the old basic feasible solution) may reappear after one or more changes of basis with no improvement in $f(X)$, thereby completing a "cycle." Recognition of the cyclical pattern and consequent discretionary change of the rule used to break ties will eventually enable us to break away from "cycling" and move on to the optimal solution. But unless this is done, the "cycle" may repeat itself indefinitely thereby preventing attainment of the optimal solution.

Fortunately, experience in the use of the simplex method to solve linear programming problems seems to suggest strongly that the probability of "cycling" is extremely small.* In fact, it is generally considered to be so small that no special counter-measures are thought necessary in computer programs for solving linear programming problems. For this reason, we shall not discuss here the special decision rule developed by Charnes for handling "cycling" problems.** Instead, when ties in the ratio $\frac{\theta_{(i)}}{y_{(i)k}}$ develop in a degenerate basic feasible solution with more than one $\theta_{(i)} = 0$, it is recommended that the basis vector (among those tied), whose $y_{(i)k}$ (must be positive) is

* So far, there exist in the linear programming literature only two theoretical examples of the 'cycling' problem, one constructed by E. M. L. Beale in 'Cycling in the Dual Simplex Algorithm,' *Naval Research Logistics Quarterly*, II, (1955), No. 4, and the other by A. Hoffman quoted by S. I. Gass in, *Linear Programming*, (New York, MacGraw-Hill, 1958).

** Cf. Charnes, Cooper and Henderson, *op. cit.*, pp. 22–24 and 62–67.

the largest, be replaced. If there are further ties in $y_{(i)k}$, arbitrarily select the one which has the lowest basis index. These should reduce the chance of "cycling" to practically nil.

4.5. Selection of the Incoming Vector—Decision Rule II.

As explained in Section 1, the underlying rationale of the simplex method is to obtain a new basic feasible solution from an old one with attendant improvement in the value of the objective function. The amount of improvement depends on (1) the value of the new basis variable just introduced into the basis, i.e., θ'_k, and (2) the *rate* of improvement in $f(X)$ per unit of θ'_k. (In the old solution, θ_k which is the old value of x_k, was, of course, zero since P_k was not in the old basis.) As we have seen, (1) is determined by Decision Rule I which is based on feasibility consideration; (2) represents an evaluation of the vectors not currently in the basis to determine whether their inclusion would improve $f(X)$, and if so, by how much. We will develop the method of evaluation in this section from two approaches, the intuitive and the analytical.

To follow the intuitive approach, let us first reconsider the meaning of (4.2.4) after replacing the specific subscript k with the general subscript j, i.e., $P_j = y_{(1)j}P_{(1)} + y_{(2)j}P_{(2)} + \ldots + y_{(m)j}P_{(m)}$. This equation states that any non-basis vector P_j can be written as a linear combination of the basis vectors $P_{(1)}, P_{(2)}, \ldots P_{(m)}$, with coefficients $y_{(1)j}, y_{(2)j}, \ldots, y_{(m)j}$ respectively. A moment's reflection will enable one to see that it means simply that one P_j is equivalent to $y_{(1)j}$ units of $P_{(1)}$ plus $y_{(2)j}$ units of $P_{(2)} \ldots$ plus finally $y_{(m)j}$ units of $P_{(m)}$. It is this equivalence that made possible the generation of new basic feasible solutions as discussed in Section 4.2. Now, since according to (4.2.2), $P_1x_1 + P_2x_2 \ldots + P_jx_j + \ldots + P_{n+m}x_{n+m} = P_0$, an increase of one P_j (i.e. increasing the value of x_j by one) must be offset by a decrease of $y_{(1)j}$ units of $P_{(1)}$ (i.e. decreasing the value of $x_{(1)}$ by $y_{(1)j}$) and a decrease of $y_{(2)j}$ units of $P_{(2)}$ (i.e. decreasing $x_{(2)}$ by $y_{(2)j}$) etc. in order that (4.2.2) will continue to be satisfied. Since $x_{(1)}, x_{(2)}, \ldots, x_{(m)}$ have coefficients $c_{(1)}, c_{(2)}, \ldots, c_{(m)}$ respectively in the objective function (4.2.1), decreases in the values of these variables will result in a loss of value of the objective function equal to

$$(4.5.1) \quad z_j = c_{(1)}y_{(1)j} + c_{(2)}y_{(2)j} + \ldots + c_{(m)}y_{(m)j} = \sum_{(i)=(1)}^{(m)} c_{(1)}y_{(i)j}.$$

On the other hand, each unit increase in the value of x_j will increase the value of the objective function by c_j. Hence, the net effect on the objective function of this exchange between P_j on the one hand, and $P_{(1)}, P_{(2)}, \ldots, P_{(m)}$ on the other is,

$$z_j - c_j = \sum_{(i)=(1)}^{(m)} c_{(i)} y_{(i)j} - c_j. \qquad \text{Decision Rule II} \tag{4.5.2}$$

We shall call $z_j - c_j$ the *evaluator* of P_j. If $z_j - c_j$ is positive, the net effect on $f(X)$ is to decrease it. Therefore, P_j should not be introduced into the basis if the objective is to maximize $f(X)$. On the other hand, if the objective is to minimize $f(X)$, then P_j could be introduced into basis with advantage. If $z_j - c_j$ is negative, the net effect on $f(X)$ is to increase it. Hence, P_j should or should not be introduced depending on whether the objective is to maximize or minimize $f(X)$. Since $z_j - c_j$ represents the net change in $f(X)$ per unit increase in x_j, if the total increase in x_j as determined by Decision Rule I (4.2.6) is θ'_j, then obviously the total change in $f(X)$, designated $\Delta f(X)$, is

$$\Delta f(X) = \theta'_j(z_j - c_j) \text{(negative means increase and positive means decrease).}$$

Analytically, the above discussion can be summarized as follows. Assume the original solution is $x_{(1)} = \theta_{(1)}, x_{(2)} = \theta_{(2)}, \ldots, x_{(m)} = \theta_{(m)}$. Its corresponding functional value, therefore, is $f(X) = c_{(1)}\theta_{(1)} + c_{(2)}\theta_{(2)} + \ldots + c_{(m)}\theta_{(m)}$. In evaluating P_j, *assume* that P_j is the new basis vector. Its new value θ'_j can then be determined by Decision Rule I. The new values of the other basis variables according to (4.2.8) are $x_1 = \theta'_{(1)} = \theta_{(1)} - \theta'_j y_{(1)j}$, $x_{(2)} = \theta'_{(2)} = \theta_{(2)} - \theta'_j y_{(2)j}$, etc., and the new value of the objective function becomes

$$\begin{aligned} f(X') &= c_{(1)}(\theta_{(1)} - \theta'_j y_{(1)j}) + c_{(2)}(\theta_{(2)} - \theta'_j y_{(2)j}) + \ldots \\ &\qquad + c_{(m)}(\theta_{(m)} - \theta'_j y_{(m)j}) + c_j\theta'_j \\ &= c_{(1)}\theta_{(1)} + c_{(2)}\theta_{(2)} + \ldots + c_{(m)}\theta_{(m)} - c_{(1)}y_{(1)j}\theta'_j - c_{(2)}y_{(2)j}\theta'_j \\ &\qquad - \ldots - c_{(m)}y_{(m)j}\theta'_j + c_j\theta'_j \\ &= f(X) - \theta'_j(c_{(1)}y_{(1)j} + c_{(2)}y_{(2)j} + \ldots + c_{(m)}y_{(m)j} - c_j). \end{aligned}$$

$$\therefore f(X') = f(X) - \theta'_j(z_j - c_j). \tag{4.5.3}$$

If $f(X')$ is to be greater than $f(X)$ as desired in the case of maximizing, then $\theta'_j(z_j - c_j)$ must be negative, which implies that $(z_j - c_j)$ must be negative since θ'_j must be positive in order to maintain feasibility. This means, therefore, that if $f(X)$ is to be maximized, we must select the incoming vector P_k among those P_j whose evaluators $(z_j - c_j)$ are negative.

We will illustrate the use of Decision Rule II to select the incoming non-basis vector by the following example.

Example. Maximize $f(X) = -x_1 - x_2 + x_3$,

$$\begin{aligned} \text{subject to:} \quad & 2x_1 - x_2 - x_3 \leq 4, \\ & x_1 + x_2 + 2x_3 \leq 6, \\ & -x_1 + x_2 \leq 3, \\ \text{and} \quad & x_1, x_2, x_3 \geq 0. \end{aligned}$$

The above is equivalent to:

$$\begin{aligned} \text{Maximize } f(X) &= -x_1 - x_2 + x_3, \\ \text{subject to:} \quad & P_1x_1 + P_2x_2 + P_3x_3 + P_4x_4 + P_5x_5 + P_6x_6 = P_0, \\ \text{and} \quad & x_j \geq 0, \quad j = 1, 2, \ldots, 6. \end{aligned}$$

$$P_1 = \begin{bmatrix} 2 \\ 1 \\ -1 \end{bmatrix}, \quad P_2 = \begin{bmatrix} -1 \\ 1 \\ 1 \end{bmatrix}, \quad P_3 = \begin{bmatrix} -1 \\ 2 \\ 0 \end{bmatrix}, \quad P_4 = \begin{bmatrix} 1 \\ 0 \\ 0 \end{bmatrix},$$

$$P_5 = \begin{bmatrix} 0 \\ 1 \\ 0 \end{bmatrix}, \quad P_6 = \begin{bmatrix} 0 \\ 0 \\ 1 \end{bmatrix}, \quad P_0 = \begin{bmatrix} 4 \\ 6 \\ 3 \end{bmatrix}.$$

Obviously P_4, P_5, P_6 represent an independent set of vectors and form a basis for the basic feasible solution, $x_1 = x_2 = x_3 = 0$, $x_4 = 4$, $x_5 = 6$, $x_6 = 3$. Since the basis $B = [P_4\, P_5\, P_6]$ is the identity matrix I, its inverse B^{-1} is also I. Consequently, the linear-combination coefficient vectors Y_1, Y_2 and Y_3 of the non-basis vectors are respectively P_1, P_2, and P_3 themselves, i.e. $Y_1 = B^{-1}P_1 = IP_1 = P_1$, etc. Applying (4.5.1), we have then

$$\begin{aligned} z_1 &= c_4y_{41} + c_5y_{51} + c_6y_{61} \\ &= c_4(2) + c_5(1) + c_6(-1) = 2c_4 + c_5 - c_6, \\ z_2 &= c_4y_{42} + c_5y_{52} + c_6y_{62} \\ &= c_4(-1) + c_5(1) + c_6(1) = -c_4 + c_5 + c_6, \\ z_3 &= c_4y_{43} + c_5y_{53} + c_6y_{63} \\ &= c_4(-1) + c_5(2) + c_6(0) = -c_4 + 2c_5. \end{aligned}$$

But, $c_4 = c_5 = c_6 = 0$ since x_4, x_5, x_6 do not appear in the objective function. Therefore, $z_1 = z_2 = z_3 = 0$. Consequently, from (4.5.2) the evaluation of P_1, P_2, P_3 results in:

$$z_1 - c_1 = 0 - (-1) = 1,$$
$$z_2 - c_2 = 0 - (-1) = 1,$$
$$z_3 - c_3 = 0 - 1 = -1.$$

Since $z_j - c_j$ has to be negative in order for $f(X)$ to increase, only P_3 is eligible. Once the incoming vector is determined as P_3, we can determine θ_3', which is the value x_3 can assume without causing other variables to become negative. This is accomplished by the use of Decision Rule I (4.2.6). Bearing in mind that $x_4 = 4$, $x_5 = 6$, $x_6 = 3$ and $y_{43} = -1$, $y_{53} = 2$ and $y_{63} = 0$, we obtain

$$x_3 = \theta_3' = \underset{y_{(i)3}>0}{\text{minimum}} \left\{ \frac{x_{(i)}}{y_{(i)3}} \right\} = \frac{x_5}{y_{53}} = 3.$$

Hence, the new solution is: $x_1 = x_2 = 0$, $x_3 = 3$, $x_4 = 4 - 3(-1) = 7$, $x_5 = 0$, $x_6 = 3 - 3(0) = 3$. The new value of the objective function is accordingly $f(X') = -1(0) - 1(0) + 1(3) = 3$ which represents an increase of 3 from the old value. This increase can be verified by the use of (4.5.3).

Two remarks are in order before we conclude this section. The first one is that although we have devised a rule (Decision Rule II) for the selection of *a* non-basis vector to enter into the basis, we have no rule for a *unique* choice. In other words, upon evaluation of all non-basis vectors by $z_j - c_j$, there may be more than one eligible for introduction into the basis. It is then not clear which is the "most" eligible—in the first place, the total improvement in $f(X)$ depends on θ_j' as well as $z_j - c_j$ (4.5.3), hence the absolute magnitude of $z_j - c_j$ is not an adequate criterion. In the second place, even the total improvement, i.e., $\theta_j'(z_j - c_j)$, is not a reliable yardstick since maximal improvement at any one stage of the solution does not necessarily mean that large improvements can continue to be made at subsequent stages of the iterative process. It is, at least, theoretically possible that a big improvement may be followed by a long series of small improvements, resulting in a slow overall convergence to the optimum. The dilemma posed here by the simplex method may be likened to the situation of a motorist who discovers that he has only compass and road signs to tell him the direction in which he should proceed and the distance to the nearest town, but no overall road map to enable him to select the shortest route to his destination. The lack of a road map is analogous to our lack of knowledge of what the ultimate destination is in a linear programming problem. In view of this, how should we select a specific incoming vector from among those eligible? The experts do not have a common answer to this question. Some advise the selection of the one with the largest rate of improvement as measured by $z_j - c_j$ (this should be negative for maximizing and positive for minimizing problems). More, however, seem to favor the selection of the first non-basis vector found to be eligible. When linear programming problems are solved by computers, the latter procedure is almost invariably adopted—due to the relatively slow speed with which non-basis vectors can be read in for evaluation.

The second remark is most important as it concerns the termination of the simplex procedure. Obviously we should continue to change the basis so long as we obtain better and better basic feasible solutions. That is to say, so long as there is some non-basis vector(s) with a favorable evaluation (i.e., negative $z_j - c_j$ for a maximizing problem and positive $z_j - c_j$ for a minimizing one), it should be brought into the basis to bring about a better solution. By the same token, if no non-basis vector exists which has a favorable evaluation, the implication is that an optimal solution has been obtained and the simplex iterative procedure should terminate. Specifically, we mean that when at any point of the simplex iterative process, the evaluators of all non-basis vectors become either positive or zero in a maximizing problem (or all become negative or zero in a minimizing problem), then no further iteration is necessary and the last basic feasible solution represents the optimal solution as defined in Chapter III Section 3.5.* The optimal value of the objective function in this case, however, is finite. It will be recalled that the terminating criterion for an infinite optimum is different and has already been discussed fully in Section 4.3 of this chapter.

4.6. Rules of Transformation.

In order to generate a new basic feasible solution from an existing one, we have seen that both Decision Rules I and II require $y_{(i)j}$ to determine which non-basis vector should come into the basis and which basis vector should go out.** There are two ways of getting Y_j.

* In the case where some $z_j - c_j$ equal zero for some non-basis vectors, there are other solutions which yield the same optimal value of $f(X)$. This problem of alternate optima will be discussed in Chapter V.

** Decision Rule II requires $y_{(i)j}$ not so much for the purpose of computing z_j according to (4.5.1) as for the direct transformation of $z_j - c_j$, which will be discussed later in this section.

One involves the use of the inverse of the basis B^{-1}. That is to say, since $P_j = BY_j$, hence $Y_j = B^{-1}P_j$ (4.2.4). When the basis is changed, the new inverse will be found and from that, the new Y_j. This approach therefore calls first of all for a transformation of the basis. Since this is the underlying concept of the so-called Revised Simplex Method (or Inverse Matrix Method) to be discussed in Chapter VI, we will not elaborate on it here. The second method involves the direct transformation of Y_j. This will be explained as follows.

Assume that we have a basis B which yields a basic feasible solution X. Assume further that by the use of Decision Rules I and II, we have selected a non-basis vector P_k to replace the basis vector $P_{(l)}$ in order to obtain the new basis B' which yields the basic feasible solution X'. According to (4.2.4), we can then write for *any* vector,

$$P_j = y_{(1)j}P_{(1)} + y_{(2)j}P_{(2)} + \dots + y_{(l)j}P_{(l)} + \dots + y_{(m)j}P_{(m)}, \tag{4.6.1}$$

and

$$P_j = y'_{(1)j}P_{(1)} + y'_{(2)j}P_{(2)} + \dots + y'_{kj}P_k + \dots + y'_{(m)j}P_{(m)}. \tag{4.6.2}$$

Since P_k, a non-basis vector relative to B (the old basis), is

$$P_k = y_{(1)k}P_{(1)} + y_{(2)k}P_{(2)} + \dots + y_{(l)k}P_{(l)} + \dots + y_{(m)k}P_{(m)}. \tag{4.6.3}$$

Substituting (4.6.3) into (4.6.2) we have

$$\begin{aligned} P_j &= y'_{(1)j}P_{(1)} + y'_{(2)j}P_{(2)} + \dots + y'_{kj}(y_{(1)k}P_{(1)} + y_{(2)k}P_{(2)} \\ &\qquad + \dots + y_{(l)k}P_{(l)} + \dots + y_{(m)k}P_{(m)}) + \dots + y'_{(m)j}P_{(m)} \\ &= (y'_{(1)j} + y'_{kj}y_{(1)k})P_{(1)} + (y'_{(2)j} + y'_{kj}y_{(2)k})P_2 \\ &\qquad + \dots + y'_{kj}y_{(l)k}P_{(l)} + \dots + (y'_{(m)j} + y'_{kj}y_{(m)k})P_{(m)}. \end{aligned} \tag{4.6.4}$$

Since both (4.6.1) and (4.6.4) are expressions for the same P_j in terms of the same set of vectors comprising the old basis B, corresponding coefficients must be equal. Hence

$$\begin{aligned} y'_{(1)j} + y'_{kj}y_{(1)k} &= y_{(1)j}, \\ y'_{(2)j} + y'_{kj}y_{(2)k} &= y_{(2)j}, \\ &\ \ \vdots \\ y'_{kj}y_{(l)k} &= y_{(l)j}, \\ &\ \ \vdots \\ y'_{(m)j} + y'_{kj}y_{(m)k} &= y_{(m)j}. \end{aligned} \tag{4.6.5}$$

From $y'_{kj}y_{(l)k} = y_{(l)j}$ above, it is clear that

$$y'_{kj} = \frac{y_{(l)j}}{y_{(l)k}}.$$

Since P_k is the lth vector in the new basis, i.e., $P_{(l)} = P_k$, therefore

$$y'_{(l)j} = y'_{kj} = \frac{y_{(l)j}}{y_{(l)k}}. \qquad \text{Transformation Rule I} \tag{4.6.6}$$

By substituting (4.6.6) into the rest of the equations in (4.6.5), we have

$$y'_{(1)j} = y_{(1)j} - y'_{kj}y_{(1)k} = y_{(1)j} - \frac{y_{(l)j}}{y_{(l)k}} y_{(1)k},$$

$$y'_{(2)j} = y_{(2)j} - y'_{kj}y_{(2)k} = y_{(2)j} - \frac{y_{(l)j}}{y_{(l)k}} y_{(2)k},$$

$$\vdots$$

$$y'_{(m)j} = y_{(m)j} - y'_{kj}y_{(m)k} = y_{(m)j} - \frac{y_{(l)j}}{y_{(l)k}} y_{(m)k}.$$

Or, in general, we can write,

$$y'_{(i)j} = y_{(i)j} - \frac{y_{(l)j}}{y_{(l)k}} y_{(i)k}, \quad \text{for} \quad (i) \neq (l), (i) = (1), \ldots, (m). \tag{4.6.7}$$

Transformation Rule II

The Transformation Rules I and II derived above may look awkward but they are really quite simple to apply as will be made clear in Chapter V. With these rules available, we can proceed from one basis to another and from one basic feasible solution to another until an optimal solution is obtained.

There remains one more detail to clean up, viz., the determination of the new evaluators $z'_j - c_j$ relative to the new basis B'. These evaluators can be found by using (4.5.2) since all necessary $y'_{(i)j}$ (not $y_{(i)j}$ as we are now engaged in evaluating the non-basis vectors relative to the new basis B') have already been determined according to Transformation Rules I and II. However, there is an easier way to obtain the $z'_j - c_j$.

From (4.5.2) we have

$$z'_j - c_j = (c_{(1)}y'_{(1)j} + c_{(2)}y'_{(2)j} + \ldots + c_k y'_{kj} + \ldots + c_{(m)}y'_{(m)j}) - c_j.$$

Substituting (4.6.6) and (4.6.7) into the above, we obtain

$$z'_j - c_j = \left[c_{(1)} \left(y_{(1)j} - \frac{y_{(l)j}}{y_{(l)k}} y_{(1)k} \right) + c_{(2)} \left(y_{(2)j} - \frac{y_{(l)j}}{y_{(l)k}} y_{(2)k} \right) \right.$$

$$\left. + \ldots + c_k \frac{y_{(l)j}}{y_{(l)k}} + \ldots + c_{(m)} \left(y_{(m)j} - \frac{y_{(l)j}}{y_{(l)k}} y_{(m)k} \right) \right] - c_j$$

$$= \left[(c_{(1)}y_{(1)j} + c_{(2)}y_{(2)j} + \ldots + c_{(m)}y_{(m)j}) - \right.$$

$$\left. \frac{y_{(l)j}}{y_{(l)k}} (c_{(1)}y_{(1)k} + c_{(2)}y_{(2)k} + \ldots - c_k + \ldots + c_{(m)}y_{(m)k}) \right] - c_j.$$

After adding and subtracting the term $c_{(l)}y_{(l)j}$, the above becomes

$$z'_j - c_j = (c_{(1)}y_{(1)j} + c_{(2)}y_{(2)j} + \ldots + c_{(l)}y_{(l)j} + \ldots + c_{(m)}y_{(m)j})$$

$$- c_j - \frac{y_{(l)j}}{y_{(l)k}} \left(c_{(1)}y_{(1)k} + c_{(2)}y_{(2)k} + \ldots + c_{(l)}y_{(l)j} \frac{y_{(l)k}}{y_{(l)j}} \right.$$

$$\left. + \ldots + c_{(m)}y_{(m)k} - c_k \right).$$

which is simply

(4.6.8) $$z'_j - c_j = (z_j - c_j) - \frac{y_{(l)j}}{y_{(l)k}} (z_k - c_k),$$

for $j = 1, 2, \ldots, n + m$. Transformation Rule III

The above transformation rule for the evaluators $z_j - c_j$ can be seen to closely resemble Transformation Rule II (4.6.7) in structure. As a matter of fact, in the next chapter, we shall see that the same procedure will be applied to the transformation of both $y_{(i)j}$ and $z_j - c_j$.

Example. We will attempt to illustrate the use of transformation rules by continuing the solution of the linear programming problem used in the example of Section 4.5.

At the end of that example, we have as a new (and better) basic feasible solution the solution vector $X^T = [0\ 0\ 3\ 7\ 0\ 3]$ or $x_1 = x_2 = 0, x_3 = 3, x_4 = 7, x_5 = 0$ and $x_6 = 3$. The logical question is: is this the optimal solution? To answer this, we have to determine the evaluators $z_j - c_j$ of all non-basis vectors P_j according to Decision Rule II (4.5.2). If they are all non-negative, then the answer is "yes" since the problem is a maximizing one. If some of them are negative, then the answer is "no" and we are off to find a still better basic feasible solution. The non-basis vectors at this stage of the solution are P_1,

P_2 and P_5 (since the current basis consists of P_3, P_4 and P_6) which have $z_1 - c_1 = 1$, $z_2 - c_2 = 1$ and $z_5 - c_5 = 0$ respectively. We know from the previous example that $(l) = 5$, $k = 3$, $y_{(l)k} = y_{53} = 2$, $y_{51} = 1$, $y_{52} = 1$, $y_{55} = 1$, and $z_3 - c_3 = -1$. Hence, using Transformation Rule III to re-evaluate these vectors, we have,

$$z_1' - c_1 = (z_1 - c_1) - \frac{y_{51}}{y_{53}}(z_3 - c_3) = 1 - \frac{1}{2}(-1) = \frac{3}{2},$$

$$z_2' - c_2 = 1 - \tfrac{1}{2}(-1) = \tfrac{3}{2},$$

$$z_5' - c_5 = 0 - \tfrac{1}{2}(-1) = \tfrac{1}{2}.*$$

Since they are all positive (therefore, non-negative), the present solution is the optimal solution and the process may be terminated here. However, for the purpose of illustrating the use of Transformation Rules I and II, we will continue to determine some y_{ij}'. In particular, we will determine Y_1, the linear-combination coefficient vector of P_1, to serve two additional purposes: (1) the verification of $z_1' - c_1$ as computed above, and (2) the verification of the Transformation Rules themselves. From the previous example, we know $y_{41} = 2$, $y_{51} = 1$ and $y_{61} = -1$. Since the new basis B' has P_3 in place of P_5, the relevant coefficients are y_{41}', y_{31}' (instead of y_{51}) and y_{61}' determined as follows:

by (4.6.7) $$y_{41}' = y_{41} - \frac{y_{51}}{y_{53}}y_{43} = 2 - \frac{1}{2}(-1) = \frac{5}{2},$$

by (4.6.7) $$y_{61}' = y_{61} - \frac{y_{51}}{y_{53}}y_{63} = (-1) - \frac{1}{2}(0) = -1,$$

by (4.6.6) $$y_{31}' = \frac{y_{51}}{y_{53}} = \frac{1}{2}.$$

These coefficients mean that P_1 can be expressed in terms of the new basis B' in the following manner:

$$P_1 = y_{31}'P_3 + y_{41}'P_4 + y_{61}'P_6 = \tfrac{1}{2}P_3 + \tfrac{5}{2}P_4 + (-1)P_6$$

Is this correct? We can verify it by substituting the values of P_3, P_4, and P_6 (given in the previous example) into the right-hand side, and obtain

* Even without calculation, we can assert on *a priori* grounds that $z_5' - c_5$ must be positive since it has *just* been displaced from the basis. However, this does not mean that P_5 once displaced will never be eligible for re-introduction again after one more change of the basis.

$$\frac{1}{2}\begin{bmatrix}-1\\2\\0\end{bmatrix}+\frac{5}{2}\begin{bmatrix}1\\0\\0\end{bmatrix}+(-1)\begin{bmatrix}0\\0\\1\end{bmatrix}=\begin{bmatrix}2\\1\\-1\end{bmatrix},$$

which is exactly the vector P_1 as expected. This should reinforce our confidence in the Transformation Rules I and II.

Next, we compute $z_1' - c_1$ by (4.5.2), since we now have the $y'_{(i)j}$:

$$z_1' - c_1 = (c_3y_{31}' + c_4y_{41}' + c_6y_{61}') - c_1 = 1(\tfrac{1}{2}) + 0(\tfrac{5}{2}) + 0(-1) - (-1) = \tfrac{1}{2} + 1 = \tfrac{3}{2},$$

which is exactly what we obtained before by Transformation Rule III. This verifies the validity of the rule.

4.7. Summary.

Let us now piece together the discussions in the foregoing sections and present an integrated view of the simplex procedure as employed in the solution of a linear programming problem. Without loss of generality we may assume that the problem is one of maximization as given by (4.2.1) and (4.2.2) in Section 4.2. It is further assumed that an initial basic feasible solution X associated with a basis B is available. (The question of obtaining an initial basic feasible solution will be discussed in detail in the next chapter.) From B we can determine B^{-1} which, in turn, enables us by $B^{-1}P_j$ to determine the various linear-combination coefficient vectors Y_j of all P_j. Now we are ready to enter into the following iterative simplex process.

Step I—Determine $z_j - c_j$ for all non-basis vectors by Decision Rule II (4.5.2) or alternatively, by Transformation Rule III (4.6.8), realizing that the evaluators $z_j - c_j$ for basis vectors are identically equal to zero and so require no computation. Two situations may arise:

(1) All computed $z_j - c_j$ are zero or positive. This means that no further improvement in $f(X)$ can be expected and the last basic feasible solution is an optimal solution. The problem is solved and computation ends.

(2) Some $z_j - c_j$ are negative. In this case proceed to Step II.

Step II—Select the P_j with the most negative $z_j - c_j$ (algebraic minimum) to be the incoming vector. Say this is P_k. Proceed to Step III.

Step III—Select the basis vector to be replaced by P_k. Call this $P_{(l)}$. Three different situations may be encountered:

(1) All $y_{(i)k}$ are non-positive. According to Section 4.3 this means that $f(X)$ has no upper bound, and that any positive value of x_k (as large as one wishes up to positive infinity) together with appropriately increased values of the existing basis variables (there are m of them) will be a solution yielding an arbitrarily large value of $f(X)$ (up to positive infinity). Here the optimal solution is not a basic feasible solution but consists of $m + 1$ variables (except in cases of degeneracy), some or all of which are infinite (positive) in value, and so yields a value of positive infinity for $f(X)$. The problem is solved and computation ends.

(2) Some $y_{(i)k}$ are positive and a unique choice of $P_{(l)}$ is possible as the result of the application of Decision Rule I (4.2.6). In this case, $P_{(l)}$ is the outgoing vector. Computation proceeds to Step IV.

(3) Some $y_{(i)k}$ are positive but ties occur as the result of the application of Decision Rule I (4.2.6). In this case, ties are broken according to the following convention and a unique choice of $P_{(l)}$ is made before proceeding to Step IV.

(a) The ratios $\dfrac{\theta_{(i)}}{y_{(i)k}}$ are zero and the relevant $y_{(i)k}$ are not equal. In this case, choose that particular basis vector in the tie which has the largest (positive) $y_{(i)k}$.

(b) In all other ties which (a) fail to resolve, choose that basis vector in the tie which has the lowest basis index (i).

Step IV—Compute the new basic feasible solution by (4.2.7) and (4.2.8). Determine the new value of $f(X)$ by (4.5.3) and proceed to Step V.

Step V—Determine the new linear-combination coefficients $y_{(i)j}$ for all non-basis vectors by Transformation Rules I (4.6.6) and II (4.6.7). Now return to Step I.

The above represents the theoretical framework upon which a practical operational procedure will be built in the next chapter for the solution of linear programming problems. If the problem has an optimal solution, whether finite or infinite, the simplex procedure will eventually reveal it—except in the very rare case of "cycling." When "cycling" occurs, the problem can be resolved quickly by the use of on-the-spot judgement to break the cyclical pattern. Thus we may consider as this chapter closes that the theoretical problems posed by linear programming problems have been completely overcome.

EXERCISES

1. Given the following system of inequalities:

$$\begin{aligned} 2x_1 - 2x_2 + x_3 &\leq 4, \\ x_2 + 2x_3 &\leq 4, \\ -x_1 + 3x_2 - x_3 &\leq 6, \\ x_j \geq 0, \quad j &= 1, 2, 3. \end{aligned}$$

   (a) Obtain the initial basic feasible solution with the slack vectors.
   (b) Bring P_1 into the basis and obtain the next basic feasible solution employing Decision Rule I.
   (c) Obtain all basic feasible solutions through repeated use of Decision Rule I.
   (d) What rule did you use to insure that the same basic feasible solution would not be repeated?

2. In generating new basic feasilble solutions such as in Exercise 1, how does Decision Rule II help avoid possible duplications?

3. Given the following inequality system:

$$P_1x_1 + P_2x_2 + P_3x_3 \leq P_0$$

where
$$P_1 = \begin{bmatrix} 2 \\ 0 \\ -1 \end{bmatrix}, \quad P_2 = \begin{bmatrix} -2 \\ 1 \\ 3 \end{bmatrix}, \quad P_3 = \begin{bmatrix} 2 \\ 2 \\ 1 \end{bmatrix}, \quad P_0 = \begin{bmatrix} 2 \\ 3 \\ 1 \end{bmatrix}.$$

Start with the initial basis $B = [P_4\,P_5\,P_6]$ where P_4, P_5, P_6 are the slack vectors associated with the first, second and third inequalities respectively. Determine whether the basis $B = [P_1\,P_2\,P_3]$ is a feasible one by successive introduction of the three desired vectors into the basis. What other method could you use to ascertain the feasibility of $[P_1\,P_2\,P_3]$ directly?

4. (a) Assuming that P_k is selected to come into a feasible basis, how does one insure that the new basis will continue to be feasible?
   (b) What would be the outcome if P_k does not form a feasible basis with any subset of $[m - 1]$ vectors out of the m vectors in the current basis?

5. Assume that P_k is selected to replace $P_{(l)}$ in the current basis $B = [P_{(1)} P_{(2)} \dots P_{(l)} \dots P_{(m)}]$ in accordance with Decision Rule I. Show that the new basis $[P_{(1)} P_{(2)} \dots P_k \dots P_{(m)}]$ is invariably made up of a linearly independent set of vectors.

6. Show that the inequality system in Exercise 1 has a degenerate solution in $x_1 = 6$ and $x_2 = 4$. What is the relationship between P_1, P_2 and P_0?

7. In Exercise 2, suppose the objective function to be optimized is $f(X) = -x_1 - x_2 + 2x_3$. Evaluate x_3, x_4 and x_6 relative to the basis $[P_1 P_2 P_5]$. According to Decisions Rule II, which variables are eligible for introduction into the basis if the objective function is to be maximized? If it is to be minimized?

8. Prove that if X_1 and X_2 are two optimal solutions, then any convex combination of them is also an optimal solution. This means that if there are more than one optimal solutions, then there are an infinite number of optimal solutions.

THE SIMPLEX METHOD—COMPUTATIONAL

5.1. Introduction.

In this chapter we shall apply the theory of linear programming developed in preceding chapters to the solution of actual problems. This involves especially the design of computational forms and procedures to facilitate the manipulation of data in accordance with theory. In other words, we might say that in this chapter, we will be "tooling up" for production.

Additionally, we will answer some specific questions concerning the determination of an initial basic feasible solution.

5.2. The Simplex Tableau—Its Structure.

The main "fixture" employed in the solution of linear programming problems is the simplex tableau. Essentially, a simplex tableau is a special tabular arrangement of all necessary information for the iterative solution of a linear programming problem with the basis vectors arranged in rows and the full set of structural and slack vectors arranged in columns. Despite variations, its basic form remains the one used by Charnes, Cooper and Henderson, which with minor modifications is reproduced in Table 5.1.

To explain the tableau, let us write the linear programming problem as follows (cf. Chapter III Section 3.1):

$$\begin{aligned} \text{Maximize } f(X) &= c_1x_1 + \ldots + c_nx_n, && (3.1.7)\\ \text{subject to:} \quad & P_1x_1 + \ldots + P_nx_n \leq P_0, && (3.1.8)\\ \text{and:} \quad & x_j \geq 0, \quad j = 1, 2, \ldots, n, && (3.1.9)\end{aligned}$$

where P_1, P_2 etc. are the m-component column vectors representing the coefficients of x_1, x_2 etc. respectively and P_0 is an m-component column vector of constants. By adding slack variables to the inequalities, the problem is rendered into the following equivalent form (cf. Chapter III Section 3.6):

$$\text{Maximize } f(X) = c_1x_1 + \ldots + c_nx_n + 0\cdot x_{n+1} + \ldots + 0\cdot x_{n+m},$$

$$\text{subject to: } P_1x_1 + \ldots + P_nx_n + P_{n+1}x_{n+1} + \ldots + P_{n+m}x_{n+m} = P_0,$$

$$\text{and} \qquad x_j \geq 0, \qquad j = 1, 2, \ldots, n, n+1, \ldots, n+m,$$

which can then be solved through the repeated application of the following rules:

$$\text{Decision Rule I:} \quad \theta'_j = \frac{\theta_{(l)}}{y_{(l)k}} = \min\left\{\frac{\theta_{(i)}}{y_{(i)k}}\right\} \quad \text{for } y_{(i)k} > 0. \qquad (4.2.6)$$

$$\text{Decision Rule II:} \quad z_j - c_j = \sum_{(i)=(1)}^{(m)} c_{(i)}y_{(i)j} - c_j,$$

$$j = 1, \ldots, n+m. \qquad (4.5.2)$$

Since both rules require explicit knowledge of $y_{(i)j}$, they are the first things that should be included in the tableau. But what are these $y_{(i)j}$? The reader would do well to pause and ponder over their meaning, perhaps in consultation with Chapter III Section 3.6 and particularly (3.6.6) and (3.6.7). The answer is, of course, that they are the linear combination coefficients of the basis vectors for P_j. That is to say, if $B = [P_{(1)}, P_{(2)}, \ldots, P_{(m)}]$ is a feasible basis such that $B\tilde{X} = P_0$, then $y_{(i)j}$ are those scalars which satisfy the following:

$$P_j = y_{(i)j}P_{(1)} + y_{(2)j}P_{(2)} + \ldots + y_{(m)j}P_{(m)}.$$

$$\text{Or,} \qquad Y_j = B^{-1}P_j, \quad \text{where } Y_j = \begin{bmatrix} y_{(1)j} \\ \cdot \\ \cdot \\ \cdot \\ y_{(m)j} \end{bmatrix}. \qquad (4.2.4)$$

These $y_{(i)j}$ for $j = 1, 2, \ldots, n + m$ are entered into the proper $P_{(i)}$ rows and P_j columns as shown in Table 5.1.*

Table 5.1

The Simplex Tableau

Based on Charnes, Cooper, and Henderson, *An Introduction to Linear Programming.* (New York: John Wiley & Sons, Inc., 1953.)

$c_{(i)}$	$P_{(i)}$ \ P_j		P_1	P_2	. . . .	P_k	. . .	P_n	P_{n+1}	. . .	P_{n+m}	P_0	
$c_{(1)}$	$P_{(1)}$	P_1	$y_{(1)1}$	$y_{(1)2}$	. . .	$y_{(1)k}$	. . .	$y_{(1)n}$	$y_{(1),n+1}$	. . .	$y_{(1),n+m}$	$\theta_{(1)}$	
$c_{(2)}$	$P_{(2)}$	P_4	$y_{(2)1}$	$y_{(2)2}$	. . .	$y_{(2)k}$	. . .	$y_{(2)n}$	$y_{(2),n+1}$	. . .	$y_{(2),n+m}$	$\theta_{(2)}$	
⋮	⋮	⋮	⋮	⋮		⋮		⋮	⋮		⋮	⋮	
$c_{(l)}$	$P_{(l)}$	P_8	$y_{(l)1}$	$y_{(l)2}$	. . .	$y_{(l)k}$	. . .	$y_{(l)n}$	$y_{(l),n+1}$	. . .	$y_{(l),n+m}$	$\theta_{(l)}$	→
⋮	⋮	⋮	⋮	⋮		⋮		⋮	⋮		⋮	⋮	
$c_{(m)}$	$P_{(m)}$	P_2	$y_{(m)1}$	$y_{(m)2}$	. . .	$y_{(m)k}$	. . .	$y_{(m)n}$	$y_{(m),n+1}$	. . .	$y_{(m),n+m}$	$\theta_{(m)}$	
z_j			z_1	z_2	. . .	z_k	. . .	z_n	z_{n+1}	. . .	z_{n+m}	—	
c_j			c_1	c_2	. . .	c_k	. . .	c_n	c_{n+1}	. . .	c_{n+m}	—	
$z_j - c_j$			$z_1 - c_1$	$z_2 - c_2$	. . .	$z_k - c_k$	. . .	$z_n - c_n$	$z_{n+1} - c_{n+1}$	. . .	$z_{n+m} - c_{n+m}$	$f(X)$	
						↑							

Secondly, we note that the Decision Rules also call for knowledge of $\theta_{(i)}$ (the solution-values of the basis variables). Since we have assumed that there is a B such that $B\tilde{X} = P_0$, these values could be obtained from $\tilde{X} = B^{-1}P_0$ and entered in the P_0 column as shown in Table 5.1. All other variables (the non-basis variables) will, of course, be zero in a basic feasible solution, hence do not have to be recorded explicitly in the tableau.

The third group of information contained in the simplex tableau is the z_j computed from the formula $z_j = \sum_{(i)=(1)}^{(m)} c_{(i)}y_{(i)j}$ (4.5.1) for $j = 1, 2, \ldots, n + m$. These are entered in the "z_j" row and the appropriate P_j columns.

* The reader is warned against the common mistake of regarding the entries under each P_j (excluding $j = 0$) in the simplex tableau as the elements of P_j itself. It will be recalled that the elements of P_j are a_{ij} and the entries in the tableau are $y_{(i)j}$. Only in the special case where the basis (usually the initial basis) is an identity matrix are the two equal. In general, however, $B \neq I$ and consequently, $y_{(i)j} \neq a_{ij}$.

Fourthly, the c_j (called price coefficients in linear programming terminology) are transcribed from the objective function to the "c_j" row and the evaluators $z_j - c_j$ are then computed and entered in the last row of the tableau. Finally, $f(X) = c_{(1)}\theta_{(1)} + c_{(2)}\theta_{(2)} + \ldots + c_{(m)}\theta_{(m)}$ is evaluated and recorded in the P_0 column as shown.

To summarize, assuming that a feasible basis $B = [P_{(1)}\, P_{(2)} \ldots P_{(m)}]$ has already been found for the type of linear programming problems described by (5.2.1), a simplex tableau can then be constructed displaying all the data, information and intermediate results necessary for the continued systematic evaluation of different basic feasible solutions until an optimal solution is reached. The main body of the tableau consists of the linear combination coefficients $y_{(i)j}$ which are obtained from $Y_j = B^{-1}P_j$. In addition, the tableau shows $\theta_{(i)}$ obtained from $\tilde{X} = B^{-1}P_0$, z_j from $z_j = \sum_{(i)=(1)}^{(m)} c_{(i)}y_{(i)j}$, c_j from the objective function, $z_j - c_j$ from the z_j and c_j just determined and finally, the value of the objective function $f(X)$ computed from $f(X) = \sum_{(i)=(1)}^{(m)} c_{(i)}\theta_{(i)}$. We shall now demonstrate the construction of such a tableau.

Example. It is known that the following linear programming problem has a feasible basis in P_3, P_4 and P_6.* Construct a corresponding simplex tableau.

$$
\begin{aligned}
\text{Maximize } f(X) &= -x_1 + x_2 + x_3, \\
\text{subject to:} \quad & 2x_1 - x_2 - x_3 + x_4 = 4, \\
& x_1 + x_2 + 2x_3 + x_5 = 6, \\
& -x_1 + x_2 + x_6 = 3, \\
\text{and} \quad & x_j \geq 0, \qquad j = 1, 2, \ldots, 6.
\end{aligned}
$$

The blank form of such a tableau is shown in Table 5.2. Certain information can immediately be entered. Since the basis vectors are given as P_3, P_4 and P_6, we can arbitrarily let $P_{(1)} = P_3$, $P_{(2)} = P_4$ and $P_{(3)} = P_6$ and write P_3, P_4 and P_6 in that order in the blank spaces under $P_{(i)}$. Similarly the price coefficients of these three basis vectors

* An obvious feasible basis is, of course, the set of vectors P_4, P_5 and P_6, which form an identity matrix. It is, however, purposely avoided here in order to stress the point that it is the $y_{(i)j}$ and not the a_{ij} that should be entered in the tableau.

Table 5.2

$c_{(i)}$	$P_{(i)}$ \ P_j	P_1	P_2	P_3	P_4	P_5	P_6	P_0
	$P_{(1)}$							
	$P_{(2)}$							
	$P_{(3)}$							
	z_j							—
	c_j							—
	$z_j - c_j$							

are put down in the same order in the column "$c_{(i)}$." Finally the price coefficients of $P_1, \ldots, P_6$ are copied down in the "c_j" row from the objective function, noting that $c_4 = c_5 = c_6 = 0$. The remaining information required by the tableau is then computed in four steps as follows:

(1) $y_{(i)j}$—To obtain $y_{(i)j}$, we must have the inverse of the basis.

$$B^{-1} = [P_3 \quad P_4 \quad P_6]^{-1} = \begin{bmatrix} -1 & 1 & 0 \\ 2 & 0 & 0 \\ 0 & 0 & 1 \end{bmatrix}^{-1}$$

$$= \frac{1}{-2}\begin{bmatrix} 0 & -1 & 0 \\ -2 & -1 & 0 \\ 0 & 0 & -2 \end{bmatrix} = \begin{bmatrix} 0 & \frac{1}{2} & 0 \\ 1 & \frac{1}{2} & 0 \\ 0 & 0 & 1 \end{bmatrix}.$$

With the inverse of the basis known, the Y_j vectors can now be found.

$$Y_1 = \begin{bmatrix} y_{(1)1} \\ y_{(2)1} \\ y_{(3)1} \end{bmatrix} = B^{-1}P_1 = \begin{bmatrix} 0 & \frac{1}{2} & 0 \\ 1 & \frac{1}{2} & 0 \\ 0 & 0 & 1 \end{bmatrix}\begin{bmatrix} 2 \\ 1 \\ -1 \end{bmatrix} = \begin{bmatrix} \frac{1}{2} \\ \frac{5}{2} \\ -1 \end{bmatrix}.$$

These coefficients are then entered in order in the column marked "P_1" in Table 5.3, which now reads, $P_1 = \frac{1}{2}P_{(1)} + \frac{5}{2}P_{(2)} - P_{(3)}$ or $P_1 = \frac{1}{2}P_3 + \frac{5}{2}P_4 - P_6$. Similar calculations for $Y_2, \ldots, Y_6$ lead to results as shown. It will be noted, however, that since the basis vectors are linearly independent, none can be expressed as a linear combination of the other two in the basis. This means $P_3 = 1 \cdot P_3 + 0 \cdot P_4 + 0 \cdot P_6$, $P_4 = 0 \cdot P_3 + 1 \cdot P_4 + 0 \cdot P_6$ and $P_6 = 0 \cdot P_3 + 0 \cdot P_4 + 1 \cdot P_6$, or

$$Y_3 = \begin{bmatrix} 1 \\ 0 \\ 0 \end{bmatrix}, \qquad Y_4 = \begin{bmatrix} 0 \\ 1 \\ 0 \end{bmatrix} \quad \text{and} \quad Y_6 = \begin{bmatrix} 0 \\ 0 \\ 1 \end{bmatrix}.$$

Therefore the Y vectors of the basis vectors are always unit vectors* and can immediately be written in the appropriate columns without any computation.

(2) $z_j - c_j$—First, compute $z_j = \sum_{(i)=(1)}^{(m)} c_{(i)}y_{(i)j}$. For instance, $z_1 = c_{(1)}y_{(1)1} + c_{(2)}y_{(2)1} + c_{(3)}y_{(3)1} = 1 \cdot \frac{1}{2} + 0 \cdot (\frac{5}{2}) + 0 \cdot (-1) = \frac{1}{2}$. Since $c_1 = -1$, $z_1 - c_1 = \frac{1}{2} - (-1) = \frac{3}{2}$ which is recorded in the last row under P_1. Similar calculations result in the evaluation of other vectors as shown in Table 5.3. It is worthwhile to point out that the evaluators

Table 5.3

$c_{(i)}$	$P_{(i)}$ \ P_j		P_1	P_2	P_3	P_4	P_5	P_6	P_0
1	$P_{(1)}$	P_3	$\frac{1}{2}$	$\frac{1}{2}$	1	0	$\frac{1}{2}$	0	3
0	$P_{(2)}$	P_4	$\frac{5}{2}$	$-\frac{1}{2}$	0	1	$\frac{1}{2}$	0	7
0	$P_{(3)}$	P_6	-1	1	0	0	0	1	3
z_j			$\frac{1}{2}$	$\frac{1}{2}$	1	0	$\frac{1}{2}$	0	—
c_j			-1	1	1	0	0	0	—
$z_j - c_j$			$\frac{3}{2}$	$-\frac{1}{2}$	0	0	$\frac{1}{2}$	0	3

of the basis vectors are necessarily zero due to the fact that their Y vectors are always unit vectors as explained earlier. This stands to reason since $z_j - c_j$ is basically an evaluation of the "economic advantage" of the jth vector relative to the existing basis. If the vector is already in the basis, its "economic advantage" relative to the basis must be zero.

(3) θ_i—To determine the solution values of the basis variables, we proceed as follows:

---

* Let $Y = [Y_{(1)}Y_{(2)} \ldots Y_{(m)}]$ and $B = [P_{(1)}P_{(2)} \ldots P_{(m)}]$. Then, $Y = B^{-1} [P_{(1)} \ldots P_{(m)}] = B^{-1}B = I$. Hence, all Y vectors of basis vectors are unit vectors of an identity matrix.

$$\tilde{X} = \begin{bmatrix} x_{(1)} \\ x_{(2)} \\ x_{(3)} \end{bmatrix} = B^{-1}P_0 = \begin{bmatrix} 0 & \frac{1}{2} & 0 \\ 1 & \frac{1}{2} & 0 \\ 0 & 0 & 1 \end{bmatrix} \begin{bmatrix} 4 \\ 6 \\ 3 \end{bmatrix} = \begin{bmatrix} 3 \\ 7 \\ 3 \end{bmatrix}.$$

Therefore, $x_{(1)} = \theta_{(1)} = 3$, $x_{(2)} = \theta_{(2)} = 7$ and $x_{(3)} = \theta_{(3)} = 3$ which are recorded in the proper rows under P_0 in the tableau. Since the first basis variable is the third variable in the problem, i.e. $x_{(1)} = x_3$, the second basis variable is the fourth and the third basis variable is the sixth, the complete basic feasible solution is $X^T = [x_1\, x_2\, x_3\, x_4\, x_5\, x_6] = [0\ 0\ 3\ 7\ 0\ 3]$. This solution, of course, can easily be verified by substituting these values into the original equations and observing whether the equations are satisfied.

(4) $f(X)$—Once the specific values of the basis variables are known, it is a simple matter to evaluate the objective function by the formula $f(X) = \sum_{(i)=(1)}^{(m)} c_{(i)}\theta_{(i)}$. In this particular case $f(X) = c_{(1)}\theta_{(1)} + c_{(2)}\theta_{(2)} + c_{(3)}\theta_{(3)} = 1\cdot 3 + 0\cdot 7 + 0\cdot 3 = 3$, which is entered in the "$z_j - c_j$" row under P_0.

With the determination of $f(X)$, the construction of the simplex tableau is now complete. The finished tableau is shown in Table 5.3. The next logical question is, how can the information contained in the tableau be used to obtain a new and better solution in accordance with the Decision and Transformation Rules derived in previous chapters? We shall turn our attention to this question in the next section.

5.3. The Simplex Tableau—Its Transformation.

To obtain a new and better solution (basic feasible), the first step is to identify that non-basis vector which will improve the objective function in accordance with Decision Rule II (4.5.2). If the function is to be maximized, it should be a vector with a negative evaluation. If the function is to be minimized, then it should be a vector with a positive evaluation. Let this vector be P_k as shown in Table 5.1. This column is then known as the *key column.* The next step is to determine the basis vector to be displaced in accordance with Decision Rule I (4.2.6). This calls for a search for positive $y_{(i)k}$ in the key column and forming ratios of $\dfrac{\theta_{(i)}}{y_{(i)k}}$ for these positive $y_{(i)k}$. The smallest of these ratios then determines the basis vector to be displaced. Let this be $P_{(l)}$ as indicated by an out-pointing arrow in Table 5.1. This row is then known as the *key row* which together with the key column determines the *pivotal element* $y_{(l)k}$. Having selected the incoming and outgoing vectors, we are ready to transform the old tableau into a new one showing the new basic feasible solution, the new value of the objective function, the new linear combination coefficients and the new evaluations of P_j. This is accomplished by applying the Transformation Rules derived in Chapter IV, which are summarized below:

(1) New linear-combination coefficients:

$$y'_{(l)j} = \frac{y_{(l)j}}{y_{(l)k}}, \qquad j = 1, 2, \ldots, n + m. \tag{4.6.6}$$

$$y'_{(i)j} = y_{(i)j} - \frac{y_{(l)j}}{y_{(l)k}} y_{(i)k}, \qquad \begin{array}{l}(i) \neq (l), (i) = (1), \ldots, (m). \\ j = 1, 2, \ldots, n + m.\end{array} \tag{4.6.7}$$

(2) New evaluations of P_j:

$$z'_j - c_j = (z_j - c_j) - \frac{y_{(l)j}}{y_{(l)k}} (z_k - c_k), \quad j = 1, 2, \ldots, n + m. \tag{4.6.8}$$

(3) New basic feasible solution:

$$\theta'_{(l)} = \frac{\theta_{(l)}}{y_{(l)k}}. \tag{4.2.7}$$

$$\theta'_{(i)} = \theta_{(i)} - \frac{\theta_{(l)}}{y_{(l)k}} y_{(i)k}, \qquad (i) \neq (l), (i) = (1), (2), \ldots, (m). \tag{4.2.8}$$

(4) New value of the objective function:

$$f(X') = f(X) - \frac{\theta_{(l)}}{y_{(l)k}} (z_k - c_k). \tag{4.5.3}$$

A close examination of the above formulas in relation to Table 5.1, however, reveals that due to the structure of the simplex tableau, the six rather complicated rules listed above can be condensed into just two working rules stated in terms of the relative positions of the data involved instead of their nature. The first one concerns the key row, i.e. the row of $P_{(l)}$. If the element to be transformed is in the key row, whether it be θ or y, the new value is always obtained by dividing the pivotal element into the old value of the element. This is true of (4.6.6) and (4.2.7) above. The second rule can be more easily grasped by reference to Figure 5.1 below. This figure shows the key row, the key column, the (i)th row and the jth column, with the latter two determined by the row and column indices of the element whose transformation is under consideration. These two rows and two columns intersect, and thus determine four elements in the tableau which are involved in the transformation. Element A is the element whose transformation is desired. Element B is fixed by the key row and the jth column just as element C is fixed by the key column and the (i)th row. Finally, element D is the pivotal element $y_{(l)k}$.

The second rule is that the transformed value of A, i.e., A′, whether it be $y_{(i)j}$, $\theta_{(i)}$, $(z_j - c_j)$ or $f(X)$, is always obtained by the following formula:

$$\text{A}' = \text{A} - \frac{\text{B}}{\text{D}}\,\text{C}$$

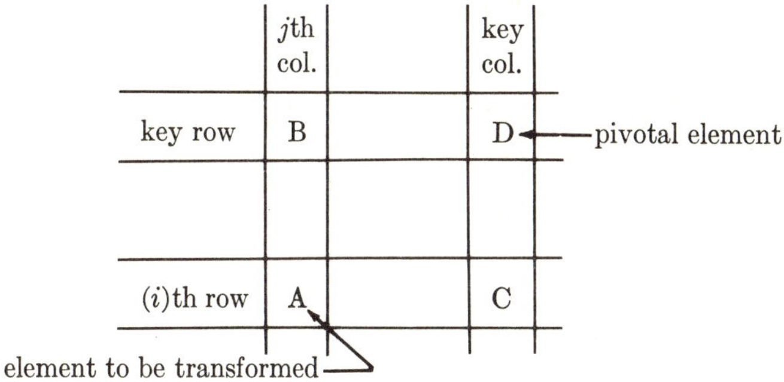

FIGURE 5.1

As intimated earlier, this rule holds true for (4.6.7), (4.6.8), (4.2.8) and (4.5.3). In fact, it is applicable to any element in the tableau except those in the key row or the z_j and c_j rows which are not transformed. It is seldom that complex theory lends itself to simple computations. The simplex method, however, is a notable exception. The structural similarity of its various transformation formulas makes it possible for the transformation of the entire tableau by the use of only two simple working rules. This fact undoubtedly has contributed significantly to the acceptance of linear programming as a practical tool of analysis.

Example. We will illustrate the discussion in this section by transforming Table 5.3 which represents the first basic feasible solution to the problem in the example of Section 5.2.

Referring to Table 5.3, we must first scan the "$z_j - c_j$" row for the vector with the most negative evaluation, since the problem is a maximizing one. This turns out to be P_2. Next, the P_2 column is scanned for positive $y_{(i)2}$. These are $y_{(1)2}$ and $y_{(3)2}$. Forming and comparing the two ratios, $\frac{\theta_{(1)}}{y_{(1)2}}$ and $\frac{\theta_{(3)}}{y_{(3)2}}$, we find that the latter is the smaller. According to Decision Rule I, $P_{(3)}$ $(= P_6)$ should be the outgoing basis vector. Thus, $P_{(3)}$ is the key row, P_2 is the key column and $y_{(3)2} = 1$ is the pivotal element. Once these identifications have been made, we can proceed to transform the tableau in accordance with the two working rules suggested in the text.

The resulting new tableau is shown in Table 5.4. It will be noticed

Table 5.4

$c_{(i)}$	$P_{(i)}$ \ P_j		P_1	P_2	P_3	P_4	P_5	P_6	P_0
1	$P_{(1)}$	P_3	1	0	1	0	$\frac{1}{2}$	$-\frac{1}{2}$	$\frac{3}{2}$
0	$P_{(2)}$	P_4	2	0	0	1	$\frac{1}{2}$	$\frac{1}{2}$	$\frac{17}{2}$
1	$P_{(3)}$	P_2	-1	1	0	0	0	1	3
	z_j								—
	c_j								—
	$z_j - c_j$		1	0	0	0	$\frac{1}{2}$	$\frac{1}{2}$	$\frac{9}{2}$

that in the $P_{(i)}$ column, P_2 is now identified as the third basis vector $(P_{(3)})$ instead of P_6 and accordingly $c_2 = 1$ is entered as $c_{(3)}$ instead of $c_6 = 0$. By the first working rule stated in the text, the new elements in the key row $(P_{(3)})$ are obtained by dividing the pivotal element $y_{(3)2} = 1$ into the old elements in that row (including $y_{(3)2}$ itself). The remaining elements in the tableau with the exception of the "z_j" and "c_j" rows are then transformed according to the second working rule described in the text, viz.,

$$A' = A - \frac{B}{D}C.$$

It will be noted that the "z_j" and the "c_j" rows of Table 5.4 are left vacant. The reason is that they do not follow the same transformation rules as the rest. Obviously the c_j remain the same as given in the objective function. The z_j should be calculated from $\sum_{(i)=(1)}^{(m)} c_{(i)}y_{(i)j}$. Although their calculations are not necessary for the derivation of new basic feasible solution (since the evaluators $z_j - c_j$ can be transformed directly), they are useful as independent checks on the validity of the transformed $y_{(i)j}$ and $z_j - c_j$.

5.4. Obtaining an Initial Basic Feasible Solution—Using Slack Vectors.

So far, we have dealt extensively with the problems of searching for better basic feasible solutions and of recognizing the end of the search. For reasons of clarity, the question of how to start such a search has been avoided until now. We have in our illustrations always "managed" somehow to find an initial basis which provided the starting point for subsequent iterative improvement towards the final optimal solution. In practice, of course, such an initial basis cannot be "guessed" correctly all the time. In this and the following sections we shall develop a standard procedure for the systematic determination of an initial feasible solution—so that wasteful guesswork may be eliminated altogether when desirable. The reader should keep in mind, however, that in solving a real industrial linear programming problem, established practices often provide significant clues for an intelligent guess at an initial basic feasible solution, or may even directly represent one themselves. In such cases it may be less time-consuming to use the existing solution as the starting solution instead of following the procedure described below.

Basically finding an initial solution means the selection of a set of m independent vectors to form a basis B so that $B^{-1}P_0$ yields a unique feasible solution. There are two complicating factors in this task. One is the independence required of the selected vectors and the other is the feasibility (or non-negativity) required of the solution. Unless the vectors are independent, there will not be an inverse of the basis. This will be revealed readily in the process of inversion or by an evaluation of $|B|$ to see if B is singular. Thus, this complicating factor does not present too much difficulty, especially in view of the fact that the probability of dependence is small among a large set of vectors selected at random. The second complicating factor, however, really presents a problem for there is no way of knowing in advance whether the solution will be feasible until the basis is inverted and the solution actually determined. Since matrix inversion is not simple, a lot of time can be wasted in trying different sets of independent vectors before a set is found that yields a feasible solution.* Therefore, it is advisable to develop a procedure which guarantees an initial basic feasible solution.

On the assumption that the right-hand-side constants of the constraints are all non-negative (if for some constraint the constant is negative, then multiply both sides by -1, and simultaneously reverse the inequality sign), the following three cases may be distinguished: (1) all constraints are of the $\leq$ type, (2) all constraints are of the $\geq$ type and (3) the constraints represent a mixture of both types. Case (1) will be examined in this section and cases (2) and (3) will be discussed in Sections 5.5 and 5.6 respectively.

Assume the system of linear constraints in a linear programming problem to be as follows:

$$P_1x_1 + P_2x_2 + \ldots + P_nx_n \leq P_0, \quad \text{where } P_0 \geq O \text{ (a zero vector)}$$

Just as in Section 5.2, we may convert the above into a system of equations by adding slack variables. Thus,

$$P_1x_1 + P_2x_2 + \ldots + P_nx_n + P_{n+1}x_{n+1} + \ldots + P_{n+m}x_{n+m} = P_0,$$

where, it will be recalled, $P_{n+1}, \ldots, P_{n+m}$ are a set of m unit vectors, called *slack vectors*, which together form an identity matrix. Let $B_0 = [P_{n+1}, \ldots, P_{n+m}] = I$ be the initial basis, then obviously,

$$\tilde{X}_0 = B_0^{-1}P_0 = I^{-1}P_0 = IP_0 = P_0,$$

$$\text{where } \tilde{X}_0 = \begin{bmatrix} x_{n+1} \\ \cdot \\ \cdot \\ \cdot \\ x_{n+m} \end{bmatrix} \quad \text{and} \quad P_0 = \begin{bmatrix} b_1 \\ \cdot \\ \cdot \\ \cdot \\ b_m \end{bmatrix}.$$

This means that the initial solution is $x_{n+1} = b_1,\ x_{n+2} = b_2, \ldots,$

* The reader will notice that the simplex method guarantees continued feasibility of successive basic solutions only if the initial solution is itself feasible. The explanation of this is left as an exercise for the curious reader who should consider carefully the effect of negative θ_i on the use of Decision Rule I (4.2.6).

$x_{n+m} = b_m$ and $x_1 = x_2 = \ldots = x_n = 0$. This solution is bound to be feasible since P_0 is assumed to be non-negative. Furthermore, it can be obtained easily without computation since the basis is I whose inverse is also I. In other words, if all constraints are of the $\leq$ type, a "natural" initial basis exists in the set of m slack vectors which yields immediately an initial basic solution that must be feasible. Furthermore, since $B_0^{-1} = I$, the linear-combination coefficient vectors Y_j are simply the corresponding coefficient vectors P_j themselves. This is so because $Y_j = B_0^{-1}P_j = IP_j = P_j$. Thus, a simplex tableau can be quickly constructed on the basis of $B_0 = I$ and iteration begun for the determination of the optimal solution. The problem of obtaining an initial basic feasible solution in case (1) is therefore extremely simple and straightforward.

Example. Let the problem be,

$$\text{Maximize } f(X) = x_1 - x_2 + 3x_3,$$

$$\begin{aligned} \text{subject to:} \quad x_1 + x_2 + x_3 &\leq 10, \\ 2x_2 - x_3 &\leq 2, \\ 2x_1 - 2x_2 + 3x_3 &\leq 0, \\ \text{and} \quad x_j \geq 0, \quad j &= 1, 2, 3. \end{aligned}$$

By adding slack variables, we have the following equivalent problem:

$$\text{Maximize } f(X) = x_1 - x_2 + 3x_3,$$

$$\begin{aligned} \text{subject to:} \quad x_1 + x_2 + x_3 + x_4 \qquad\qquad &= 10, \\ 2x_2 - x_3 \qquad + x_5 \qquad &= 2, \\ 2x_1 - 2x_2 + 3x_3 \qquad\qquad + x_6 &= 0, \\ \text{and} \quad x_j \geq 0, \quad j = 1, 2, \ldots&, 6, \end{aligned}$$

which suggests immediately as discussed in Section 5.4 that the initial basis B_0 could be composed of the three unit vectors P_4, P_5 and P_6. Hence

$$\tilde{X} = B_0^{-1}P_0 = IP_0 = P_0 = \begin{bmatrix} 10 \\ 2 \\ 0 \end{bmatrix},$$

and similarly

$$Y_1 = B_0^{-1}P_1 = IP_1 = P_1 = \begin{bmatrix} 1 \\ 0 \\ 2 \end{bmatrix},$$

$$Y_2 = B_0^{-1}P_2 = IP_2 = P_2 = \begin{bmatrix} 1 \\ 2 \\ -2 \end{bmatrix},$$

and
$$Y_3 = B_0^{-1}P_3 = IP_3 = P_3 = \begin{bmatrix} 1 \\ -1 \\ 3 \end{bmatrix}.$$

This means an initial basic feasible solution that is $X^T = [x_1\ x_2\ x_3\ x_4\ x_5\ x_6] = [0\ 0\ 0\ 10\ 2\ 0]$ and an initial simplex tableau as shown in Table 5.5. This tableau will serve as the starting point of

Table 5.5

$c_{(i)}$	$P_{(i)}$ \ P_j		P_1	P_2	P_3	P_4	P_5	P_6	P_0
0	$P_{(1)}$	P_4	1	1	1	1	0	0	10
0	$P_{(2)}$	P_5	0	2	−1	0	1	0	2
0	$P_{(3)}$	P_6	2	−2	3	0	0	1	0
z_j			0	0	0	0	0	0	—
c_j			1	−1	3	0	0	0	—
$z_j - c_j$			−1	1	−3	0	0	0	0

an iterative process leading to the determination of the optimal solution.

In conclusion, it should be noted that although the entries under P_1, P_2, etc. in Table 5.5 are exactly the same as the components of P_1, P_2, etc., they have completely different meanings. Under P_1 in Table 5.5, for instance, the entries really do not represent $P_1 = \begin{bmatrix} 1 \\ 0 \\ 2 \end{bmatrix}$, but rather signify that $P_1 = 1 \cdot P_4 + 0 \cdot P_5 + 2P_6$. To underscore the point, the reader should try to set up a tableau for the same solution but with the basis vectors arranged in an order different from P_4, P_5 and P_6; for instance, in the order of P_6, P_5 and P_4.

5.5. Obtaining an Initial Basic Feasible Solution—Using Artificial Vectors.

If the constraints are all of the $\geq$ type such as represented by the following inequality:

$$P_1x_1 + P_2x_2 + \ldots + P_nx_n \geq P_0, \quad \text{where } P_0 \geq 0, \tag{5.5.1}$$

how then can we insure an initial basic feasible solution without resorting to the haphazard method of selecting at random a set of independent P_j vectors to form a basis?

To answer this question, we first note that the inequalities (5.5.1) can be converted to the following equivalent equation by subtracting non-negative slack variables $x_{n+1}, x_{n+2}, \ldots, x_{n+m}$.

$$P_1x_1 + P_2x_2 + \ldots + P_nx_n + P_{n+1}x_{n+1} + \ldots + P_{n+m}x_{n+m} = P_0, \tag{5.5.2}$$

$$\text{where } x_j \geq 0, \qquad j = 1, \ldots, n, \ldots, n+m,$$

$$P_{n+1} = \begin{bmatrix} -1 \\ 0 \\ \cdot \\ \cdot \\ \cdot \\ 0 \end{bmatrix}, \ldots, P_{n+m} = \begin{bmatrix} 0 \\ 0 \\ \cdot \\ \cdot \\ \cdot \\ -1 \end{bmatrix} \quad \text{and} \quad P_0 \geq 0.$$

Although the slack vectors here are still unit vectors and linearly independent, if we let the initial basis $B_0 = [P_{n+1}\, P_{n+2} \ldots P_{n+m}]$, we would get an initial basic solution $\tilde{X}_0 = B_0^{-1}P_0 = [P_{n+1}\, P_{n+2} \ldots P_{n+m}]^{-1}P_0 = (-I)^{-1}P_0 = -IP_0 = -P_0$ which is unfortunately not feasible, and therefore unacceptable. In this section we shall discuss a method by which the above difficulty can be overcome.

Suppose we were to add an *artificial variable* to each of the m equations implied in (5.5.2) we should obtain the following expanded system of constraints:

$$\begin{array}{llllll}
a_{11}x_1 + \ldots + a_{1n}x_n & - x_{n+1} & & + x_{n+m+1} & & = b_1, \\
a_{21}x_1 + \ldots + a_{2n}x_n & & - x_{n+2} & & + x_{n+m+2} & = b_2, \\
\cdot & \cdot & \cdot & \cdot & & \cdot \\
\cdot & \cdot & \cdot & \cdot & & \cdot \\
\cdot & \cdot & \cdot & \cdot & & \cdot \\
a_{m1}x_1 + \ldots + a_{mn}x_n & & - x_{n+m} & & + x_{n+2m} & = b_m.
\end{array}$$

This can be written as

$$(5.5.3) \qquad P_1x_1 + \ldots + P_nx_n - P_{n+1}x_{n+1} - \ldots - P_{n+m}x_{n+m} + P_{n+m+1}x_{n+m+1} + \ldots + P_{n+2m}x_{n+2m} = P_0,$$

where $P_1, \ldots, P_{n+m}$ are coefficient vectors of $x_1, \ldots, x_{n+m}$

As defined in (5.5.2), $P_0 \geq 0$ and

$$P_{n+m+1} = \begin{bmatrix} 1 \\ 0 \\ \cdot \\ \cdot \\ \cdot \\ 0 \end{bmatrix}, P_{n+m+2} = \begin{bmatrix} 0 \\ 1 \\ \cdot \\ \cdot \\ \cdot \\ 0 \end{bmatrix}, \ldots, P_{n+2m} = \begin{bmatrix} 0 \\ 0 \\ \cdot \\ \cdot \\ \cdot \\ 1 \end{bmatrix},$$

Clearly, for the reason discussed in Section 5.4, the set of m positive unit vectors $P_{n+m+1}, \ldots, P_{n+2m}$ form a basis that is an identity matrix, and so yield a non-negative basic solution which is $x_{n+m+1} = b_1$, $x_{n+m+2} = b_2, \ldots, x_{n+2m} = b_m$. To be sure, this is not a solution to the original set of constraints (5.5.2) since $x_{n+m+1}, \ldots, x_{n+2m}$ are not in the original equations. But it is non-negative, hence can be used as a starting solution for the simplex iterative procedure. Since the simplex procedure guarantees continued feasibility of solutions, we may eventually arrive at a basic solution that is feasible and at the same time satisfies the original constraints—assuming, of course, that such a solution exists in the first place. A moment's reflection makes it clear that this could be achieved if we succeed by simplex iterations in reducing all artificial variables to zero. This in turn means that all artificial vectors $P_{n+m+1}, \ldots, P_{n+2m}$ must be displaced from the basis. Thus we conclude that the way to obtain a legitimate basic feasible solution from an initial artificial one is to assign such grossly unfavorable price coefficients to the artificial variables that in the normal process of optimization they will inevitably be replaced by the regular variables in the basis. In other words, if the original problem calls for the maximization of $f(X) = \sum_{j=1}^{n} c_jx_j$ subject to constraints (5.5.1) and $x_j \geq 0$, we will now maximize $f'(X) = \sum_{j=1}^{n} c_jx_j + 0\left(\sum_{i=1}^{m} x_{n+i}\right) - \text{M}\left(\sum_{i=1}^{m} x_{n+m+i}\right)$ subject to (5.5.3) and $x_j \geq 0$, $j = 1, \ldots, n + 2m$, where $-\text{M}$ symbolizes the price coefficient of negative infinity. Since $-\text{M}$ is infinitely more unfavorable than any c_j or combination of c_j, any specific $f(X)$ will be better than a $f'(X)$ in which x_{n+m+i} are not entirely zero. Thus, so long as such a $f(X)$ exists (i.e. the original problem has a solution), the maximization of $f'(X)$ will eventually lead to it through the complete elimination of all artificial vectors from the basis. Contrarily, this also implies that if not all of them can be eliminated, then the original problem must have no solution. In the case of minimization, it is clear that the same guarantee of proper optimal solution can be obtained by assigning $+\mathrm{M}$ (instead of $-\mathrm{M}$) to be the price coefficients of the artificial variables.

Before we proceed with a numerical illustration of the above approach, let us make two more related observations. The first is that occasionally a problem may contain a constraint that is strictly an equation rather than an inequality. It should be obvious that such constraints could be handled in exactly the same way by adding artificial variables. The only difference would be that slack variables would be neither necessary nor appropriate. The second observation is that since the fundamental reason for adding artificial variables is to develop an identity matrix to serve as the initial basis, it becomes clear that if some of the original P_j are already unit vectors, then to the extent that they could be used to form an identity matrix, the number of additional artificial vectors required would be reduced. This point is illustrated in the Example of Section 5.6 below.*

Example. Maximize $f(X) = 2x_1 - x_2 + 2x_3$,

subject to:
$$\begin{aligned} x_1 + x_2 + x_3 &\geq 6, \\ -2x_1 + x_3 &\geq 2, \\ 2x_2 - x_3 &\geq 0, \end{aligned}$$
and
$$x_1, x_2, x_3 \geq 0.$$

First, we convert and modify the problem into the following form by adding and subtracting artificial and slack variables with appropriate price coefficients.

Maximize $f'(X)$
$$= 2x_1 - x_2 + 2x_3 + 0x_4 + 0x_5 + 0x_6 - \mathrm{M}x_7 - \mathrm{M}x_8 - \mathrm{M}x_9,$$

* By a simple linear transformation of the system of constraints, the number of artificial vectors in the system can be reduced to only one, which will materially shorten the number of iterations required to reach optimal solution. See S. I. Gass, *op. cit.*, pp. 120–122.

subject to:

$$\begin{aligned}
x_1 + x_2 + x_3 - x_4 \quad + x_7 \quad &= 6,\\
-2x_1 \quad + x_3 \quad - x_5 \quad + x_8 \quad &= 2,\\
2x_2 - x_3 \quad - x_6 \quad + x_9 &= 0,
\end{aligned}$$

and $\qquad x_j \geq 0, \quad j = 1, 2, \ldots, 9.$

In the above formulation, x_4, x_5, x_6 are slack variables and x_7, x_8, x_9 are artificial variables. The modified objective function shows that the slack variables are given a price coefficient of zero, which is equivalent to their absence in the original objective function. The artificial variables are given a price coefficient of $-$M to insure their ultimate elimination from the basis, if that is possible.

Using the artificial vectors which form an identity matrix as the initial basis and following procedure shown in the Example of Section 5.4 we obtain the initial basic feasible solution of $X^T = [x_1\ x_2\ x_3\ x_4\ x_5\ x_6\ x_7\ x_8\ x_9] = [0\ 0\ 0\ 0\ 0\ 0\ 6\ 2\ 0]$. In setting up the initial simplex tableau we note that the two different types of price coefficients, $-$M associated with the artificial vectors and the usual numerical c_j associated with the other vectors make it desirable to split the z_j, c_j and $z_j - c_j$ rows into two parts; one (the upper half) to record the coefficient of M and the other (the lower half) to record the numerical component in the expression. For instance, relative to the initial basis

Table 5.6a

$c_{(i)}$	$P_{(i)}$ \ P_j		P_1	P_2	P_3	P_4	P_5	P_6	P_7	P_8	P_9	P_0
$-$M	$P_{(1)}$	P_7	1	1	1	-1	0	0	1	0	0	6
$-$M	$P_{(2)}$	P_8	-2	0	1	0	-1	0	0	1	0	2
$-$M	$P_{(3)}$	P_9	0	(2)	-1	0	0	-1	0	0	1	0
z_j		M	1	-3	-1	1	1	1	-1	-1	-1	—
		N	0	0	0	0	0	0	0	0	0	—
c_j		M	0	0	0	0	0	0	-1	-1	-1	—
		N	2	-1	2	0	0	0	0	0	0	—
$z_j - c_j$		M	1	-3	-1	1	1	1	0	0	0	-8
		N	-2	1	-2	0	0	0	0	0	0	0

$B_0 = [P_7\, P_8\, P_9]$, $z_1 = -\text{M}(1) - \text{M}(-2) - \text{M}(0) = \text{M}$. Hence, 1 is written in the upper half of the z_j row (marked "M" in the stub) and 0 in the lower half (marked "N") as shown in Table 5.6a. Since $c_1 = 2 = 0 \cdot \text{M} + 2$, 0 and 2 are entered in the "M" and "N" parts of the c_j row respectively. Finally, $z_1 - c_1 = \text{M} - 2$. Accordingly, 1 and -2 are written into the "M" and "N" parts of the $z_j - c_j$ row respectively. With all other data computed as explained in the previous sections, we obtain the initial tableau as shown in Table 5.6a.*

In choosing the incoming vector for the next basic feasible solution, it must be borne in mind that since M represents infinity, the decision is always dominated by the sign of M alone if the evaluation is composite, i.e., if $z_j - c_j$ contains both an M component and a numerical value. Scanning the upper part of the $z_j - c_j$ row, we find

Table 5.6b

$c_{(i)}$	$P_{(i)}$ \ P_j		P_1	P_2	P_3	P_4	P_5	P_6	P_7	P_8	P_0
$-$M	$P_{(1)}$	P_7	1	0	$\frac{3}{2}$	-1	0	$\frac{1}{2}$	1	0	6
$-$M	$P_{(2)}$	P_8	-2	0	(1)	0	-1	0	0	1	2
-1	$P_{(3)}$	P_2	0	1	$-\frac{1}{2}$	0	0	$-\frac{1}{2}$	0	0	0
z_j		M	1	0	$-\frac{5}{2}$	1	1	$-\frac{1}{2}$	-1	-1	—
		N	0	-1	$\frac{1}{2}$	0	0	$\frac{1}{2}$	0	0	—
c_j		M	0	0	0	0	0	0	-1	-1	—
		N	2	-1	2	0	0	0	0	0	—
$z_j - c_j$		M	1	0	$-\frac{5}{2}$	1	1	$-\frac{1}{2}$	0	0	-8
		N	-2	0	$-\frac{3}{2}$	0	0	$\frac{1}{2}$	0	0	0

that P_2 has a negative coefficient of M, so we choose it to be the incoming vector. Applying Decision Rule I, we further discover that P_9 is the basis vector that should be replaced. Since it is an artificial vector that does not belong to the original problem, it will be elim-

* The M part of the c_j row corresponds to the auxiliary objective function, $Q(X) = 0(x_1 + x_2 + \ldots + x_{n+m}) + (-1)(x_{n+m+1} + \ldots + x_{n+2m})$ proposed by Harvey Wagner in "A Two Phase Method for the Simplex Tableau," *Operations Research*, IV, (1956), No. 4, pp. 443–447.

inated from the next tableau altogether. The pivotal element is now fixed and encircled in Table 5.6a. In the ensuing transformation it can be shown readily that the "M" part of $z_j - c_j$ follows Transformation Rule III just as the "N" part does separately. We obtain therefore the second simplex tableau as shown in Table 5.6b.

The solution is now $X^T = [x_1\, x_2\, x_3\, x_4\, x_5\, x_6\, x_7\, x_8] = [0\,0\,0\,0\,0\,0\,6\,2]$ and $f'(X) = -8\text{M}$ as in the first solution. No improvement in $f'(X)$ is achieved, but a new basis is established which opens ways to possible future improvement, as we shall see presently.

Repeated applications of the same criteria reveal that P_3 can be introduced into the basis in place of P_8 with the result shown in Table 5.6c. In the meantime, P_8 now displaced, is also removed from the tableau.

Table 5.6c

$c_{(i)}$	$P_{(i)}$ \ P_j		P_1	P_2	P_3	P_4	P_5	P_6	P_7	P_0
$-$M	$P_{(1)}$	P_7	(4)	0	0	-1	$\frac{3}{2}$	$\frac{1}{2}$	1	3
2	$P_{(2)}$	P_3	-2	0	1	0	-1	0	0	2
-1	$P_{(3)}$	P_2	-1	1	0	0	$-\frac{1}{2}$	$-\frac{1}{2}$	0	1
z_j		M	-4	0	0	1	$-\frac{3}{2}$	$-\frac{1}{2}$	-1	—
		N	-3	-1	2	0	$-\frac{3}{2}$	$\frac{1}{2}$	0	—
c_j		M	0	0	0	0	0	0	-1	—
		N	2	-1	2	0	0	0	0	—
$z_j - c_j$		M	-4	0	0	1	$-\frac{3}{2}$	$-\frac{1}{2}$	0	-3
		N	-5	0	0	0	$-\frac{3}{2}$	$\frac{1}{2}$	0	3

We have then the solution, $X^T = [x_1\, x_2\, x_3\, x_4\, x_5\, x_6\, x_7] = [0\,1\,2\,0\,0\,0\,3]$, with $f'(X) = -3\text{M} + 3$ which is greater than the previous value of -8M.

Continuing the iteration, we select P_1 to replace P_7, which happens to be the last artificial vector remaining. Its removal firmly establishes a basic feasible solution to the original problem and at the same time reduces the tableau to its simpler conventional form, i.e., without split rows for z_j, c_j and $z_j - c_j$. Table 5.6d shows the results of the third transformation.

Table 5.6d

$c_{(i)}$	$P_{(i)}$ \ P_j		P_1	P_2	P_3	P_4	P_5	P_6	P_0
2	$P_{(1)}$	P_1	1	0	0	$-\frac{1}{4}$	$\frac{3}{8}$	$\frac{1}{8}$	$\frac{3}{4}$
2	$P_{(2)}$	P_3	0	0	1	$-\frac{1}{2}$	$-\frac{1}{4}$	$\frac{1}{4}$	$\frac{7}{2}$
-1	$P_{(3)}$	P_2	0	1	0	$-\frac{1}{4}$	$-\frac{1}{8}$	$-\frac{3}{8}$	$\frac{7}{4}$
z_j			2	-1	2	$-\frac{5}{4}$	$\frac{3}{8}$	$\frac{9}{8}$	—
c_j			2	-1	2	0	0	0	—
$z_j - c_j$			0	0	0	$-\frac{5}{4}$	$\frac{3}{8}$	$\frac{9}{8}$	$\frac{27}{4}$

The solution at this stage is $X^T = [x_1\, x_2\, x_3\, x_4\, x_5\, x_6] = [\frac{3}{4}\, \frac{7}{4}\, \frac{7}{2}\, 0\, 0\, 0]$ with $f'(X) = f(X) = \frac{27}{4}$.

Looking for further improvement, we notice that P_4 should be introduced. However, since all $y_{(i)4} \leq 0$, according to Chapter IV Section 4.3, we conclude that this problem has no upper bound. In other words, we could select any large, positive, arbitrary value for x_4, thereby increasing x_1, x_2 and x_3 correspondingly, and obtain a large, arbitrary value of the objective function. This means that if we let $x_4 = \infty$, x_1, x_2 and x_3 will also equal infinity and the value of the objective function will increase beyond all bounds. An incidental point made clear by this example is that maximizing a linear function whose variables are subject to linear constraints of the $\geq$ type, i.e., bounded from below, usually lead to an optimal solution which is unbounded from above. An exception is, of course, when the c_j in the objective function are all negative, in which case the largest $f(X)$ will be a finite value. Stripped of mathematical terminologies, this "revelation" is, of course, nothing more than simple common sense. (Why?).

5.6. Obtaining an Initial Basic Feasible Solution—Using Mixed Slack and Artificial Vectors.

In this section, we will deal with the general case of mixed constraints, i.e., constraints of both the $\leq$ and the $\geq$ types with the same assumption of $P_0 \geq 0$ as before.

The theoretical considerations involved in linear programming problems having both types of constraints are quite straightforward. Under the assumption of $P_0 \geq 0$, an initial feasible basis can evidently be formed with the unit slack vectors in the $\leq$ type of constraints and with as many artificial vectors as there are constraints of the $\geq$ (or the $=$) type remaining. The objective function, to be sure, should be modified as suggested in Section 5.5 in order to insure the displacement of the artificial vectors from the basis as soon as possible. But once these changes have been made, routine application of the standard simplex procedure will take the solution to its logical end.

A minor complication arises when some constraints have $b_i < 0$. In such cases, however, b_i can be made positive by multiplying the whole inequality by -1 and simultaneously reversing the direction of its inequality. For instance, the constraint $2x_1 - 3x_2 + 3x_3 \leq -5$ is equivalent to $-2x_1 + 3x_2 - 3x_3 \geq 5$ and the constraint $-x_1 - x_2 \geq -2$ is equivalent to $x_1 + x_2 \leq 2$. After all negative b_i have been so converted into positive, an initial feasible basis can be constructed as discussed earlier in this section.

In summary, we may set forth a general procedure for handling all kinds of linear constraints encountered in a linear programming problem. This consists of the following steps:

(a) Check the signs of b_i. Convert all negative b_i to positive by multiplying through the relevant constraints by -1, at the same time reversing the inequality signs.

(b) Convert all inequality constraints into equations by:

(1) adding non-negative slack variables to constraints of the $\leq$ type;

(2) subtracting non-negative slack variables to constraints of the $\geq$ type.

(c) Add non-negative artificial variables *as needed* to constraints of the $\geq$ and the $=$ types to develop an identity matrix for use as the initial basis.

Once the initial basis is established (which will be feasible according to the procedure described above), the solution of the problem follows the method discussed in Sections 5.2 and 5.3. In the following, two examples will be provided to illustrate in order (1) the procedure for handling mixed constraints, and (2) the case of a solution-less linear programming problem.

Example 1. Minimize $f(X) = 3x_1 - 2x_2 + x_3$,

$$\begin{aligned} \text{subject to:} \quad & x_1 + x_3 \geq 6, \\ & 2x_1 - x_3 \leq 4, \\ & x_1 - x_2 \leq -1, \\ \text{and} \quad & x_1, x_2, x_3 \geq 0 \end{aligned}$$

Since the third constraint has a negative constant, we first multiply it through by -1 and obtain $-x_1 + x_2 \geq 1$. Then adding the slack variable x_5 to the fourth constraint and subtracting the slack variables x_4 and x_6 from the first and third constraints respectively, we arrive at the following:

$$\text{Minimize } f(X) = 3x_1 - 2x_2 + x_3,$$

$$\begin{aligned} \text{subject to:} \quad & x_1 + x_3 - x_4 = 6, \\ & 2x_1 - x_3 + x_5 = 4, \\ & -x_1 + x_2 - x_6 = 1, \\ \text{and} \quad & x_j \geq 0, \quad j = 1, 2, \ldots, 6. \end{aligned}$$

It is obvious from the above that only one additional artificial variable is needed to make up the identity matrix since P_2 and P_5 are independent unit vectors already in the problem. This artificial variable, designated x_7, must be added to the first constraint so that its coefficient vector $P_7 = \begin{bmatrix} 1 \\ 0 \\ 0 \end{bmatrix}$ is of the necessary form to make up an identity matrix with P_2 and P_5. Thus, the problem is finally modified into the following form:

Minimize $f'(X) =$

$$3x_1 - 2x_2 + x_3 + 0 \cdot x_4 + 0 \cdot x_5 + 0 \cdot x_6 + Mx_7$$

$$\begin{aligned} \text{subject to:} \quad & x_1 + x_3 - x_4 + x_7 = 6, \\ & 2x_1 - x_3 + x_5 = 4, \\ & -x_1 + x_2 - x_6 = 1, \\ \text{and} \quad & x_j \geq 0, \quad j = 1, 2, \ldots, 7. \end{aligned}$$

Using P_2, P_5 and P_7 as basis vectors, we construct the initial simplex tableau as shown in Table 5.7a.

Table 5.7a

$c_{(i)}$	$P_{(i)}$ \ P_j		P_1	P_2	P_3	P_4	P_5	P_6	P_7	P_0
−2	$P_{(1)}$	P_2	−1	1	0	0	0	−1	0	1
0	$P_{(2)}$	P_5	2	0	−1	0	1	0	0	4
M	$P_{(3)}$	P_7	1	0	(1)	−1	0	0	1	6
z_j		M	1	0	1	−1	0	0	1	—
		N	2	−2	0	0	0	2	0	—
c_j		M	0	0	0	0	0	0	1	—
		N	3	−2	1	0	0	0	0	—
$z_j - c_j$		M	1	0	1	−1	0	0	0	6
		N	−1	0	−1	0	0	2	0	−2

Since $f'(X)$ is to be minimized, the incoming vector should have a positive evaluation (Chapter IV Section 4.5). This points to either P_1 or P_3. Looking ahead, however, we notice that if P_3 is introduced into the basis, the artificial vector P_7 will be displaced. Therefore, we select P_3 to go into the basis in place of P_7 which is presently dropped from the tableau altogether. The second tableau, after proper transformation to reflect the new basis, appears as follows:

Table 5.7b

$c_{(i)}$	$P_{(i)}$ \ P_j		P_1	P_2	P_3	P_4	P_5	P_6	P_0
−2	$P_{(1)}$	P_2	−1	1	0	0	0	−1	1
0	$P_{(2)}$	P_5	3	0	0	−1	1	0	10
1	$P_{(3)}$	P_3	1	0	1	−1	0	0	6
z_j			3	−2	1	−1	0	2	—
c_j			3	−2	1	0	0	0	—
$z_j - c_j$			0	0	0	−1	0	2	4

At this juncture, it becomes clear that only P_6 can bring about a further decrease in $f(X)$ [it will be recalled that when $x_7 = 0$, as now,

$f'(X) = f(X)$]. But since all $y_{(i)6} \leq 0$, a solution with no lower bound is indicated. More specifically, $f(X)$ can be decreased to negative infinity by letting x_6 be positive infinity. As a result, we adjust the other variables to the following values, $x_2 = \infty$, $x_3 = 6$, $x_5 = 10$, and $x_1 = x_4 = 0$.

Example 2. Minimize $f(X) - 3x_1 + 2x_2 - x_3$,

$$\begin{aligned} \text{subject to:} \quad & x_1 + x_2 + x_3 \leq 6, \\ & x_1 \qquad - x_3 \geq 4, \\ & \qquad x_2 - x_3 \geq 3, \\ \text{and,} \quad & x_1, x_2, x_3 \geq 0. \end{aligned}$$

Modifying the problem along the same lines as in Example 1, we have:

Minimize $f'(X) = 3x_1 + 2x_2 - x_3 + \text{M}x_7 + \text{M}x_8$

$$\begin{aligned} \text{subject to:} \quad & x_1 + x_2 + x_3 + x_4 = 6, \\ & x_1 - x_3 - x_5 + x_7 = 4, \\ & x_2 - x_3 - x_6 + x_8 = 3, \\ \text{and,} \quad & x_j \geq 0, \quad j = 1, 2, \ldots, 8. \end{aligned}$$

The ensuing sequence of tableaus (Tables 5.8a–5.8c) traces the course of its solution.

Table 5.8a

$c_{(i)}$	$P_{(i)}$	P_j	P_1	P_2	P_3	P_4	P_5	P_6	P_7	P_8	P_0
0	$P_{(1)}$	P_4	1	1	1	1	0	0	0	0	6
M	$P_{(2)}$	P_7	(1)	0	−1	0	−1	0	1	0	4
M	$P_{(3)}$	P_8	0	1	−1	0	0	−1	0	1	3
z_j		M	1	1	−2	0	−1	−1	1	1	—
		N	0	0	0	0	0	0	0	0	—
c_j		M	0	0	0	0	0	0	1	1	—
		N	3	2	−1	0	0	0	0	0	—
$z_j - c_j$		M	1	1	−2	0	−1	−1	0	0	7
		N	−3	−2	1	0	0	0	0	0	0

Table 5.8b

$c_{(i)}$	$P_{(i)}$ \ P_j		P_1	P_2	P_3	P_4	P_5	P_6	P_8	P_0
0	$P_{(1)}$	P_4	0	(1)	2	1	1	0	0	2
3	$P_{(2)}$	P_1	1	0	−1	0	−1	0	0	4
M	$P_{(3)}$	P_8	0	1	−1	0	0	−1	1	3
z_j		M	0	1	−1	0	0	−1	1	—
		N	3	0	−3	0	−3	0	0	—
c_j		M	0	0	0	0	0	0	1	—
		N	3	2	−1	0	0	0	0	—
$z_j - c_j$		M	0	1	−1	0	0	−1	0	3
		N	0	−2	−2	0	−3	0	0	12

Table 5.8c

$c_{(i)}$	$P_{(i)}$ \ P_j		P_1	P_2	P_3	P_4	P_5	P_6	P_8	P_0
2	$P_{(1)}$	P_2	0	1	2	1	1	0	0	2
3	$P_{(2)}$	P_1	1	0	−1	0	−1	0	0	4
M	$P_{(3)}$	P_8	0	0	−3	−1	−1	−1	1	1
z_j		M	0	0	−3	−1	−1	−1	1	—
		N	3	2	1	2	−1	0	0	—
c_j		M	0	0	0	0	0	0	1	—
		N	3	2	−1	0	0	0	0	—
$z_j - c_j$		M	0	0	−3	−1	−1	−1	0	1
		N	0	0	2	2	−1	0	0	16

Since the coefficients of M in the evaluation of all non-basis vectors are negative, no further improvement of $f'(X)$ is possible.* This

* If some of these coefficients of M in $z_j - c_j$ are zero instead of negative, then whether $f'(X)$ can be further improved or not depends on the numerical part of

means that there is no basic feasible solution that does not include the artificial vector in the basis. Therefore the original problem has no solution. An examination of the original constraints quickly reveals the cause: Since the minimum value of x_3 is zero (due to the non-negativity requirement), according to the second and the third constraints, x_1 and x_2 cannot be smaller than 4 and 3 respectively. But these values will violate the first constraint. Therefore the constraints are contradictory and no solution to the problem exists.

5.7. Alternative Optimal Basic Feasible Solutions.

Often in a linear programming problem, the optimal solution is not unique in the sense that other basic feasible solutions exist which yield the same optimal value of the objective function. When this situation obtains, we refer to those equally optimal solutions as *alternative optima*.

Alternate optima can be readily detected from the final tableau of the simplex process. Recalling from Chapter IV Section 4.4 that

$$f(X') = f(X) - \theta_k'(z_k - c_k) \qquad (4.5.3)$$

$$\text{where } \theta_k' = \frac{\theta_{(l)}}{y_{(l)k}}, \qquad y_{(l)k} > 0, \qquad (4.2.6)$$

we can see clearly that $f(X')$ will be equal to $f(X)$ if either θ_k' is zero which implies $\theta_{(l)} = 0$, or $(z_k - c_k)$ is zero. If $f(X)$ is optimal, then the substitution of x_k for $x_{(l)}$ in the basis will yield an equally optimal solution. As a matter of fact, if $\theta_{(l)}$ is zero, then its substitution by any x_j whose $y_{(l)j}$ is not zero, regardless of the sign of $z_j - c_j$, will yield an alternative optimum. This type of alternative optimum has little applied significance since obviously in an industrial or business problem a change of variables with the same value of zero calls for no overt change of policy, program or action. On the other hand, if $z_k - c_k$ is zero, the substitution of x_k for $x_{(l)}$ to obtain an alternative optimal solution will necessitate adjustments of all other basis variables (unless $x_{(l)}$ happens to be zero also) and will represent a totally different policy, program or action on the part of the decision maker.

---

these $z_j - c_j$ (positive for decrease and negative for increase). In either case no displacement of artificial vector from the basis will result. Hence one needs only to consult the "M" part of $z_j - c_j$ to determine whether the original problem has a basic feasible solution. If the coefficients of M are all zero or negative (for minimizing $f(X)$), then no feasible solution exists for the original problem.

Presumably this range of alternatives should result in better decision since the decision maker is allowed a greater degree of freedom in his choice.

Example. In Table 5.7b, the solution is $X^T = [x_1\ x_2\ x_3\ x_4\ x_5\ x_6] = [0\ 1\ 6\ 0\ 10\ 0]$ with $f(X) = 4$. Since $z_1 - c_1$ is zero, an equally desirable solution is $X'^T = [\frac{10}{3}\ \frac{13}{3}\ \frac{8}{3}\ 0\ 0\ 0]$, obtained by substituting P_1 for P_5 in the basis. It can quickly be verified that $f(X) = f(X') = 4$.

There remains one final comment before we close this chapter. Since, as we have seen in Chapter III Section 3.4, the solutions of a linear programming problem form a convex set, it follows that if there are two basic feasible solutions that are equally optimal, any convex combination of the two solutions is also an optimal solution, although it will be non-basic. The proof of this statement is left as an exercise for the reader.

EXERCISES

1. Solve the following linear programming problems by the simplex method:

(a) Maximize $f(X) = 2x_1 - x_2 + x_4$,

subject to:
$$\begin{aligned} x_1 + x_2 + x_3 + x_4 &\le 8,\\ 2x_2 - x_3 &\le 4,\\ 2x_1 - x_2 - x_4 &\le 2, \end{aligned}$$
and $x_j \ge 0, \quad j = 1, 2, 3, 4.$

(b) Minimize $f(X) = -x_1 + 2x_2 - x_3$,

subject to:
$$\begin{aligned} -3x_1 - x_2 + x_3 &\ge -6,\\ x_1 + x_3 &\le 3,\\ x_1 + x_2 - x_3 &\le 2,\\ 2x_2 + 3x_3 &\ge 6, \end{aligned}$$
and $x_j \ge 0, \quad j = 1, 2, 3.$

(c) Minimize $f(X) = x_1 + x_2$,

subject to:
$$\begin{aligned} x_1 + x_2 + x_3 + x_4 &\ge 10,\\ 2x_2 - x_3 - x_4 &\ge 2,\\ x_2 - 2x_3 + 2x_4 &\ge 6, \end{aligned}$$
and $x_j \ge 0, \quad j = 1, \ldots, 4.$

(d) Maximize $f(X) = 400x_1 + 250x_2$,

subject to:
$$\begin{aligned} x_1 + x_2 + x_3 + x_4 &= 10,\\ -2x_1 + x_3 - 2x_4 + x_5 &= -2,\\ 2x_1 + 2x_2 - 3x_3 + x_5 &= 4, \end{aligned}$$
and $x_j \ge 0, \quad j = 1, \ldots, 5.$

(e) Maximize $f(X) = -2x_4 - x_5 + x_6,$
subject to:
$$-x_2 + 2x_3 + x_4 = 5,$$
$$x_1 + 2x_2 - 2x_3 + x_5 = 4,$$
$$-x_1 + x_2 + x_6 = 6,$$
and $x_j \geq 0, \quad j = 1, \ldots, 6.$

(f) Minimize $f(X) = x_1 + 3x_2 - x_3,$
subject to:
$$-x_1 - x_2 + 2x_3 \leq 5,$$
$$2x_1 - x_2 + x_3 \leq 6,$$
$$-x_1 - 3x_2 + 5x_3 \geq 20,$$
and $x_j \geq 0, \quad j = 1, 2, 3.$

(g) Maximize $f(X) = x_1 - x_2 + 2x_3,$
subject to:
$$-x_1 + x_2 + 2x_3 \leq 5,$$
$$-2x_1 + 5x_2 - x_3 \geq 10,$$
$$2x_1 - x_2 + x_3 \geq 4,$$
and $x_j \geq 0, \quad j = 1, 2, 3.$

2. The following simplex tableau shows the intermediate results of solving a linear programming problem. It is known additionally that $c_3 = 4$ and $c_5 = 0$. Is the solution in the tableau optimal? If not, find the maximal solution.

$c_{(i)}$	$P_{(i)}$ \ P_j		P_1	P_2	P_3	P_4	P_5	P_0
4	$P_{(1)}$	P_2	0	1	2	0	-1	4
0	$P_{(2)}$	P_4	0	0	3	1	-1	2
1	$P_{(3)}$	P_1	1	0	-4	0	2	6

3. Solve the nut-mix problem of Chapter I Exercise 1.

4. Suppose the specifications of the nut mixtures described in Exercise 1, Chapter I are changed to the following:

	Peanuts	Cashew	Hazel
Mixture A	at most $\frac{1}{4}$	at least $\frac{1}{2}$	any proportion
Mixture B	any proportion	at least $\frac{1}{2}$	at most $\frac{1}{2}$
Supply (lbs)	1,000	400	400

What then would be the vendor's best product-mix? Is there any other solution that is equally profitable? What is the next best product-mix?

5. Prescribe a procedure for obtaining the next best basic feasible solution. (Hint: there may be more than one optimal solution).

6. In connection with the freight-movement problem described in Chapter I, Section 1.3, let the number of trains required to be moved be $t_{ab} = 4$, $t_{ac} = 8$, $t_{ba} = 5$, $t_{bc} = 10$, $t_{ca} = 2$, $t_{cb} = 9$. Further, let the engine-and-crew costs be \$250 for route A-B-A, \$280 for route B-C-B, \$200 for route C-A-C, \$400 for routes A-B-C-A and A-C-B-A. Determine the optimal schedule of engine-and-crew assignments.

7. A factory is engaged in producing three products, X, Y and Z which involve lathe work, grinding and assembling. The cutting, grinding and assembling time required for one unit of X are 2, 1 and 1 hours respectively. Similarly, they are 3, 1 and 3 hours for one unit of Y, and 1, 3 and 1 hours for one unit of Z. The profits of X, Y and Z are \$2, \$2 and \$4 per unit respectively. Assuming that there are available 300 hours of lathe time, 300 hours of grinder time and 240 hours of assembly time, how many units of each product should be produced to maximize profits? If the factory receives an inquiry about the possibility of producing a fourth product which calls for 1 hour each of cutting, grinding and assembling per unit, and would contribute a profit of \$1 per unit, should the order be sought after? If so, how much of the additional product should the factory promise to deliver?

8. Determine the optimal solution to Exercise 1 (a) without the non-negative constraints.

VARIATIONS OF THE SIMPLEX METHOD

6.1. Introduction.

In this chapter, two variations of the basic simplex method will be discussed. The first is commonly known as the revised simplex method (or the inverse matrix method). For reasons which will become clear in the following sections, this method requires less computation and record keeping, and is generally regarded as faster and more efficient than the regular simplex method, particularly for large problems. An industrial linear programming problem having 50 constraints and 200 variables is not considered inordinately large today. Problems of this size simply have to be solved by a computer. Due to the high cost of computer time and the limited internal memory capacity of computers it has become almost a rule that the revised simplex be used instead of the regular simplex. For this reason, some basic knowledge of the revised simplex is considered essential. Sections 6.2, 6.3, and 6.4 will be devoted to it.

The second variation which will be discussed in Sections 6.5 and 6.6 is called the dual simplex method. It is essentially a computational scheme based on the regular simplex but modified to handle non-feasible basic initial solutions. It will be recalled that up to now our standard approach to linear programming problems with the "$\geq$" type of constraints is to add artificial variables in order to obtain an initial basic feasible solution (cf. Chapter V, Sections 5.5 and 5.6). This not only enlarges the problem, making it more difficult to handle, but also prolongs the solution process since all artificial variables in the basis must eventually be replaced in order to arrive at the optimal solution to the original problem. It would be highly desirable, therefore, if a technique could be devised to circumvent the need for artificial variables by accepting a *non-feasible* solution as a starting solution for subsequent iteration towards the optimum. The dual simplex method is such a technique. Although it does not eliminate the need for artificial variables in all circumstances—e.g., when unit vectors are lacking due to the presence of the "=" type of constraints, it does represent a significant enough extension of the regular simplex that special treatment is justified.

6.2. The Revised Simplex Method—Its Theoretical Foundation.

It will be recalled that the preliminary steps in solving a linear programming problem are to modify the constraints in such a way as to convert them into equations and to develop (preferably) an identity matrix to serve as the initial basis. In the following discussion *we will assume that these modifications have already been made* so that we can in general write a linear programming problem as follows:

(6.2.1) $$\text{Maximize } f(X) = CX$$
(6.2.2) $$\text{subject to: } AX = P_0$$
(6.2.3) $$\text{and } X \geq 0$$

where:

$$A = [P_1 P_2 \ldots P_n] = \begin{bmatrix} a_{11} & \ldots & a_{1n} \\ a_{21} & \ldots & a_{2n} \\ \cdot & & \cdot \\ \cdot & & \cdot \\ \cdot & & \cdot \\ a_{m1} & \ldots & a_{mn} \end{bmatrix}, \quad X = \begin{bmatrix} x_1 \\ x_2 \\ \cdot \\ \cdot \\ \cdot \\ x_n \end{bmatrix},$$

$$P_0 = \begin{bmatrix} b_1 \\ b_2 \\ \cdot \\ \cdot \\ \cdot \\ b_m \end{bmatrix} \geq O \quad \text{and} \quad C = [c_1 c_2 \ldots c_n].$$

Let there be an initial basis $B = [P_{(1)} P_{(2)} \ldots P_{(m)}]$ such that

(6.2.4)
$$\tilde{X} = B^{-1}P_0, \qquad \text{where } \tilde{X} = \begin{bmatrix} x_{(1)} \\ \cdot \\ \cdot \\ \cdot \\ x_{(m)} \end{bmatrix}.$$

(6.2.5)
$$\begin{aligned} Y &= [Y_1\, Y_2 \ldots Y_n] \\ &= [B^{-1}P_1 \quad B^{-1}P_2 \ldots B^{-1}P_n] = B^{-1}[P_1\, P_2 \ldots P_n] \\ &= B^{-1}A; \end{aligned}$$

(6.2.6)
$$\begin{aligned} Z - C &= [z_1 - c_1 \quad z_2 - c_2 \ldots z_n - c_n] \\ &= \left[\sum_{(i)=(1)}^{(n)} c_{(i)}y_{(i)1} - c_1 \quad \sum_{(i)=(1)}^{(m)} c_{(i)}y_{(i)2} - c_2 \ldots \sum_{(i)=(1)}^{(m)} c_{(i)}y_{(i)n} - c_n\right] \\ &= [\tilde{C}Y_1 - c_1 \quad \tilde{C}Y_2 - c_2 \ldots \tilde{C}Y_n - c_n], \\ &\qquad \text{where } \tilde{C} = [c_{(1)} \ldots c_{(m)}] \\ &\qquad Y_1 = \begin{bmatrix} y_{(1)1} \\ \cdot \\ \cdot \\ \cdot \\ y_{(m)1} \end{bmatrix}, \text{ etc.} \\ &= [\tilde{C}B^{-1}P_1 - c_1 \quad \tilde{C}B^{-1}P_2 - c_2 \ldots \tilde{C}B^{-1}P_n - c_n] \\ &= \tilde{C}B^{-1}[P_1\, P_2 \ldots P_n] - [c_1\, c_2 \ldots c_n] \\ &= \tilde{C}B^{-1}A - C; \end{aligned}$$

and finally

(6.2.7)
$$\begin{aligned} f(X) &= CX = \tilde{C}\tilde{X} \\ &= \tilde{C}B^{-1}P_0. \end{aligned}$$

Let us recall further the content of the simplex tableau (Figure 5.1 in Chapter V) and denote it by T. Then with the exclusion of the z_j and c_j rows,*

(6.2.8)
$$T = \left[\begin{array}{cccc|c} y_{(1)1} & y_{(1)2} & \cdots & y_{(1)n} & x_{(1)} \\ y_{(2)1} & y_{(2)2} & \cdots & y_{(2)n} & x_{(2)} \\ \cdot & \cdot & & \cdot & \cdot \\ \cdot & \cdot & & \cdot & \cdot \\ \cdot & \cdot & & \cdot & \cdot \\ y_{(m)1} & y_{(m)2} & \cdots & y_{(m)n} & x_{(m)} \\ \hline z_1 - c_1 & z_2 - c_2 & \cdots & z_n - c_n & f(X) \end{array}\right]$$

* These two rows are not essential since the new $z_j - c_j$ may be directly obtained from the old by Transformation Rule III (4.6.8).

$$= \begin{bmatrix} Y & X \\ Z - C & f(X) \end{bmatrix}.$$

Substituting (6.2.4), (6.2.5), (6.2.6) and (6.2.7) for X, Y, $Z - C$ and $f(X)$ respectively in (6.2.8) we have

(6.2.9) $$T = \begin{bmatrix} B^{-1}A & B^{-1}P_0 \\ \tilde{C}B^{-1}A - C & \tilde{C}B^{-1}P_0 \end{bmatrix}.$$

This expression can be rewritten as the product of two matrices as follows:

(6.2.10) $$T = \begin{bmatrix} B^{-1} & O \\ \tilde{C}B^{-1} & I \end{bmatrix} \begin{bmatrix} A & P_0 \\ -C & O \end{bmatrix}$$

$$= \begin{bmatrix} B & O \\ -\tilde{C} & I \end{bmatrix}^{-1} \begin{bmatrix} A & P_0 \\ -C & O \end{bmatrix}$$

$$= \hat{B}^{-1}\hat{A},$$

where:

(6.2.11) $$\hat{B} = \begin{bmatrix} B & O \\ -\tilde{C} & I \end{bmatrix},$$

and

(6.2.12) $$\hat{A} = \begin{bmatrix} A & P_0 \\ -C & O \end{bmatrix}.$$

Thus, we see that each simplex tableau is the product of two matrices. The first matrix ($\hat{B}^{-1}$) is essentially the inverse of the basis (B) and the second ($\hat{A}$) is nothing more than a complete tabulation of the *original* coefficients in the objective function and the system of constraints [cf. (6.2.11) and (6.2.12)]. So constituted, it is obvious that $\hat{A}$ is not affected by change of basis. Herein lies the basic difference between the regular and the revised simplex methods. In the former the entire simplex tableau is transformed as we move from one basic feasible solution to another; in the latter only the transformation of $\hat{B}^{-1}$ is required. Since the tableau usually is much larger in dimensions $[(m + 1) \times (n + 1)$ with $n > m]$ than the inverse of $\hat{B}$ $[(m + 1) \times (m + 1)]$, it is apparent that the revised simple method offers important advantages in speed over the regular simplex. In addition, the smaller matrix stored in the internal memory of the computer now enables it to handle larger problems. These are the reasons why computer programs for the solution of linear programming problems are more commonly based on the revised than the regular simplex method.

All of these advantages would, of course, come to naught if the transformation of $\hat{B}^{-1}$ were much more complicated and time-consuming than the transformation of the tableau (T) itself. In Section 6.3, however, we shall derive rules for the transformation of $\hat{B}^{-1}$ and show that this is not so. In the meantime we shall demonstrate the representation of T by $\hat{B}^{-1}$ and A with the following example:

Example. Maximize $f(X) = -x_1 + x_2 + x_3,$

$$\begin{aligned} \text{subject to:} \quad & 2x_1 - x_2 - x_3 + x_4 = 4, \\ & x_1 + x_2 + 2x_3 + x_5 = 6, \\ & -x_1 + x_2 + x_6 = 3, \end{aligned}$$

$$\text{and} \quad x_j \geq 0, \quad j = 1, \ldots, 6.$$

This problem is the same one used in Chapter V Section 5.2 where it was found that one of the feasible bases is $B = [P_3 \quad P_4 \quad P_6]$. From this, we can write immediately

$$B = [P_3 \quad P_4 \quad P_6] = \begin{bmatrix} -1 & 1 & 0 \\ 2 & 0 & 0 \\ 0 & 0 & 1 \end{bmatrix}$$

and

$$\tilde{C} = [c_3 \quad c_4 \quad c_6] = [1 \quad 0 \quad 0].$$

From which it can be easily shown that

$$B^{-1} = \begin{bmatrix} 0 & \frac{1}{2} & 0 \\ 1 & \frac{1}{2} & 0 \\ 0 & 0 & 1 \end{bmatrix},$$

and

$$\tilde{C}B^{-1} = [1 \quad 0 \quad 0] \begin{bmatrix} 0 & \frac{1}{2} & 0 \\ 1 & \frac{1}{2} & 0 \\ 0 & 0 & 1 \end{bmatrix} = [0 \quad \tfrac{1}{2} \quad 0].$$

Hence, according to (6.2.11),

$$\hat{B}^{-1} = \begin{bmatrix} B^{-1} & O \\ \tilde{C}B^{-1} & I \end{bmatrix} = \begin{bmatrix} 0 & \frac{1}{2} & 0 & 0 \\ 1 & \frac{1}{2} & 0 & 0 \\ 0 & 0 & 1 & 0 \\ 0 & \frac{1}{2} & 0 & 1 \end{bmatrix}$$

Further, by the definition of $\hat{A}$ in (6.2.12),

$$\hat{A} = \begin{bmatrix} A & P_0 \\ -C & O \end{bmatrix} = \begin{bmatrix} 2 & -1 & -1 & 1 & 0 & 0 & 4 \\ 1 & 1 & 2 & 0 & 1 & 0 & 6 \\ -1 & 1 & 0 & 0 & 0 & 1 & 3 \\ 1 & -1 & -1 & 0 & 0 & 0 & 0 \end{bmatrix}.$$

Now, in accordance with (6.2.10) the simplex tableau (T) can be computed.

$$T = \hat{B}^{-1}\hat{A} = \begin{bmatrix} 0 & \frac{1}{2} & 0 & 0 \\ 1 & \frac{1}{2} & 0 & 0 \\ 0 & 0 & 1 & 0 \\ 0 & \frac{1}{2} & 0 & 1 \end{bmatrix} \begin{bmatrix} 2 & -1 & -1 & 1 & 0 & 0 & 4 \\ 1 & 1 & 2 & 0 & 1 & 0 & 6 \\ -1 & 1 & 0 & 0 & 0 & 1 & 3 \\ 1 & -1 & -1 & 0 & 0 & 0 & 0 \end{bmatrix}$$

$$= \begin{bmatrix} \frac{1}{2} & \frac{1}{2} & 1 & 0 & \frac{1}{2} & 0 & 3 \\ \frac{5}{2} & \frac{1}{2} & 0 & 1 & \frac{1}{2} & 0 & 7 \\ -1 & 1 & 0 & 0 & 0 & 1 & 3 \\ \frac{3}{2} & -\frac{1}{2} & 0 & 0 & \frac{1}{2} & 0 & 3 \end{bmatrix}.$$

This is identical with the results obtained in Chapter V Section 5.2 and shown in Table 5.3 except for the omission of the z_j and c_j rows. Since T stands for the simplex tableau, the elements of T are subject to the same interpretation. In other words, the values of the basis variables and of the objective function are found in the last column of T; the evaluators, in the last row. The rest represents simply the linear-combination coefficients ($y_{(i)j}$) of the vectors (P_j).

All this may appear to be merely a circuitous way of obtaining the simplex tableau, with little practical advantage. This impression is entirely correct, but it must be emphasized that our example was not meant to demonstrate the computational efficiency of the revised simplex method but only its feasibility. But even at this point, not much imagination is required to realize the advantage of the revised simplex. If we have an efficient way of transforming $\hat{B}^{-1}$, i.e., obtaining the new inverse from the old, there is no need to transform all the $y_{(i)j}$ in accordance with Transformation Rules I and II. In short, referring to the present example, we would not have to complete the computation of T. The computation could have been ended with the finding of the first vector with a favorable evaluation. In this example, if we proceed to evaluate the vectors in T from left to right, P_2 will be the first vector to show a favorable evaluation ($z_2 - c_2 = -\frac{1}{2}$). Therefore no additional computation for P_3, P_4, P_5 and P_6 will be necessary.

6.3. The Transformation of $\hat{B}^{-1}$.

As intimated in the previous section, the purpose of this section is to derive rules by which the elements of the new $\hat{B}^{-1}$ may be obtained from the elements of the old.

Let us assume that we have at the beginning:

$$\hat{B} = [B_1\, B_2 \ldots B_l \ldots B_{m+1}] \tag{6.3.1}$$

where
$$B_1 = \begin{bmatrix} P_{(1)} \\ -c_{(1)} \end{bmatrix},\; B_2 = \begin{bmatrix} P_{(2)} \\ -c_{(2)} \end{bmatrix} \text{ etc.}$$

and
$$B_l = \begin{bmatrix} P_{(l)} \\ -c_{(l)} \end{bmatrix},\; B_{m+1} = \begin{bmatrix} O \\ 1 \end{bmatrix}.$$

By substituting P_k for $P_{(l)}$, we obtain a new basis,

$$\hat{B}' = [B_1\, B_2 \ldots B_l' \ldots B_{m+1}] \quad \text{where } B_l' = \begin{bmatrix} P_k \\ -c_k \end{bmatrix}. \tag{6.3.2}$$

Since $\hat{B}$ is a basis, called *augmented basis*, B_l' can be expressed as a linear combination of the vectors in $\hat{B}$, i.e.,

$$B_l' = y_{(1)k}B_1 + y_{(2)k}B_2 + \ldots + y_{(l)k}B_l + \ldots + y_{(m+1)k}B_{m+1}{}^*$$

Thus, as can be easily verified,

$$[B_1\, B_2 \ldots B_l' \ldots B_{m+1}] = [B_1\, B_2 \ldots B_l \ldots B_{m+1}] \begin{bmatrix} 1 & 0 & \ldots & y_{(1)k} & \ldots & 0 \\ 0 & 1 & \ldots & y_{(2)k} & \ldots & 0 \\ \cdot & \cdot & & \cdot & & \cdot \\ \cdot & \cdot & & \cdot & & \cdot \\ \cdot & \cdot & & \cdot & & \cdot \\ 0 & 0 & \ldots & y_{(l)k} & \ldots & 0 \\ \cdot & \cdot & & \cdot & & \cdot \\ \cdot & \cdot & & \cdot & & \cdot \\ \cdot & \cdot & & \cdot & & \cdot \\ 0 & 0 & \ldots & y_{(m+1)k} & \ldots & 1 \end{bmatrix}.$$

---

* Due to the definition of $\hat{B}$, the $y_{(i)k}$ in this expression are identical with their counterparts in (4.2.9) [except for the lack of $y_{(m+1)k}$ in (4.2.9)].

The second matrix on the right-hand side, which looks like an identity matrix with the exception of one column of $y_{(i)k}$, is called an *elementary matrix*. Let it be denoted by E. Then we have from above

$$\hat{B}' = \hat{B}E, \tag{6.3.3}$$

which shows the relationship between two successive bases when one is obtained from the other by substituting P_k for P_l as suggested in (6.3.1) and (6.3.2).

Taking the inverse of both sides of (6.3.3) we have

$$(\hat{B}')^{-1} = (\hat{B}E)^{-1} \quad \text{or}$$

$$(\hat{B}')^{-1} = E^{-1}\hat{B}^{-1}. \tag{6.3.4}$$

But it can be easily verified that

$$E^{-1} = \begin{bmatrix} 1 & 0 \ldots & y_{(1)k} & \ldots 0 \\ 0 & 1 \ldots & y_{(2)k} & \ldots 0 \\ \cdot & \cdot\;\cdot & \cdot & \cdot \\ \cdot & \cdot\;\;\cdot & \cdot & \cdot \\ \cdot & \cdot\;\;\;\cdot & \cdot\,\cdot & \cdot \\ 0 & 0 \ldots & y_{(l)k} & \ldots 0 \\ \cdot & \cdot & \cdot & \cdot\;\;\cdot \\ \cdot & \cdot & \cdot & \cdot\;\cdot \\ \cdot & \cdot & \cdot & \cdot\cdot \\ 0 & 0 \ldots & y_{(m+1)k} & \ldots 1 \end{bmatrix}^{-1} \tag{6.3.5}$$

$$= \begin{bmatrix} 1 & 0 \ldots & -y_{(1)k}/y_{(l)k} & \ldots 0 \\ 0 & 1 \ldots & -y_{(2)k}/y_{(l)k} & \ldots 0 \\ \cdot & \cdot\;\cdot & \cdot & \cdot \\ \cdot & \cdot\;\;\cdot & \cdot & \cdot \\ \cdot & \cdot\;\;\;\cdot & \cdot & \cdot \\ 0 & 0 \ldots & 1/y_{(l)k} & \ldots 0 \\ \cdot & \cdot & \cdot & \cdot\;\;\cdot \\ \cdot & \cdot & \cdot & \cdot\;\cdot \\ \cdot & \cdot & \cdot & \cdot\cdot \\ 0 & 0 \ldots & -y_{(m+1)k}/y_{(l)k} & \ldots 1 \end{bmatrix}.$$

By defining $\hat{B}^{-1} = [d_{ij}]$ and $(\hat{B}')^{-1} = [d'_{ij}]$, and by substituting (6.3.5) into (6.3.4), we obtain

$$\begin{bmatrix} d'_{11} & d'_{12} & \dots d'_{1,m+1} \\ d'_{21} & d'_{22} & \dots d'_{2,m+1} \\ \cdot & \cdot & \cdot \\ \cdot & \cdot & \cdot \\ \cdot & \cdot & \cdot \\ d'_{m+1,1} & d'_{m+1,2} & \dots d'_{m+1,m+1} \end{bmatrix}$$

$$= \begin{bmatrix} 1 & 0 \dots -y_{(1)k}/y_{(l)k} & \dots 0 \\ 0 & 1 \dots -y_{(2)k}/y_{(l)k} & \dots 0 \\ \cdot & \cdot\cdot \qquad \cdot & \cdot \\ \cdot & \cdot \quad \cdot \qquad \cdot & \cdot \\ \cdot & \cdot \quad \cdot \qquad \cdot & \cdot \\ 0 & 0 \dots \qquad 1/y_{(l)k} & \dots 0 \\ \cdot & \cdot \qquad \cdot & \cdot \; \cdot \\ \cdot & \cdot \qquad \cdot & \cdot \; \cdot \\ \cdot & \cdot \qquad \cdot & \cdot \; \cdot \\ 0 & 0 \dots -y_{(m+1)k}/y_{(l)k} & \dots 1 \end{bmatrix} \begin{bmatrix} d_{11} & d_{12} & \dots d_{1,m+1} \\ d_{21} & d_{22} & \dots d_{2,m+1} \\ \cdot & \cdot & \cdot \\ \cdot & \cdot & \cdot \\ \cdot & \cdot & \cdot \\ d_{m+1,1} & d_{m+1,2} & \dots d_{m+1,m+1} \end{bmatrix},$$

which means according to ordinary matrix multiplication:

$$d'_{ij} = \frac{d_{ij}}{y_{(l)k}} \qquad \text{for } i = (l) \quad \text{and all } j, \tag{6.3.6}$$

$$d'_{ij} = d_{ij} - \frac{y_{(i)k}}{y_{(l)k}} d_{lj} \qquad \text{for } i \neq (l),\ i = 1, 2, \dots, m+1, \text{ and all } j. \tag{6.3.7}$$

Thus we see that the elements of the new inverse may be obtained directly from the corresponding elements of the old inverse by using (6.3.6) and (6.3.7). What is most interesting about these two rules is their striking similarity to (4.6.6) and (4.6.7) which are as we recall the transformation rules of the regular simplex method. Structurally speaking, these two sets of transformation rules are identical—the only difference being in the elements to be transformed, viz., d_{ij} in the case of the revised simplex and $y_{(i)j}$ in the case of the regular simplex. The following illustration will make clear the direct transformation of the inverse of $\hat{B}$ by the two rules, (6.3.6) and (6.3.7), derived in this section.

Example. Using the problem in the preceding example, we note that for

$$\hat{B} = \begin{bmatrix} B & O \\ -\tilde{C} & I \end{bmatrix} = \begin{bmatrix} P_3 & P_4 & P_6 & O \\ -c_3 & -c_4 & -c_6 & 1 \end{bmatrix}$$

the inverse has already been determined to be

$$\hat{B}^{-1} = \begin{bmatrix} 0 & \frac{1}{2} & 0 & 0 \\ 1 & \frac{1}{2} & 0 & 0 \\ 0 & 0 & 1 & 0 \\ 0 & \frac{1}{2} & 0 & 1 \end{bmatrix}.$$

Suppose we want to introduce P_2 into the basis to replace P_6, to determine the inverse of the new basis by transformation rules (6.3.6) and (6.3.7). We must first obtain the linear-combination coefficients $y_{(i)j}$ for P_2 in terms of $\hat{B}$. This is accomplished as follows:

$$\hat{Y}_2 = \hat{B}^{-1}\hat{P}_2 = \begin{bmatrix} B & O \\ -\tilde{C} & I \end{bmatrix}^{-1} \begin{bmatrix} P_2 \\ -c_2 \end{bmatrix}$$

$$= \begin{bmatrix} 0 & \frac{1}{2} & 0 & 0 \\ 1 & \frac{1}{2} & 0 & 0 \\ 0 & 0 & 1 & 0 \\ 0 & \frac{1}{2} & 0 & 1 \end{bmatrix} \begin{bmatrix} -1 \\ 1 \\ 1 \\ -1 \end{bmatrix} = \begin{bmatrix} \frac{1}{2} \\ -\frac{1}{2} \\ 1 \\ -\frac{1}{2} \end{bmatrix},$$

which means that $y_{(1)2} = y_{32} = \frac{1}{2}$, $y_{(2)2} = y_{42} = -\frac{1}{2}$, $y_{(3)2} = y_{62} = 1$ and $y_{(4)2} = z_2 - c_2 = -\frac{1}{2}$. Bearing in mind that $k = 2$ and $(l) = (3)$, we have from (6.3.6)

$$d'_{31} = \frac{d_{31}}{y_{(3)2}} = \frac{0}{1} = 0,$$

$$d'_{32} = \frac{d_{32}}{y_{(3)2}} = \frac{0}{1} = 0, \text{ etc.,}$$

and from (6.3.7)

$$d'_{11} = d_{11} - \frac{y_{(1)2}}{y_{(3)2}} d_{31} = 0 - \frac{\frac{1}{2}}{1}(0) = 0$$

$$d'_{12} = d_{12} - \frac{y_{(1)2}}{y_{(3)2}} d_{32} = \frac{1}{2} - \frac{\frac{1}{2}}{1}(0) = \frac{1}{2}, \text{ etc.}$$

The complete new inverse is therefore

$$(\hat{B}')^{-1} = \begin{bmatrix} 0 & \frac{1}{2} & -\frac{1}{2} & 0 \\ 1 & \frac{1}{2} & \frac{1}{2} & 0 \\ 0 & 0 & 1 & 0 \\ 0 & \frac{1}{2} & \frac{1}{2} & 1 \end{bmatrix}.$$

The validity of this inverse can easily be verified by multiplying it by $\hat{B}' = \begin{bmatrix} P_3 & P_4 & P_2 & O \\ -c_3 & -c_4 & -c_2 & 1 \end{bmatrix}$ to see if the result is an identity matrix.

6.4. Revised Simplex Algorithm.

In this section we will give close attention to the computational details of the revised simplex method and conclude with a description of the step-by-step algorithm involved.

First there is the question of the initial basis and its inverse. There are two answers to this question. As in the case of the regular simplex method, we may begin with an identity matrix by adding appropriate slack and artificial vectors, or, if a known solution exists, we may begin with the basis implied by that solution. In the first case no calculation would be required to obtain the inverse of $\hat{B}$ since the inverse of an identity matrix is also an identity matrix. But the solution of the problem may be prolonged due to the inclusion of slack and artificial vectors in the initial basis. In the second case, although the solution process may be shortened, the inversion of a non-identity basis may be so complicated as to nullify its advantage. Clearly then, the particular approach to be followed depends on one's knowledge of the problem at hand and the computational device available. In some computer codes an identity basis is automatically provided by the machine at the start so that the advantage of using any other basis is open to serious question.

Secondly, to reiterate a point already made in Section 6.2, the computation of $y_{(i)j}$ need not be complete for all P_j. Since the link between successive solutions is no longer $Y = [y_{(i)j}]$ but the inverse, $\hat{B}^{-1}$, we compute only $y_{(i)k}$ after P_k has been selected to come into the basis, and only for the purpose of applying Decision Rule I to select the basis vector to be replaced. This means that we will compute the linear-combination coefficients of only one vector instead of n vectors, which will result in a significant saving of time, particularly if the number of vectors (n) is very much larger than the number of constraints (m).

Thirdly, to determine the vector to be introduced into the basis, the evaluator $z_j - c_j$ can be obtained by a simple multiplication of the jth column of $\hat{A}$ by the last row of $\hat{B}^{-1}$ [cf. (6.2.6), (6.2.8), (6.2.9) and (6.2.10)]. This should be done immediately after the inverse of the basis has been transformed for all P_j (omitting those which are already in the basis) until the *first* one with a favorable evaluation is found.* Once this vector, i.e., P_k is determined, we cease to evaluate the remaining vectors and proceed immediately to determine $y_{(i)k}$ as indicated in the previous paragraph.

Except for the points observed above, the procedure of the revised simplex method remains the same as that of the regular simplex, particularly with respect to the selection of the outgoing vector, the handling of degenerate basic feasible solution and the recognition of unbounded problems. We are now in a position to set forth the precise computational algorithm of the revised simplex method.

Assuming that the problem is a maximizing one and that an initial basis $\hat{B}$ together with its inverse $\hat{B}^{-1}$ is already known, the revised simplex algorithm is as follows:

Step I—Determine the $z_j - c_j$ for $j = 1, 2, \ldots n$ by multiplying the successive columns of $\hat{A}$ by the last row of $\hat{B}^{-1}$ until one of the following two conditions is met:

(1) the first negative $z_j - c_j$ is found (first positive $z_j - c_j$ for minimizing problems), say $z_k - c_k$ for P_k. Proceed to Step II.

(2) no negative (positive for minimizing problems) $z_j - c_j$ is found for $j = 1, 2, \ldots, n$. The current solution is optimal and computation ends.

Step II—Determine the $y_{(i)k}$ by multiplying the kth column of $\hat{A}$ by the ith row of $\hat{B}^{-1}$ for $i = 1, 2, \ldots, m$.

Step III—Determine $x_{(i)} = \theta_{(i)}$ by multiplying the P_0 column of $\hat{A}$ by the ith row of $\hat{B}^{-1}$ for $i = 1, 2, \ldots, m$, and determine the value of $f(X)$ by multiplying the P_0 column of $\hat{A}$ by the last or $(m + 1)$th row of $\hat{B}^{-1}$.

Step IV—Identify the positive $y_{(i)k}$. The following three situations may be encountered:

(1) no positive $y_{(i)k}$ is found, which means that the problem has no upper (lower for minimizing problems) bound and computation ends.

(2) some $y_{(i)k}$ are positive. Form ratios of $\dfrac{\theta_{(i)}}{y_{(i)k}}$ for all $y_{(i)k} > 0$ and select the basis vector, say $P_{(l)}$, with the minimum ratio to

---

* It will be noticed that this is at variance with the regular simplex algorithm (Chapter IV, Section 4.7) where it was stated that all $z_j - c_j$ have to be computed, and then the vector with the most favorable evaluation chosen to be P_k. This is changed to the 'first favorable evaluation' in the revised simplex in the interest of saving time.

be the outgoing vector in accordance with Decision Rule I (4.2.6). Proceed to Step V.

(3) same as (2) above except that more than one basis vectors have the same minimum ratios. In such cases, resolve the tie as follows, and then proceed to Step V:

(a) if the ratio is zero, choose the basis vector among those in tie, which has the largest $y_{(i)k}$;

(b) in all other cases, chose among those in tie the basis vector with the lowest (i) index.

Step V—Transform $\hat{B}^{-1}$ by (6.3.6) and (6.3.7) and then return to Step I.

Example. In the interest of facilitating the comparison of the revised simplex with the regular simplex, the example of Chapter V Section 5.5 will be used here. The problem is:

$$\begin{aligned}
\text{Maximize } f(X) &= 2x_1 - x_2 + 2x_3,\\
\text{subject to:}\quad & x_1 + x_2 + x_3 \geq 6,\\
& -2x_1 + 3x_3 \geq 2,\\
& 2x_2 - x_3 \geq 0,\\
\text{and}\quad & x_1, x_2, x_3 \geq 0.
\end{aligned}$$

By introducing slack and artificial variables with proper price coefficients, the problem is modified into the following equivalent one. Maximize $f'(X)$

$$= 2x_1 - x_2 + 2x_3 + 0x_4 + 0x_5 + 0x_6 - \mathrm{M}x_7 - \mathrm{M}x_8 - \mathrm{M}x_9,$$

subject to:

$$\begin{aligned}
x_1 + x_2 + x_3 - x_4 \qquad\qquad + x_7 \qquad\qquad &= 6,\\
-2x_1 \qquad + x_3 \qquad - x_5 \qquad\qquad + x_8 \qquad &= 2,\\
2x_2 - x_3 \qquad\qquad - x_6 \qquad\qquad + x_9 &= 0,\\
\text{and}\qquad x_j \geq 0, \qquad j = 1, 2, \ldots, 9.&
\end{aligned}$$

The above formulation immediately suggests the following initial basis and its inverse.

$$\hat{B} = \begin{bmatrix} B & O \\ -\tilde{C} & I \end{bmatrix} = \begin{bmatrix} 1 & 0 & 0 & 0 \\ 0 & 1 & 0 & 0 \\ 0 & 0 & 1 & 0 \\ \mathrm{M} & \mathrm{M} & \mathrm{M} & 1 \end{bmatrix}.$$

$$\hat{B}^{-1} = \begin{bmatrix} B^{-1} & O \\ \tilde{C}B^{-1} & I \end{bmatrix} = \begin{bmatrix} 1 & 0 & 0 & 0 \\ 0 & 1 & 0 & 0 \\ 0 & 0 & 1 & 0 \\ -M & -M & -M & 1 \end{bmatrix}. \tag{6.4.3}$$

Table 6.2a*

	P_1	P_2	P_3	P_4	P_5	P_6	P_0
$\hat{A}$	1	1	1	−1	0	0	6
	−2	0	1	0	−1	0	2
	0	2	−1	0	0	−1	0
	−2	1	−2	0	0	0	0

Table 6.2b

Basis vectors	$\hat{B}^{-1}$				P_1	P_2	P_3	P_4	P_5	P_6	P_0
P_7	1	0	0	0		1					6
P_8	0	1	0	0		0					2
P_9	0	0	1	0		②					0
$z_j - c_j$	−M	−M	−M	1	M−2	−3M +1					−8M

Table 6.2c

P_7	1	0	$-\frac{1}{2}$	0			$\frac{3}{2}$				6
P_8	0	1	0	0			①				2
P_2	0	0	$\frac{1}{2}$	0			$-\frac{1}{2}$				0
$z_j - c_j$	−M	−M	$\frac{1}{2}$M $-\frac{1}{2}$	1	M−2	0	$-\frac{5}{2}$M $-\frac{3}{2}$				−8M

Table 6.2d

P_7	1	$-\frac{3}{2}$	$-\frac{1}{2}$	0	④						3
P_3	0	1	0	0	−2						2
P_2	0	$\frac{1}{2}$	$\frac{1}{2}$	0	−1						1
$z_j - c_j$	−M	$\frac{3}{2}$M $+\frac{3}{2}$	$\frac{1}{2}$M $-\frac{1}{2}$	1	−4M −5						−3M +3

Table 6.2e

P_1	$\frac{1}{4}$	$-\frac{3}{8}$	$-\frac{1}{8}$	0				$-\frac{1}{4}$			$\frac{3}{4}$
P_3	$\frac{1}{2}$	$\frac{1}{4}$	$-\frac{1}{4}$	0				$-\frac{1}{2}$			$\frac{7}{2}$
P_2	$\frac{1}{4}$	$\frac{1}{8}$	$\frac{3}{8}$	0				$-\frac{1}{4}$			$\frac{7}{4}$
$z_j - c_j$	$\frac{5}{4}$	$-\frac{3}{8}$	$-\frac{9}{8}$	1	0	0	0	$-\frac{5}{4}$			$\frac{27}{4}$

* Tables 6.2a through 6.2e do not include P_7, P_8 and P_9 because they are artificial vectors and as such will not be allowed to reenter the basis once they are removed.

There is no standard format for the revised simplex method. The one used here is merely a suggestion. First of all, the matrix $\hat{A}$ (6.2.10) is written down as in Table 6.2a for constant future reference. In the tableaux that follow, the inverse of $\hat{B}$ is always kept in full view on the left while space is allowed in the middle for the computation and recording of $z_j - c_j$, Y_k, X and $f(X)$ as needed. Thus, in Table 6.2b we show (6.4.3) on the left and mark the rows P_7, P_8 and P_9 to help identify the vectors in that basis.

To carry out Step I in the revised simplex algorithm, we multiply the successive columns of $\hat{A}$ in Table 6.2a by the last row of $\hat{B}^{-1}$ in Table 6.2b to obtain the $z_j - c_j$. Thus, for instance,

$$z_1 - c_1 = [-\mathrm{M} \quad -\mathrm{M} \quad -\mathrm{M} \quad 1] \begin{bmatrix} 1 \\ -2 \\ 0 \\ -2 \end{bmatrix} = \mathrm{M} - 2$$

which is recorded in the last row under P_1 in Table 6.2b. Since it is positive, P_1 is rejected and $z_2 - c_2$ is then computed in a similar manner. This is found, however, to be $-3\mathrm{M} + 1$ which is an overall negative number due to the predominance of M. Therefore, P_2 is selected to be the incoming vector and computation progresses to Step II. In Step II, the linear-combination coefficients are determined for P_2 by multiplying the P_2 column by the first, second and third rows of $\hat{B}^{-1}$. For instance,

$$y_{(1)2} = [1 \quad 0 \quad 0 \quad 0] \begin{bmatrix} 1 \\ 0 \\ 2 \\ 1 \end{bmatrix} = 1,$$

which is entered in the first row under P_2 in Table 6.2b. After $y_{(2)2}$ and $y_{(3)2}$ have been similarly determined and recorded, we move on to Step III which calls for the determination of $\theta_{(i)}$ and $f(X)$. These are accomplished as suggested earlier by multiplying the P_0 column of $\hat{A}$ by the successive rows of $\hat{B}^{-1}$. Hence

$$x_{(1)} = x_7 = \theta_7 = [1 \quad 0 \quad 0 \quad 0] \begin{bmatrix} 6 \\ 2 \\ 0 \\ 0 \end{bmatrix} = 6, \text{ etc.}$$

and

$$f(X) = [-M \quad -M \quad -M \quad 1]\begin{bmatrix}6\\2\\0\\0\end{bmatrix} = -8M,$$

all of which are recorded in the proper places as shown in Table 6.2b. Next in order are the forming and comparison of the ratios $\dfrac{\theta_{(i)}}{y_{(i)k}}$ as called for by Step IV. With $\theta_{(i)}$ and $y_{(i)k}$ recorded respectively in the columns P_0 and P_2 in Table 6.2b, this is easily accomplished. It is found that $P_{(3)}$ $(= P_9)$ has the smallest positive ratio and hence is to be removed from the basis. P_2 and P_9 together fix the pivotal element, $y_{(3)2} = 2$, which is circled in Table 6.2b. We now proceed with Step V which calls for the transformation of $\hat{B}^{-1}$ in much the same manner as the transformation of the simplex tableau in the regular simplex method. The new $\hat{B}^{-1}$ is recorded in Table 6.2c, providing a start for a second iteration, and a third, etc. These subsequent calculations are shown in Tables 6.2c and 6.2d, terminating in Table 6.2e when the solution is identified as without upper bound. In other words, if x_4 which has a favorable evaluation is assigned any positive value θ_4, x_1, x_2 and x_3 will all increase in value as follows:

$$x_1 = \tfrac{3}{4} - (-\tfrac{1}{4})\theta_4 = \tfrac{3}{4} + \tfrac{1}{4}\theta_4.$$
$$x_2 = \tfrac{7}{4} - (-\tfrac{1}{4})\theta_4 = \tfrac{7}{4} + \tfrac{1}{4}\theta_4,$$
$$x_3 = \tfrac{7}{2} - (-\tfrac{1}{2})\theta_4 = \tfrac{7}{2} + \tfrac{1}{2}\theta_4.$$

The value of $f(X)$ will also increase as

$$f(X) = \tfrac{27}{4} - (-\tfrac{5}{4})\theta_4 = \tfrac{27}{4} + (\tfrac{5}{4})\theta_4.$$

Thus, if θ_4 is allowed to approach positive infinity, x_1, x_2 and x_3 will all approach positive infinity also, resulting in a value of $f(X)$ which rapidly exceeds all finite bounds.

6.5. Dual Simplex Method—Theoretical Considerations.*

The dual simplex method as indicated in Section 6.1 of this chapter is a partial answer to the objection of prolonged computations due to the use of artificial vectors in the initial basis. The answer is only

* Although generically the dual simplex is a derivative of the duality theorem which will not be discussed until Chapter VII, it is felt that its *modus operandi* is so similar to the regular simplex that there is considerable pedagogical merit in its inclusion here as a refinement of the latter.

partial because it eliminates the need for artificial vectors only when some unit (or slack) vectors in the problem have the wrong signs for yielding a feasible initial solution. If the full set of linearly independent unit vectors (regardless of their signs) does not exist in the first place, then artificial unit vectors normally will have to be added to obtain a convenient initial basic feasible solution in spite of the dual simplex method. Furthermore, the dual simplex method is advantageous only when certain conditions concerning the evaluators are satisfied. These will be specified in the ensuing discussion.

Let us now concentrate on the former type of problems wherein a full set of linearly independent unit vectors exists (or can be generated within the logical interpretation of the problem) but with some of the vectors bearing the wrong signs. Such a problem is shown below.

$$
\begin{array}{llllll}
\text{Minimize} & f(X) = c_1x_1 & + c_2x_2 & + \ldots + c_nx_n, & \\
\text{subject to:} & a_{11}x_1 & + a_{12}x_2 & + \ldots + a_{1n}x_n & \geq b_1, \\
& \cdot & \cdot & \cdot & \cdot \\
& \cdot & \cdot & \cdot & \cdot \\
& \cdot & \cdot & \cdot & \cdot \\
& a_{f1}\,x_1 & + a_{f2}\,x_2 & + \ldots + a_{fn}\,x_n & \geq b_f, \\
& a_{f+1,1}x_1 & + a_{f+1,2}x_2 & + \ldots + a_{f+1,n}x_n & \leq b_{f+1}, \\
& \cdot & \cdot & \cdot & \cdot \\
& \cdot & \cdot & \cdot & \cdot \\
& \cdot & \cdot & \cdot & \cdot \\
& a_{m1}x_1 & + a_{m2}x_2 & + \ldots + a_{mn}x_n & \leq b_m,
\end{array}
$$

and $$x_1, x_2, \ldots, x_n \geq 0,$$

where we stipulate further,

$$b_1, b_2, \ldots, b_m \geq 0,$$
$$c_1, c_2, \ldots, c_n \geq 0.$$

This is a familiar linear programming problem with mixed constraints (f of the "$\geq$" type and $(m - f)$ of the "$\leq$" type). The additional restrictions that $c_1, c_2, \ldots, c_n \geq 0$ are inserted for heuristic purposes and will be removed in Section 6.7. By adding or subtracting nonnegative slack variables in the m inequality constraints, we can develop a full set of linearly independent unit vectors, $P_{n+1}, P_{n+2}, \ldots,$ P_{n+m} to form a basis. This basis then yields a basic solution, viz., $x_{n+1} = -b_1, \ldots, x_{n+f} = -b_f, x_{n+f+1} = b_{f+1}, \ldots, x_{n+m} = b_m$, which is of course not feasible. Let us summarize the present state of the solution by the simplex tableau shown in Table 6.3. It will be seen that because the basis is

$$B = [P_{n+1} \ldots P_{n+f}\, P_{n+f+1} \ldots P_{n+m}] = \begin{bmatrix} -1 & \ldots & 0 & 0 & \ldots & 0 \\ & \cdot\cdot & \cdot & \cdot & & \cdot \\ \cdot & \cdot & \cdot & \cdot & & \cdot \\ \cdot & \cdot & \cdot & \cdot & & \cdot \\ 0 & \ldots & -1 & 0 & \ldots & 0 \\ 0 & \ldots & 0 & 1 & \ldots & 0 \\ \cdot & & \cdot & \cdot\cdot & & \cdot \\ \cdot & & \cdot & \cdot & \cdot & \cdot \\ \cdot & & \cdot & \cdot & & \cdot\cdot \\ 0 & \ldots & 0 & 0 & \ldots & 1 \end{bmatrix},$$

therefore,

$$y_{(i)j} = -a_{ij} \qquad \text{for } i = 1, \ldots, f; \qquad j = 1, \ldots, n,$$
$$y_{(i)j} = a_{ij} \qquad \text{for } i = f+1, \ldots, m; \quad j = 1, \ldots, n,$$

and $\qquad z_j - c_j = 0 - c_j = -c_j, \quad \text{for } j = 1, \ldots, n.$

Let us now raise this question: What should be the signs of the $z_j - c_j$ if and when the solution is already a minimum? Clearly, accord-

Table 6.3

$c_{(i)}$	$P_{(i)}$ \ P_j		P_1	$\ldots P_k$	$\ldots P_n$	$P_{n+1} \ldots P_{n+m}$	P_0
$c_{(1)}$	$P_{(1)}$	P_{n+1}	$y_{(1)1}$	$\ldots y_{(1)k}$	$\ldots y_{(1)n}$	$1 \ldots\ldots 0$	$-b_1$
$\cdot$	$\cdot$	$\cdot$	$\cdot$	$\cdot$	$\cdot$	$\cdot\cdot \quad \cdot$	$\cdot$
$\cdot$	$\cdot$	$\cdot$	$\cdot$	$\cdot$	$\cdot$	$\cdot \; \cdot \quad \cdot$	$\cdot$
$\cdot$	$\cdot$	$\cdot$	$\cdot$	$\cdot$	$\cdot$	$\cdot \; \cdot \; \cdot \; \cdot$	$\cdot$
$c_{(f)}$	$P_{(f)}$	P_{n+f}	$y_{(f)1}$	$\ldots y_{(f)k}$	$\ldots y_{(f)n}$	$0 \ldots 1 \ldots 0$	$-b_f$
$c_{(f+1)}$	$P_{(f+1)}$	P_{n+f+1}	$y_{(f+1)1}$	$\ldots y_{(f+1)k}$	$\ldots y_{(f+1)n}$	$0 \ldots 1 \ldots 0$	b_{f+1}
$\cdot$	$\cdot$	$\cdot$	$\cdot$	$\cdot$	$\cdot$	$\cdot \quad \cdot \; \cdot$	$\cdot$
$\cdot$	$\cdot$	$\cdot$	$\cdot$	$\cdot$	$\cdot$	$\cdot \quad \cdot\cdot$	$\cdot$
$\cdot$	$\cdot$	$\cdot$	$\cdot$	$\cdot$	$\cdot$	$\cdot \quad \cdot\cdot$	$\cdot$
$c_{(m)}$	$P_{(m)}$	P_{n+m}	$y_{(m)1}$	$\ldots y_{(m)k}$	$\ldots y_{(m)n}$	$0 \ldots\ldots 1$	b_m
	$z_j - c_j$		$-c_1$	$\ldots -c_k$	$\ldots -c_n$	$0 \ldots\ldots 0$	$f(X)$

ing to the discussion in Chapter IV, they should all be non-positive (either negative or zero). Referring back to the last row of Table 6.3, one can see that this condition is fully met already by the initial solution since c_j have previously been assumed to be non-negative. The only reason the initial solution is not also the optimal solution is that the solution is not feasible, i.e., some variables assume negative values. This immediately suggests a new approach to solving linear programming problems. Starting with an initial basic solution which has all evaluators with the right signs but some variables with the wrong signs, if some method could be found to substitute these variables with others having the right signs while maintaining the correct signs of all evaluators, then we would arrive at the solution which is both optimal and feasible. This is the central idea that guides the development and understanding of the dual simplex method as treated in the remainder of this chapter.

The motif of the dual simplex method is the same as that of the regular simplex, i.e., a systematic evaluation of basic (but not feasible) solutions obtained by successive replacement of basis vectors by non-basic vectors until the desired optimal feasible solution is reached. In the regular simplex, it will be recalled that this systematic evaluation is achieved through the use of a pair of decision rules which insures that the value of the objective function will steadily press forward toward the optimum while feasibility is maintained. In the dual simplex, the same decision rules with minor modifications again provide the essential guidance, but in an opposite direction. This will be more easily understood in the light of the following theorem:

Given a linear function $f(X) = CX$ where X is constrained by $AX = P_0$ with $P_0 \geq 0$; let X_M be the optimal basic feasible solution, and X_A, any basic non-feasible solution for which $Z_A - C \leq O$ (≥ 0), then $f(X_A) \leq f(X_M)$ $[\geq f(X_M)]$.* In words, this theorem states that

* The proof for the minimizing case is as follows: since X_M is the optimal feasible solution, we infer:

$$AX_M = P_0. \tag{1}$$
$$X_M \geq 0, \tag{2}$$
$$Z_M - C \leq 0. \tag{3}$$

Since X_A is a basic solution also,

$$AX_A = P_0. \tag{4}$$

Furthermore, we are given $\quad Z_A - C \leq 0. \quad (5)$

Now, according to (6.2.6), $Z = \tilde{C}B^{-1}A$ where $\tilde{C}$ is the price vector of the basis vectors and B^{-1} is the inverse of the basis. Let the $\tilde{C}$ and B^{-1} of X_A be $\tilde{C}_A$, B_A^{-1}. Then, from (5) we have

$$\tilde{C}_A B_A^{-1} A \leq C. \tag{6}$$

So long as $X_M \geq 0$, we can multiply (6) by X_M on both sides and obtain

$$\tilde{C}_A B_A^{-1} A X_M \leq C X_M. \tag{7}$$

Substituting (1) into (7),

$$\tilde{C}_A B_A^{-1} P_0 \leq C X_M. \tag{8}$$

Substituting $B_A^{-1} P_0 = \tilde{X}_A$ (the basis components of X_A) into (8), we obtain finally

$$\tilde{C}_A \tilde{X}_A \leq C X_M. \tag{9}$$

given a minimizing (or maximizing) linear programming problem, any non-feasible basic solution always yields a lower (or higher) value of the objective function than the minimal (or maximal) basic feasible solution does *provided* all evaluations of the vectors $(z_j - c_j)$ associated with the non-feasible solution are non-positive (or non-negative). The implication for a minimizing problem is: if we can conveniently obtain a non-feasible basic solution with all evaluations non-positive, we will be able to arrive at the optimal solution by trying to remove the non-feasible basis variables while at the same time increasing the value of the objective function. Similarly, for a maximizing problem, if we can start conveniently with a basic non-feasible solution for which all evaluations are non-negative, we will also be able to arrive at the optimal basic feasible solution by trying to remove the non-feasible basis variables while at the same time decreasing the value of the objective function. By simple deduction, in a minimizing problem where the optimal value of the objective function is negative infinity, then there is no basic non-feasible solution for which all evaluations are non-positive. On the other hand, if the minimizing problem has no basic feasible solution at all, then we expect the set of basic non-feasible solutions with non-positive evaluations to have no upper bound so that the value of the objective function can be increased to positive infinity.

6.6. Dual Simplex Method—Computational Procedure.

This section will illustrate the computational details of the dual simplex method by means of two numerical examples. In the first example we deal with the simpler case where a convenient basis exists, relative to which all evaluations possess the proper signs (≥ 0 or ≤ 0). Consequently, efforts will be concentrated on the explanation of computational rules. The second example, on the other hand, illustrates a more complicated situation where although a convenient basis exists

---

Due to the definition of a basic solution

$$\tilde{C}_A \tilde{X}_A = CX_A. \tag{10}$$

Substituting (10) into (9), thus,

$$CX_A \leq CX_M$$

or

$$f(X_A) \leq f(X_M). \qquad \text{Q.E.D.}$$

For a maximizing problem, suitable changes in the direction of inequality in (3), (5), (6), (7), (8) and (9) will lead to the opposite conclusion, viz., $f(X_A) \geq f(X_M)$.

and a basic non-feasible solution can readily be obtained, the relevant evaluations do not all possess the proper signs. A method originated by Vajda will then be explained which brings about the right signs for the application of the dual simplex procedure.*

Example 1. Suppose we are to minimize $f(X) = x_1 + 2x_2 + 3x_3$,

subject to:
$$2x_1 - x_2 + x_3 \geq 4,$$
$$x_1 + x_2 + 2x_3 \leq 8,$$
$$x_2 - x_3 \geq 2,$$
and
$$x_1, x_2, x_3 \geq 0.$$

The problem is, of course, equivalent to the following:

$$\begin{aligned} \text{Minimize } f(X) &= x_1 + 2x_2 + 3x_3, \\ \text{subject to:} \quad & -2x_1 + x_2 - x_3 + x_4 = -4, \\ & x_1 + x_2 + 2x_3 + x_5 = 8, \\ & -x_2 + x_3 + x_6 = -2, \\ \text{and} \quad & x_j \geq 0, \quad j = 1, 2, \ldots, 6. \end{aligned}$$

Obviously a convenient basis for an initial basic solution is $B = [P_4 \quad P_5 \quad P_6]$. Table 6.4a below shows the simplex tableau that

Table 6.4a

$c_{(i)}$	$P_{(i)}$ \ P_j		P_1	P_2	P_3	P_4	P_5	P_6	P_0
0	$P_{(1)}$	P_4	(−2)	1	−1	1	0	0	−4
0	$P_{(2)}$	P_5	1	1	2	0	1	0	8
0	$P_{(3)}$	P_6	0	−1	1	0	0	1	−2
	$z_j - c_j$		−1	−2	−3	0	0	0	0

corresponds to this basis. The solution is clearly a non-feasible one but the evaluations of all the vectors are non-positive. Therefore, according to the theorem of Section 6.5, the corresponding value of the objective function (which is zero) must be below the minimum for the feasible solutions. If we try to remove the non-feasible variables from the basis while simultaneously increasing the value of the ob-

---

* S. Vajda, *The Theory of Games and Linear Programming* (New York: John Wiley & Sons, Inc., 1957), pp. 79–82.

jective function, we will eventually obtain the minimal basic feasible solution.

There are two basis variables in the present solution which are eligible for elimination: one is $x_4 = -4$ and the other is $x_6 = -2$. Let us arbitrarily choose x_4 to go out. The next question is then which variable should come into the basis to take its place. To answer this question, let us recall from (4.2.6) and (4.5.3) that,

$$f(X') = f(X) - \frac{\theta_{(l)}}{y_{(l)k}} (z_k - c_k)$$

Since $\theta_{(l)}$ and $(z_k - c_k)$ are both negative, the former by choice and the latter by previous calculations, we must select P_k whose $y_{(l)k}$ is negative so that the value of the objective function will increase. Should there be more than one P_j whose $y_{(l)j} \leq 0$, then we must select among them that vector P_k which satisfies the condition below to be the incoming vector.

$$\text{(6.6.1)} \qquad \frac{z_k - c_k}{y_{(l)k}} = \text{minimum} \left\{ \frac{z_j - c_j}{y_{(l)j}} \right\} \quad \text{for all } y_{(l)j} < 0.$$

If any vector other than the one with the minimum ratio is selected, then the new evaluations which can be determined by Transformation Rule III (4.6.8) will no longer be all non-positive (which would destroy the underlying condition for the application of the dual simplex method). Along the same line of reasoning it can be stated that in a maximizing problem such as *Example 2* which follows, the rule for the selection of P_k should be

$$\text{(6.6.2)} \qquad \frac{z_k - c_k}{y_{(l)k}} = \text{maximum} \left\{ \frac{z_j - c_j}{y_{(l)j}} \right\} \quad \text{for all } y_{(l)j} < 0.$$

It may be pointed out in passing that the use of the above rules to select P_k also permits x_k to enter the basis with a positive value—since according to (4.2.7) $x_k = \theta'_k = \dfrac{\theta_{(l)}}{y_{(l)k}}$ and, in dual simplex, both $\theta_{(l)}$ and $y_{(l)k}$ are always negative. This is of course essential if the ultimate solution is to be a feasible one.

Returning to our present problem we notice that in the row of the outgoing basis vector $P_{(l)} = P_4$ (Table 6.4a), there are two P_j whose $y_{(l)j}$ are negative. These are $y_{(1)1} = -2$ and $y_{(1)3} = -1$. Following (6.6.1) we find the ratio for P_1 to be $\frac{1}{2}$ and the smaller of the two.

Hence P_k is chosen to be P_1 and $y_{(1)1}$ circled as the pivotal element. Once the incoming and out-going vectors have been properly selected, what remains is a complete transformation of the simplex tableau by exactly the same rules used in the regular simplex method. The results of the current transformation are shown in Table 6.4b. Note that the

Table 6.4b

$c_{(i)}$	$P_{(i)}$	P_j	P_1	P_2	P_3	P_4	P_5	P_6	P_0
1	$P_{(1)}$	P_1	1	$-\frac{1}{2}$	$\frac{1}{2}$	$-\frac{1}{2}$	0	0	2
0	$P_{(2)}$	P_5	0	$\frac{3}{2}$	$\frac{3}{2}$	$\frac{1}{2}$	1	0	6
0	$P_{(3)}$	P_6	0	(−1)	1	0	0	1	−2
$z_j - c_j$			0	$-\frac{5}{2}$	$-\frac{5}{2}$	$-\frac{1}{2}$	0	0	2

solution is still not feasible but the value of the objective function is now increased and presumably closer to the minimum of the problem. Note further how all of the evaluations have continued to be kept non-positive.

Now that the only negative variable is x_6, $P_{(3)} = P_6$ must be replaced. Looking across the row of $P_{(3)}$, however, we see only one negative $y_{(3)j}$, which is $y_{(3)2}$. P_2 must then be the incoming vector and $y_{(3)2}$ is circled accordingly as the pivotal element. In Table 6.4c the

Table 6.4c

$c_{(i)}$	$P_{(i)}$	P_j	P_1	P_2	P_3	P_4	P_5	P_6	P_0
1	$P_{(1)}$	P_1							3
0	$P_{(2)}$	P_5							3
2	$P_{(3)}$	P_2							2
$z_j - c_j$			0	0	−5	$-\frac{1}{2}$	0	$-\frac{5}{2}$	7

new solution, evaluations and value of the objective function are displayed as a result of transformation. Since the solution is now a feasible one and the evaluations are all non-positive, the optimal solution to the original problem is reached and no $y_{(i)j}$ are transformed. To repeat in words, the optimal and minimizing solution is: $x_1 = 3$, $x_2 = 2$, $x_5 = 3$ and $x_3 = x_4 = x_6 = 0$, yielding a minimal value of the objective function equal to 7. The reader is advised to solve the foregoing problem by the regular simplex method using two artificial vectors in order to make a comparison between the two methods with respect to the path and speed of convergence towards the optimal solution.

Example 2. Consider the following slightly more complicated problem:

$$\begin{aligned} \text{Maximize } f(X) &= 3x_1 - 2x_2 - x_3, \\ \text{subject to:} \quad & x_1 - x_2 - x_3 = 4, \\ & x_2 + 2x_3 \leq 8, \\ & x_2 - x_3 \geq 2, \\ \text{and} \quad & x_1, x_2, x_3 \geq 0. \end{aligned}$$

The usual addition and subtraction of slack variable convert this problem into the following:

$$\begin{aligned} \text{Maximizing } f(X) &= 3x_1 - 2x_2 - x_3, \\ \text{subject to:} \quad & x_1 - x_2 - x_3 = 4, \\ & x_2 + 2x_3 + x_4 = 8, \\ & - x_2 + x_3 + x_5 = -2, \\ \text{and} \quad & x_j \geq 0, \quad j = 1, \ldots, 5. \end{aligned}$$

An identity basis can obviously be formed with P_1, P_4 and P_5 yielding a non-feasible basic solution. This, however, would not present any new problem if all evaluations were non-negative, for then the dual simplex procedure could be immediately applied. Unfortunately, the evaluations are not all non-negative as we find that

$$z_2 - c_2 = \tilde{C}B^{-1}P_2 - c_2 = [c_1 \quad c_4 \quad c_5]IP_2 - c_2$$

$$= [3 \quad 0 \quad 0]\begin{bmatrix} 1 & 0 & 0 \\ 0 & 1 & 0 \\ 0 & 0 & 1 \end{bmatrix}\begin{bmatrix} -1 \\ 1 \\ -1 \end{bmatrix} - (-2) = -3 + 2 = -1;$$

$$z_3 - c_3 = \tilde{C}B^{-1}P_3 - c_3 = -3 - (-1) = -2.$$

Since both are not non-negative, the underlying condition for the application of the dual simplex method is not met. Therefore, some adjustment must be made to bring all evaluations to non-negativity before the method can be applied.

Let us denote by t the set of variables (necessarily non-basis) whose evaluations are of the wrong sign.* Let these variables in t be represented by $x_{t_1}, x_{t_2}, \ldots, x_{t_r}$. Define a new variable x_a as follows:

$$x_a + x_{t_1} + x_{t_2} + \ldots + x_{t_r} = \text{M}, \tag{6.6.3}$$

where M is an arbitrarily large positive number.
Now select the variable in t, say x_{t_p} which has the largest *absolute* evaluation, i.e., $z_{t_p} - c_{t_p}$, and write from (6.6.3)

$$x_{t_p} = \text{M} - (x_a + x_{t_1} + x_{t_2} + \ldots + x_{t_r}). \tag{6.6.4}$$

Substituting (6.6.4) into the original objective function and structural constraints, we obtain a modified but equivalent problem with the following changes:

$(c_{t_1} - c_{t_p})$, $(c_{t_2} - c_{t_p})$ etc. are now the price coefficients of x_{t_1}, x_{t_2}, etc. instead of simply c_{t_1}, c_{t_2} etc. as before;

$(P_{t_1} - P_{t_p})$, $(P_{t_2} - P_{t_p})$ etc. are now the coefficient vectors of x_{t_1}, x_{t_2}, etc. instead of simply P_{t_1}, P_{t_2} etc. as before;

$(z_{t_1} - z_{t_p}) - (c_{t_1} - c_{t_p})$, $(z_{t_2} - z_{t_p}) - (c_{t_2} - c_{t_p})$, etc. are now the evaluations of x_{t_1}, x_{t_2}, etc. instead of simply $(z_{t_1} - c_{t_1})$, $(z_{t_2} - c_{t_2})$, etc. as before; and finally

x_a replaces x_{t_p}; $P_a = -P_{t_p}$; $c_a = -c_{t_p}$; $z_a - c_a = -(z_{t_p} - c_{t_p})$; and $(P_0 - \text{M}P_{t_p})$ replaces P_0.

After these changes have been made, all evaluations in the modified problem will bear the right signs and with (6.6.3) added to the system of constraints to insure the non-negativity of x_{t_p}, the dual simplex method can be applied to obtain the optimal solution.

---

* Vajda (*op. cit.*, p. 79), implied that it was the price coefficients rather than the evaluations of the variables that made necessary the modification of the problem. He writes, "If all coefficients in the linear form have the correct sign, then we proceed by the dual simplex method without any difficulty. But if there are wrong signs in the linear form, . . . then we introduce an artificial equation. . . ." While this is true for problems which happen to have a full set of slack variables with zero price coefficients to make up the initial basis, in general it is quite possible to have price coefficients of the wrong signs while all evaluations are of the right signs. In the latter case, the dual simplex method can still be applied regardless of the signs of the price coefficients. This point must be emphasized. Unfortunately Gass, *op. cit.*, p. 128, repeats the error in discussing briefly Vajda's approach to this type of difficulty encountered in the application of the dual simplex.

Returning to our example, since both x_2 and x_3 have wrong evaluations (of negative sign), we define the new variable x_a by $x_a + x_2 + x_3 = \text{M}$. From previous calculations, we know that x_3 has the larger absolute evaluation of the two. Therefore, substituting $x_3 = \text{M} - x_a - x_2$ into the objective function and the original constraints, and adding the additional constraint to the problem, we arrive at the following modified problem:

$$
\begin{aligned}
\text{Maximize } f(X) &= x_a + 3x_1 - x_2 - \text{M}, \\
\text{subject to:} \quad & x_a + x_1 = \text{M} + 4, \\
& -2x_a - x_2 + x_4 = -2\text{M} + 8, \\
& -x_a - 2x_2 + x_5 = -\text{M} - 2, \\
& x_a + x_2 + x_3 = \text{M}, \\
\text{and} \quad & x_a, x_1, \ldots, x_5 \geq 0.
\end{aligned}
$$

Clearly a convenient basis is $B = [P_1 \;\; P_4 \;\; P_5 \;\; P_3]$. Using this basis we obtain the initial simplex tableau as shown in Table 6.5a.

Table 6.5a

$c_{(i)}$	$P_{(i)}$	P_j	P_a	P_1	P_2	P_3	P_4	P_5	P_0
3	$P_{(1)}$	P_1	1	1	0	0	0	0	M + 4
0	$P_{(2)}$	P_4	(−2)	0	−1	0	1	0	−2M + 8
0	$P_{(3)}$	P_5	−1	0	−2	0	0	1	−M − 2
0	$P_{(4)}$	P_3	1	0	1	1	0	0	M
$z_j - c_j$			2	0	1	0	0	0	2M + 12

As expected all evaluations are now non-negative, indicating that the dual simplex method can now be applied to decrease the value of the objective function from 2M + 12 to the desired optimum. Following the rules discussed in the previous section, we select P_4 to leave the basis which in turn fixes P_a to come in.* The results of subsequent transformations are shown in Table 6.5b. Note especially the effects of the inclusion of P_a in the basis on the values of the other basis vari-

* There is actually a tie between P_a and P_2 here based on ratios computed according to (6.6.2). In general, ties should be resolved by the same procedure prescribed for the regular simplex. In this particular case, however, P_a is selected because its inclusion in the basis makes possible its elimination from further consideration.

Table 6.5b

$c_{(i)}$	$P_{(i)}$ \ P_j		P_a	P_1	P_2	P_3	P_4	P_5	P_0
3	$P_{(1)}$	P_1	0	1	$-\frac{1}{2}$	0	$\frac{1}{2}$	0	8
1	$P_{(2)}$	P_a	1	0	$\frac{1}{2}$	0	$-\frac{1}{2}$	0	M − 4
0	$P_{(3)}$	P_5	0	0	$\circled{-\frac{3}{2}}$	0	$-\frac{1}{2}$	1	−6
0	$P_{(4)}$	P_3	0	0	$\frac{1}{2}$	1	$\frac{1}{2}$	0	4
$z_j - c_j$			0	0	0	0	1	0	20

ables as well as on that of the objective function. Due to the auxiliary constraint $x_a + x_2 + x_3 = \text{M}$, so long as $x_a = 0$ (out of the basis), all basis variables must contain an M component making the value of the objective function also a function of M such as shown in Table 6.5a. But as soon as x_a is put into the basis as a result of reducing $f(X)$ to its maximal value for feasible solutions, all M components in the other basis variables as well as in $f(X)$ disappear (Table 6.5b). Consequently we have a basic solution (not necessarily feasible yet) to the *original* problem. From this point on, since P_a is not a part of the original problem and can never be displaced from the basis (Why?) it can be left in the basis in subsequent iterations without any further attention while the rest of the tableau is transformed successively towards the optimal solution. Or, it can be discarded altogether in subsequent transformations as is done here.

Referring to Table 6.5b we find that $x_5 = -6$ is still non-feasible. Following the rules of the dual simplex we select P_2 to replace P_5 with the results shown in Table 6.5c, giving the final optimal (maximizing) basic feasible solution.

Table 6.5c

$c_{(i)}$	$P_{(i)}$ \ P_j		P_1	P_2	P_3	P_4	P_5	P_0
3	$P_{(1)}$	P_1	1	0	0	$\frac{2}{3}$	$-\frac{1}{3}$	10
−1	$P_{(2)}$	P_2	0	1	0	$\frac{1}{3}$	$-\frac{2}{3}$	4
0	$P_{(3)}$	P_3	0	0	1	$\frac{1}{3}$	$\frac{1}{3}$	2
$z_j - c_j$			0	0	0	1	0	20

EXERCISES

1. Given the following two bases:

$$A = \begin{bmatrix} 2 & 3 & 0 \\ 0 & 1 & 0 \\ -1 & 1 & 1 \end{bmatrix}, \qquad B = \begin{bmatrix} 2 & 3 & 0 \\ 0 & 1 & 2 \\ -1 & 1 & -2 \end{bmatrix},$$

   (a) Find the elementary matrix E such that $AE = B$.
   (b) Find a series of three elementary matrices, E_1, E_2 and E_3, which transforms an identity matrix to B in three steps. In other words, find E_1, E_2 and E_3 such that $IE_1E_2E_3 = B$.
   (c) It has been suggested that elementary matrices provide a new method of inverting a basis. Do you see how? Do you see any advantage?

2. Solve the following linear programming problems by the revised simplex method.

   (a) Maximize $f(X) = 2x_1 - x_2 + x_3$,
   subject to:

$$\begin{aligned} x_1 + x_2 + x_3 + x_4 &\leq 8, \\ 2x_2 - x_3 &\leq 4, \\ 2x_1 - x_2 - x_4 &\leq 2, \end{aligned}$$

   and $x_j \geq 0, \quad j = 1, \ldots, 4.$

   (b) Maximize $f(X) = -2x_4 - x_5 + x_6$,
   subject to:

$$\begin{aligned} - x_2 + 2x_3 + x_4 &= 5, \\ x_1 + 2x_2 - 2x_3 + x_5 &= 4, \\ -x_1 + x_2 + x_6 &= 6, \end{aligned}$$

   and $x_j \geq 0, \quad j = 1, \ldots, 6.$

   (c) Minimize $f(X) = x_1 + x_2$,
   subject to:

$$\begin{aligned} x_1 + x_2 + x_3 + x_4 &\geq 10, \\ 2x_2 - x_3 - x_4 &\geq 2, \\ x_2 - 2x_3 + x_4 &\geq 6, \end{aligned}$$

   and $x_j \geq 0, \quad j = 1, \ldots, 4.$

3. Suppose that we are to maximize $f(X) = CX$ subject to $AX = P_0$. Let X_A be a basic but non-feasible solution, and X_0 be the optimal basic feasible solution. Prove that if relative to X_A, the evaluations of all vectors are non-negative, then $f(X_A) \geq f(X_0)$.

4. Solve Exercise 2 (a), (b) and (c) by the dual simplex method.

5. Solve the following linear programming problem first by the revised simplex method, and then by the dual simplex method.

$$\begin{aligned} \text{Minimize } f(X) &= -x_1 + 2x_2 - x_3, \\ \text{subject to:} \quad & 3x_1 + x_2 - x_3 \le 10, \\ & -x_1 + 4x_2 + x_3 \ge 6, \\ & x_2 + x_3 \ge 4, \\ \text{and} \quad & x_j \ge 0, \quad j = 1, 2, 3. \end{aligned}$$

6. The following is a simplex tableau as found at some intermediate stage in the solution of a minimizing problem.

$c_{(i)}$	$P_{(i)}$ \ P_j		P_1	P_2	P_3	P_4	P_5	P_6	P_0
-1	$P_{(1)}$	P_1	1	$\frac{2}{3}$	0	0	$\frac{4}{3}$	0	4
-3	$P_{(2)}$	P_4	0	$-\frac{7}{3}$	3	1	$-\frac{2}{3}$	0	2
1	$P_{(3)}$	P_6	0	$-\frac{2}{3}$	-2	0	$\frac{2}{3}$	1	2
$z_j - c_j$			0	$\frac{8}{3}$	-11	0	$\frac{4}{3}$	0	-8

Additionally, it is known that the inverse of the current basis is

$$B^{-1} = [P_1\, P_4\, P_6] = \begin{bmatrix} \frac{1}{3} & \frac{1}{3} & -\frac{1}{3} \\ \frac{1}{3} & -\frac{2}{3} & \frac{2}{3} \\ -\frac{1}{3} & \frac{2}{3} & \frac{1}{3} \end{bmatrix}.$$

Find the original problem.

7. A branch warehouse manager has the responsibility of allocating a capital fund of \$500,000 and warehouse space of 100,000 cubic feet to the purchase of three commodities A, B and C which cost \$4, \$1.50 and \$5, and require 10, 1 and 5 cubic feet of storage space per unit respectively. All other costs are assumed to be either fixed or negligible. The prices of A, B and C are respectively \$5, \$2 and \$7 per unit. The general manager of the concern has previously issued a directive to all branch managers to the effect that priority should be given to C, A and B in that order, presumably reflecting their relative profit margins. Moreover, because of the relatedness of their demands, there should be at least one A for every two C purchased. Under this directive it is clear that B would never be stored because either capital or warehouse space would be completely tied up in purchasing

and storing C and A. The market potentials for all three commodities are unlimited so far as this warehouse is concerned.

(a) In following the directive, the branch manager needs to know in advance whether capital or space would be the first limiting constraint in a program of purchases to maximize profit.

(b) He also wants to find out, to his own satisfaction at least, whether the priority system specified in the directive is actually consistent with his responsibility of maximizing the profit of this branch.

Find the answers to the two questions above by testing the optimality of the implied basis.

OPTIMALITY ANALYSIS

7.1. Introduction.

It should not be difficult to see from our previous discussions that an optimal linear programming solution has two different aspects—one numerical and the other structural. By the former we mean the specific values of the objective function and the basis variables, and by the latter, the "character" or composition of the solution as represented by the set of variables comprising the basis. Although in mathematics this distinction is still primarily one of magnitude rather than of kind, i.e., one of having a positive value (greater than zero) if the variable is in basis and a zero value if it is not, in business and industry a structural change in the optimal solution often signifies a change in policy, which is, of course, a matter of considerable importance. For instance, in the simple production planning example discussed in Chapter III Section 3.2, when the profit margins of A and B are respectively \$30 and \$20, the optimal solution (point E in Figure 3.1) indicates that both products should be produced. But if B should become more profitable relative to A at some point (e.g. when profit margins are \$20 and \$30 for A and B respectively), the optimal solution would change structurally to production of B alone and no A at all (point F in Figure 3.1). In normal business processes this would constitute a change in policy, requiring serious consideration by high-level management of such matters as dealer relationship, consumer good-will and, in some cases, legal implications. Since in terms of linear programming such a structural change in the optimal solution is always the result of a change of the basis, we are interested in learning the limiting conditions of a particular optimal basis. The analysis devoted to this end comprises the first important part of what we shall refer to in general as optimality analysis.

From a practical standpoint, it is important to know not only what program (or solution) optimizes the objective function under the prevailing constraints, but also how the objective could best be advanced by removing one or several of these constraints. In the simple illustration we used above, for example, it is desirable to ask such questions as: By how much could profit be increased if the supply of labor were increased by a certain amount in the foundry department? In the machine department? The answers to these questions often form intelligent bases for the costing of input factors and the formulation of expansion policies. Similarly, inasmuch as technological advances result in changes in the structural coefficients (a_{ij}), by analyzing their effects on the objective function, (i.e., by studying their profit implications), it is at least conceptually possible to develop some logical criteria for the evaluation of research projects. These are all studies of the sensitivity of the objective function to changes in particular parameters of the problem. They constitute the second important area of optimality analysis.

In this chapter, the effects of changes in the parameters of a linear programming problem, i.e., the c_j, b_i, a_{ij}, m and n, will be discussed in detail from the two standpoints described above. In addition, whenever the optimality of a basis is destroyed as a result of these changes, method and procedure for obtaining the new optimal basis will be described and explained. There is a third aspect of optimality analysis which ought to be pointed out. It is the determination of sub-optima, i.e., the determination of the next best solution(s). However, since this problem closely parallels the determination of alternative optima already discussed in Chapter V Section 5.7, it is left to the reader as an exercise without further elaboration.

7.2. Effects of Changes in c_j on Optimality.

The simplest question that may be asked under this heading is: Would the same optimal solution still prevail if the price coefficient of some *non-basis* variable x_j is changed from c_j to c_j'? Since the evaluation of x_j is $z_j - c_j$, we simply compute $z_j - c_j'$. If the new evaluation is now favorable, i.e., ≤ 0 for maximizing problems or ≥ 0 for minimizing problems, then the optimality of the old basis is destroyed, and a new optimal solution has to be established by routine simplex iterations.

From the above it is clear that for any optimal solution there is a corresponding set of critical values of the non-basis price coefficients, viz., $c_j' = z_j$ where j = index of non-basis variables, which define the boundaries of that particular optimal solution. Any change in c_j which causes it to exceed the corresponding z_j will result in a new optimal solution.

The next logical question is: What effects does a change in the price coefficient of a *basis* variable have on the optimality of the solution? Let us denote in general a basis variable by x_i, where i represents the index of a basis variable.* Since all z_j include a term $c_i y_{ij}$ (4.5.1), the evaluations of all non-basis variables are affected by a change in c_i, unless y_{ij} happens to be zero. If the new price coefficient of x_i is c_i', then the optimality of the existing basis will be destroyed when:

$$(z_j - c_j) - (c_i - c_i')y_{ij} < 0 \quad \text{for a maximizing problem,}$$
$$(z_j - c_j) - (c_i - c_i')y_{ij} > 0 \quad \text{for a minimizing problem.}$$

The condition given above for a maximizing problem implies that:

$$\text{if } y_{ij} > 0 \qquad \text{then} \qquad c_i' < c_i - \frac{z_j - c_j}{y_{ij}};$$

$$\text{if } y_{ij} < 0 \qquad \text{then} \qquad c_i' > c_i - \frac{z_j - c_j}{y_{ij}}.$$

Thus, there are two critical values of c_i, which together define the range of c_i that is compatible with the existing optimal basis. Designating the upper and lower bounds of this range by $\bar{c}_i$ and $\underline{c}_i$ respectively, this optimal range of c_i is given by:

$$\text{(7.2.1)} \qquad \bar{c}_i = \text{Min}\left\{c_i - \frac{z_j - c_j}{y_{ij}}\right\} \quad \text{for } y_{ij} < 0,$$

* In this and subsequent chapters, the notation (i) used previously to denote order in the basis is replaced by a simple i. By now, the reader should realize fully that the ith basis variable (x_i) may be any x_j, $j = 1, \ldots, n$.

$$\underline{c}_i = \text{Max}\left\{c_i - \frac{z_j - c_j}{y_{ij}}\right\} \quad \text{for } y_{ij} > 0, \tag{7.2.2}$$

where i = index of basis variables,
and j = index of non-basis variables.

Should the new price coefficient c_i' be either greater than $\bar{c}_i$ or smaller than $\underline{c}_i$ *algebraically*, a new optimal solution would have to be determined by further iterations.

In a similar manner but with the signs of inequality reversed at appropriate places, we arrive at the corresponding range of c_i for a minimizing problem, which is defined by:

$$\bar{c}_i = \text{Min}\left\{c_i - \frac{z_j - c_j}{y_{ij}}\right\} \quad \text{for } y_{ij} > 0, \tag{7.2.3}$$

$$\underline{c}_i = \text{Max}\left\{c_i - \frac{z_j - c_j}{y_{ij}}\right\} \quad \text{for } y_{ij} < 0, \tag{7.2.4}$$

where i = index of basis variables,
and j = index of non-basis variables.

Table 7.1 below summarizes these results in tabular form for convenient reference in the analysis of optimality from the standpoint of changes in c_i.

Table 7.1

Problem	Range of c_i	
	$\bar{c}_i$	$\underline{c}_i$
Maximizing	$\text{Min}_{y_{ij}<0}\left\{c_i - \frac{z_j - c_j}{y_{ij}}\right\}$	$\text{Max}_{y_{ij}>0}\left\{c_i - \frac{z_j - c_j}{y_{ij}}\right\}$
Minimizing	$\text{Min}_{y_{ij}>0}\left\{c_i - \frac{z_j - c_j}{y_{ij}}\right\}$	$\text{Max}_{y_{ij}<0}\left\{c_i - \frac{z_j - c_j}{y_{ij}}\right\}$

An extension of this type of analysis is the so-called *parametric programming* first investigated by Gass and Saaty.* They discovered that in certain industrial linear programming problems, the price coefficients are related by a one-parameter linear form, e.g., $c_j = d_j + f_j t$ with the parameter t varying over a certain finite (or infinite) range from, say, t_L to t_U. Obviously, the value of t importantly affects the

* S. I. Gass, *op. cit.*, Chapter 8.

optimality of a particular basis. In the following paragraph we shall briefly trace the steps involved in solving this parametric programming problem.

Let us begin by setting $t = t_L$ (the lower limit of t) and determine the corresponding optimal solution by the usual simplex routine. Since the price coefficients are linear functions of t, the evaluators $(z_j - c_j)$ are also linear functions of t, say, $g_j + h_j t$, where g_j and h_j are two coefficients recorded separately in the simplex tableau. If the problem is a maximizing one, with $t = t_L$ as previously assumed, a specific optimal solution can be obtained for which all $g_j + h_j t_L \geq 0$. Next we raise the question: What value of t greater than t_L will cause some $g_j + h_j t$ to *first* become negative? This means that $h_j t$ must be decreased. Since $t > t_L$, this could be brought about only for those h_j which are negative; therefore we concentrate on all $h_j < 0$ and determine

$$t_1 = \underset{h_j < 0}{\text{Minimum}} \left\{ -\frac{g_j}{h_j} \right\} = -\frac{g_k}{h_k}, \tag{7.2.5}$$

which then marks the upper boundary of the first optimal solution. Since x_k is the non-basis variable identified in (7.2.5), we next bring x_k into the basis and recompute all evaluations relative to the new optimal basis in terms of $t = t_1$. This enables us then to reapply (7.2.5) to determine the next critical value of t, and so on. In this manner, the entire range of $t_L \leq t \leq t_U$ can be segmented into $t_L < t_1 < t_2 < \ldots < t_U$, where there exists a different optimal solution corresponding to each interval.

7.3. Effects of Changes in b_i on Optimality.

It is relatively easy to appraise the effects on optimality of changes in b_i. They are, it will be recalled, the constants on the right-hand side of the constraints. In general, such changes affect only the solution values of the basis variables (x_i) and the value of the objective function $f(X)$ through the equations $X = B^{-1}P_0$ (6.2.4) and $f(X) = \tilde{C}B^{-1}P_0$ (6.2.7) respectively.* Since the evaluations of the non-basis variables are not at all affected, the same basis remains

* As the basis inverse is needed for the computation of the new solution, clearly the revised simplex method should be used whenever studies on the effects of changes in b_i are contemplated.

optimal unless and until some x_i are rendered negative, thereby making the basis infeasible. Thus, we have come to an interesting discovery, viz., when b_i are changed, it is the destruction of the feasibility rather than the optimality of the current basis that creates the need for re-optimization. It is noteworthy, however, that re-optimization can be quickly accomplished by means of the dual simplex method discussed in Chapter VI, Sections 6.5 and 6.6 because all evaluations are already of the proper sign by virtue of the optimality of the previous basis.

The feasibility range of a specific b_i, say b_k, for a given optimal solution can be determined in much the same way as the optimality range of c_i is determined in the previous section. Equations (7.2.1) and (7.2.2) become directly applicable after the following substitutions are made: (1) θ_i (the solution-value of the basis variable x_i) for $z_j - c_j$, (2) d_{ik} (the element in the ith row and kth column of B^{-1}) for y_{ij}, (3) b_k for c_i, and finally (4) $\bar{b}_k$, $\underline{b}_k$ for $\bar{c}_i$ and $\underline{c}_i$ respectively. Thus we have

(7.3.1) $$\bar{b}_k = \operatorname*{Min}_{d_{ik}<0} \left\{ b_k - \frac{\theta_i}{d_{ik}} \right\},$$

(7.3.2) $$\underline{b}_k = \operatorname*{Max}_{d_{ik}>0} \left\{ b_k - \frac{\theta_i}{d_{ik}} \right\},$$

where i = index of basis variables,
k = index of the constraint whose b_k has been changed.

Equations (7.2.3) and (7.2.4) are not relevant since unlike $z_j - c_j$, θ_i are always positive, or rather, never negative. In other words, the process of analyzing the effects of a change in b_k is the same for both maximizing and minimizing problems.

A moment's reflection will also reveal that the change in b_k does not mean that the kth basis variable is necessarily the one that will be rendered infeasible first. The fact that there is a feasibility range of b_k is sufficient indication that there are two different basis variables involved, one associated with the upper limit $\bar{b}_k$, and the other with the lower limit $\underline{b}_k$. Although x_k may be one of them, it certainly cannot be both. Then again, it may not be either of them.

Let us hark back for a moment to the old equation (6.2.7) which says

$$f(X) = \tilde{C}B^{-1}P_0.$$

Let us designate $\tilde{C}B^{-1}$ which is a row vector of m components, by

$W = [w_1 \ldots w_m]$. Clearly we can write the objective function in the following manner:

$$f(X) = WP_0 = w_1b_1 + w_2b_2 + \ldots + w_mb_m, \tag{7.3.3}$$

which shows that the value of the objective function associated with a given basis B is a linear function of the constants b_i. Assuming that B is the optimal basis, then w_i represents the increment (or decrement if w_i is negative) in the value of $f(X)$ due to a unit increase in b_i. Hence, w_i may be regarded as the economic value of b_i, or its *implicit* (opportunity) *cost* in the language of economics.* It provides a highly rational basis for evaluating b_i in terms of the effect of its change on the objective being sought. We will discuss this point more thoroughly in Section 8.7 of the next chapter.

7.4. Effects of Changes in a_{ij} on Optimality.

A change in the structural coefficient a_{ij} may have one of two kinds of effects depending on whether the vector P_j to which it belongs is in the optimal basis or not. If it is *not* in the basis, then the optimality of the present basis is affected only when the new evaluation of x_j is made favorable by this change. Otherwise, the present basis remains optimal and no other effects need to be considered. The formula $z_j - c_j = \tilde{C}B^{-1}P_j$ (6.2.6) is pertinent to the purpose just described.

If, on the other hand, P_j *is* in the basis, then the basis is inevitably affected and its optimality has to be re-appraised. This can be accomplished quite simply by the same process which underlies the revised simplex method discussed in Chapter VI Section 6.3. Let P_k be the basis vector with elements a_{ik} and P'_k be the new form of P_k with elements a'_{ik}. Let B^{-1} and B'^{-1} be the inverses of the old and the new bases respectively. We define an elementary matrix E which has $B^{-1}P'_k$ imbedded as the kth column in what is otherwise an identity matrix. Then,

$$B'^{-1} = E^{-1}B^{-1}, \tag{6.3.4}$$

and can be computed by (6.3.6) and (6.3.7). Once the new basis inverse is obtained, the evaluators and solution can be re-computed. If the new solution is feasible but not optimal, we resume the revised simplex iteration to obtain a new optimum. If, on the other hand, the

* The w_i are also called "shadow price", "efficiency price" and "incremental value."

new solution turns out to be infeasible, we may switch to the dual simplex method to re-establish feasibility and, eventually, optimality. It is possible, however, that the new evaluations do not all have the proper sign, which is a prerequisite for the application of the dual simplex method. In that event, the procedure explained in Example 2 of Section 6.6, Chapter VI may be employed to overcome the difficulty.

A different approach has been suggested by Garvin,* which essentially treats P'_k as an additional non-basis vector with, however, the same price coefficient c_k. The old P_k in the basis is now given a highly unfavorable price coefficient, i.e., $-M$ for maximizing problems and M for minimizing ones. All evaluators are then recomputed and the simplex routine is resumed until the new optimal solution is reached. Since P_k is no longer valid, it should be dropped altogether as soon as it is driven out of the basis due to its unfavorable price coefficient. The new optimal basis, however, may not contain P'_k again. The disadvantage of this approach is that simplex iteration has to begin anew from a highly unfavorable level, which prolongs computation. On the other hand, its simpler concept and procedure do argue in its favor, especially when the changes in a_{ij} are not confined to a single coefficient vector.

7.5. Effects of Adding or Subtracting Variables.

Occasionally changes in the content of a linear programming problem call for the addition or subtraction of new variables to or from the original formulation. For instance, in the simple production planning problem cited earlier, a new product may have been perfected for production. Or, in a transportation problem, a new market may have been penetrated successfully. The effects of these new developments on the optimality of an existing solution can be effectively investigated by the introduction of new variables without reworking the entire problem from the very beginning.

Assume that a new variable x_{n+1} with coefficient vector P_{n+1} is introduced into the problem. Whether the existing optimal solution is affected depends quite obviously on whether its evaluation relative to the current basis, viz., $z_{n+1} - c_{n+1} = \tilde{C}B^{-1}P_{n+1}$ is favorable or unfavorable. If the evaluation is favorable, it should be put into the

* W. W. Garvin, *Introduction to Linear Programming* (New York: McGraw-Hill, 1960), pp. 57–58.

basis and the simplex iteration restarted to determine a new optimum. Otherwise, nothing has to be changed and the old optimum prevails.

Should the reverse be called for, i.e., should it be necessary to remove any variable from the problem, the procedure again depends on whether the variable is or is not in the basis. If it is not, then the existing optimal solution is in no way affected. If it is, we may follow the procedure suggested in Section 7.4, i.e., change its price coefficient to +M or −M as the case may be, so as to force it out of the basis eventually when a new optimum is obtained. If this cannot be done, then the implication is that the constraints are inconsistent without this variable and no solution is possible.

7.6. Effects of Adding or Subtracting Constraints.*

An important part of planning in business and industry is to study the alternatives available in the absence or presence of a particular constraint. In terms of linear programming, this problem is equivalent to considering the effect of adding or subtracting a constraint on the optimality of the existing solution.

Let us first discuss the case of the addition of a constraint. Assume that the original problem has $(m - 1)$ constraints and an optimal solution whose associated basis is

$$B = \begin{bmatrix} a_{11} & \cdots & a_{1,m-1} \\ \cdot & & \cdot \\ \cdot & & \cdot \\ \cdot & & \cdot \\ a_{m-1,1} & \cdots & a_{m-1,m-1} \end{bmatrix}.$$

Now, the constraint $a_{m1}x_1 + \ldots + a_{mn}x_n \leq b_m$ has to be added to the system. The immediate effect of this addition is to increase the dimensions of the basis from $(m - 1) \times (m - 1)$ to $m \times m$. Call this new and enlarged basis $\hat{B}$. We will see that $\hat{B}$ can be constructed with ease from the known basis B so as to provide a basic feasible solution to the enlarged problem. From there on it is a routine matter to either reaffirm its optimality, or failing that, to proceed to determine the new optimal solution.

Obviously if the original basis is retained in the new basis, $(m - 1)$ of the m required basis variables are immediately fixed. Because of the

* An alternative approach to this problem is to solve the dual of the linear programming problem. This will be taken up in Chapter VIII Section 8.5.

additional constraint, the coefficient vectors of these variables all have m components with the last ones being $a_{m1}, a_{m2}, \ldots, a_{m,m-1}$ taken from the new constraint. To complete the new basis, we need one more basis variable whose coefficient vector meets with two conditions, viz., it must be independent of the other $(m - 1)$ vectors already in the basis and it must be such that the inverse of the new basis can easily be obtained. The one variable that satisfies these conditions is either the slack variable of the new constraint or any other variable which appears *only* in the new constraint. For in either case the mth basis vector is going to be a unit vector with a 1 in the mth or new row. If this 1 is positive and b_m (the constant of the new constraint) is also positive, or if both are negative, the new basis $\hat{B}$ will be a feasible basis additionally, which is desirable. Otherwise, it would be advisable to introduce an artificial variable with a +M or −M price coefficient—as the situation calls for—to the new constraint so as to complete a feasible basis, even though in so doing, the optimality of the original basis is automatically destroyed. In this manner, we obtain the new basis $\hat{B}$ which is related to the old optimal basis B as follows:

$$\text{(7.6.1)} \qquad \hat{B} = \begin{bmatrix} a_{11} & \ldots a_{1,m-1} & 0 \\ \cdot & \cdot & \cdot \\ \cdot & \cdot & \cdot \\ \cdot & \cdot & \cdot \\ a_{m-1,1} & \ldots a_{m-1,m-1} & 0 \\ a_{m1} & \ldots a_{mm} & 1 \end{bmatrix}$$

$$= \begin{bmatrix} B & O \\ U & I \end{bmatrix}, \quad \text{where } U = [a_{m1} \ldots a_{m,m-1}].$$

Recalling (6.2.10) in Chapter VI Section 6.2 where a structurally similar matrix was inverted, we can immediately obtain

$$\text{(7.6.2)} \qquad \hat{B}^{-1} = \begin{bmatrix} B^{-1} & O \\ -UB^{-1} & I \end{bmatrix}.$$

If B^{-1} is already available, as normally would have been the case had the revised simplex method been employed, all that remains to be done is to complete the simple matrix multiplication $(-UB^{-1})$ and the new inverse $\hat{B}^{-1}$ is determined. The new solution can then be obtained from $\hat{B}^{-1}\hat{P}_0$ and its optimality investigated by $\hat{\hat{C}}\hat{B}^{-1}\hat{P}_j$, j = index of non-basis variables, where $\hat{P}_0$, $\hat{P}_j$ and $\hat{\hat{C}}$ represent their unhatted counterparts augmented by the proper coefficients from the new constraint.

In the interest of completeness, the approach to the opposite question, viz., the effects of removing a constraint on optimality, will be briefly indicated here without detailed discussion. Assume that the mth constraint is to be removed. Identify the variable which appears in the mth constraint only. Check and see if it is already in the basis. If it is a basis variable, say the kth, then the mth *column* and the kth *row* can be deleted from the *inverse* of the current basis. The remaining $(m - 1) \times (m - 1)$ matrix becomes the inverse of a basis, necessarily feasible, for the reduced problem with $(m - 1)$ constraints. If the variable is not a basis variable, bring it into the basis and transform the inverse according to (6.3.6) and (6.3.7). Note its position in the basis, say kth column. The kth row and the mth column of the transformed inverse may now be deleted to yield the inverse of a basis for the reduced problem. This basis is necessarily feasible and would normally be non-optimal so that re-optimization is required. If no such variable exists, an artificial variable could be introduced temporarily into the mth constraint only to be dropped together with the mth constraint as soon as it is brought into the basis for the purpose of effecting the desired transformation of the basis inverse.*

We will now illustrate the theoretical discussions in the foregoing sections with two well-known applications of linear programming, namely, the gasoline blending problem and the production smoothing problem.

7.7. The Aviation Gasoline Blending Problem.**

The refining of crude oil yields a wide range of petroleum products including various grades of gasoline, kerosene, light and heavy lubricating oil, tar, etc. Because of the technical difficulties in controlling the properties of gasoline in the refining operation, refined gasolines

---

* If a computer is used, the insertion of a desired variable into the basis can be insured by giving it a highly favorable price coefficient. This procedure is desirable because it takes maximum advantage of the simplex program stored in the machine and requires minimum changes.

** The materials in this section are primarily based on A. Charnes and W. W. Cooper's, "Blending Aviation Gasolines" *Econometrica,* XX, (1952), 2, pp. 135–159, to which the reader is referred for details.

must be blended afterwards with other refined products, notably alkylate and isopentane, to produce aviation gasoline that meets specifications. These specifications include (1) a maximum allowable level of volatility as measured by RVP (Reid Vapor Pressure) index, (2) a minimal level of the required anti-knock property as measured by the PN (Performance Number) which is closely related to octane rating, and (3) a maximal safe level of tetraethyl lead which is a poisonous additive purchased from outside for the purpose of improving the octane rating of gasoline.

In order to keep the problem down to a manageable size for hand computation (this study was made in 1951), Charnes and Cooper selected only four of the outputs of the refining operation as inputs of the blending operation. These inputs together with their physical and chemical properties and their daily production rates are listed in Table 7.2 below.

Table 7.2

Blending input	Input index (j)	RVP index	PN at TEL = 0.5 cc/gal.	PN at TEL = 4.0 cc/gal.	Daily rate of production (bbl)
Alkylate I	1	5.0	94.0	107.5	3,800
Catalytic gasoline I	2	8.0	83.0	93.0	2,652
Straight-run gasoline I	3	4.0	74.0	87.0	4,081
Isopentane	4	20.5	95.0	108.0	1,300
					11,833 Total

The range of aviation gasolines normally obtained from blending these inputs includes six different grades of which three (all non-military) were chosen for study. These are grades 80L, 91C and 100C whose specifications are shown in Table 7.3. The TEL levels of these

Table 7.3

Grade	Index (i)	Specifications: max. RVP	Specifications: min. PN	Specifications: TEL (cc/gal.)	Net revenue ($/bbl)
80L	1	6.9	80	0.5	4.908823
91C	2	6.9	91	4.0	5.436584
100C	3	6.9	100	4.0	6.041584
Residue	4	—	—	—	4.548138

blended gasolines are specified at their maximal allowable limits because it is generally accepted that adding tetraethyl lead is the most economical way of raising PN. This makes it possible to predetermine the TEL cost per barrel, which is subtracted from unit price to yield the net revenue per barrel of each grade of the aviation gasolines listed in Table 7.3. The "Residue" shown in the table is actually the total excess inputs left over from blending. It is piped into another installation and blended into a premium grade of motor gasoline which contributes a net revenue of \$4.548138 per barrel as shown. The specifications of the motor gasoline are obviously unimportant to our problem here.

Our objective is to maximize profits in the blending phase of the operation with the output rates of the refining phase assumed constant. Hence, the costs of producing the inputs for blending do not enter into consideration and the objective is simply one of maximizing the total net revenue from the blended gasolines. This can be written

$$f(X) = 4.908823\text{M} + 5.436584\text{N} + 6.041584\text{Q} + 4.548138\text{Z},$$

where M, N, Q and Z represent respectively the blending outputs of aviation gasolines 80L, 91C, 100C and the residue. Since the total daily refining output is 11,833 barrels, we may substitute $\text{Z} = 11{,}833 - \text{M} - \text{N} - \text{Q}$ into the above and, after rounding off the less significant digits, obtain

$$f(X) = 0.36\text{M} + 0.889\text{N} + 1.494\text{Q} + 53{,}818. \tag{7.7.1}$$

Now, the volume of output of any aviation gasoline is obviously the sum of the inputs used in blending it. In general, if we define x_{ij} to be the unknown volume of input j used in blending output i (see Tables 7.2 and 7.3 for the identities of i and j), clearly then

$$\begin{aligned} \text{M} &= x_{11} + x_{12} + x_{13} + x_{14} = \text{output volume of grade 80L}, \\ \text{N} &= x_{21} + x_{22} + x_{23} + x_{24} = \text{output volume of grade 91C}, \\ \text{Q} &= x_{31} + x_{32} + x_{33} + x_{34} = \text{output volume of grade 100C}, \\ \text{Z} &= x_{41} + x_{42} + x_{43} + x_{44} = \text{output volume of motor gasoline (residue)}. \end{aligned} \tag{7.7.2}$$

Having taken care of the objective function, let us now develop the relevant constraints. First of all, the variables x_{ij} are subject to the constraints representing specifications. Fortunately the physical and chemical properties of the various inputs blend with each other in an essentially linear manner. In other words, a specific characteristic of the output is basically the weighted average of the same characteristic possessed by the inputs of the blend. For instance, product 1 (grade 80L aviation gasoline) is blended from the four inputs in amounts x_{11}, x_{12}, x_{13} and x_{14}. Since the four inputs have RVP indices of 5, 8, 4 and 20.5 (Table 7.2) respectively, the RVP index of 80L aviation gasoline can be expected to be

$$\frac{5x_{11} + 8x_{12} + 4x_{13} + 20.5x_{14}}{x_{11} + x_{12} + x_{13} + x_{14}}.$$

According to the specifications shown in Table 7.3, 80L aviation gasoline cannot have a RVP index greater than 6.9. Therefore, we can write the constraint which insures the RVP specification of 80L aviation gasoline as follows:

$$5x_{11} + 8x_{12} + 4x_{13} + 20.5x_{14} \leq 6.9(x_{11} + x_{12} + x_{13} + x_{14}).$$

Similarly for the PN specification of 80L aviation gasoline which is leaded at 0.5cc/gal., we can write

$$94x_{11} + 83x_{12} + 74x_{13} + 95x_{14} \geq 80(x_{11} + x_{12} + x_{13} + x_{14}).$$

In this manner, one pair of RVP and PN constraints can be set down for each of the two remaining grades of aviation gasoline (note that both are leaded at the higher level of 4cc/gal.). There is no need to write additional constraints for the maximum allowable TEL levels because in practice inputs are leaded in accordance with the TEL requirement of the specific output for which they are intended. No high-TEL input is used for low-TEL output for obvious economic reasons. Therefore, we have a total of six specification constraints as shown in (7.7.3).

$$\begin{aligned}
-1.9x_{11} + 1.1x_{12} - 2.9x_{13} + 13.6x_{14} + x_1 &= 0,\\
-14x_{11} - 3x_{12} + 6x_{13} - 15x_{14} + x_2 &= 0,\\
-1.9x_{21} + 1.1x_{22} - 2.9x_{23} + 13.6x_{24} + x_3 &= 0,\\
-16.5x_{21} - 2x_{22} + 4x_{23} - 17x_{24} + x_4 &= 0,\\
-1.9x_{31} + 1.1x_{32} - 2.9x_{33} + 13.6x_{34} + x_5 &= 0,\\
-7.5x_{31} + 7x_{32} + 13x_{33} - 8x_{34} + x_6 &= 0.
\end{aligned} \tag{7.7.3}$$

In addition to the above specification constraints, there are four capacity constraints imposed by the daily output rates of the four inputs used in blending. These would be the usual "$\leq$" type of constraints if it were not for the fact that the slack variables (excess inputs) are themselves a distinct product (premium motor gasoline). Thus, in a sense there is no slack and all inputs are fully utilized. Consequently, with data obtained from Table 7.2, the capacity constraints are expressed in equation form as follows:

(7.7.4)
$$x_{11} + x_{21} + x_{31} + x_{41} = 3{,}800 \quad \text{(daily output rate of alkylate I)},$$
$$x_{12} + x_{22} + x_{32} + x_{42} = 2{,}652 \quad \text{(daily output of catalytic gasoline I)},$$
$$x_{13} + x_{23} + x_{33} + x_{43} = 4{,}081 \quad \text{(daily output of straight-run gasoline I)},$$
$$x_{14} + x_{24} + x_{34} + x_{44} = 1{,}300 \quad \text{(daily output of isopentane)}.$$

Finally, we must not forget the non-negativity constraints.

(7.7.5)
$$x_{ij} \geq 0 \qquad i, j = 1, 2, 3, 4,$$
$$x_j \geq 0 \qquad j = 1, \ldots, 6.$$

In summary, the gasoline blending problem formulated by Charnes and Cooper consists of maximizing (7.7.1) whose variables are defined in (7.7.2), subject to the specification constraints in (7.7.3), the capacity constraints in (7.7.4) and the non-negativity constraints in (7.7.5). It has altogether 22 variables and 10 structural constraints. Although extremely small as modern linear programming problems in the oil industry go, it embodies sufficient computational work to keep a man busy on a desk calculator for days. However, the actual solution-process need not detain us unduly here. We will only display the initial simplex tableau in Table 7.4, and the final optimal basis together with its inverse in Table 7.5. Our main task is to illustrate by this example some aspects of optimality analysis. Incidentally, the optimal solution as can readily be determined by $B^{-1}P_0$ is $x_{11} = x_{12} = 0$, $x_{21} = 60$, $x_{23} = 3{,}025$, $x_{24} = 653$, $x_{31} = 3{,}740$, $x_{32} = 2{,}650$, $x_{33} = 1{,}056$, $x_{34} = 534$, $x_{44} = 113$ with a maximal net revenue of $69,163 per day. This indicates that 3,738 barrels of Grade 91C and 7,980 barrels of Grade 100C aviation gasoline together with 113 barrels of motor gasoline should be blended daily, which seems to make sense. Grade 80L aviation gasoline would not be produced at all, having the lowest net revenue per barrel of the three. Inputs with high PN, viz., alkylate and catalytic gasoline, would be used mostly in blending output with high PN requirement, viz., Grade 100C. Due to its high RVP index, however, isopentane would not be completely

Table 7.4

Initial Tableau of the Gasoline Blending Problem

c_i	Basis variable	x_{11}	x_{12}	x_{13}	x_{14}	x_{21}	x_{22}	x_{23}	x_{24}	x_{31}	x_{32}	x_{33}	x_{34}	x_1	x_2	x_3	x_4	x_5	x_6	x_{41}	x_{42}	x_{43}	x_{44}	θ_i
0	x_1	−1.9	1.1	−2.9	13.6									1										0
0	x_2	−14.0	−3.0	6.0	−15.0										1									0
0	x_3					−1.9	1.1	−2.9	13.6							1								0
0	x_4					−16.5	−2.0	4.0	−17.0								1							0
0	x_5									−1.9	1.1	−2.9	13.6					1						0
0	x_6									−7.5	7.0	13.0	−8.0						1					0
0	x_{41}	1				1				1										1				3800
0	x_{42}		1				1				1										1			2652
0	x_{43}			1				1				1										1		4081
0	x_{44}				1				1				1										1	1300
	z_j	0	0	0	0	0	0	0	0	0	0	0	0	0	0	0	0	0	0	0	0	0	0	
	c_j	.36	.36	.36	.36	.889	.889	.889	.889	1.494	1.494	1.494	1.494	0	0	0	0	0	0	0	0	0	0	53818
	$z_j - c_j$	−.36	−.36	−.36	−.36	−.889	−.889	−.889	−.889	−1.494	−1.494	−1.494	−1.494	0	0	0	0	0	0	0	0	0	0	

Table 7.5a

Final Optimal Basis of Gasoline Blending Problem

$$\hat{B} = \begin{array}{c} \\ \end{array}\begin{bmatrix} x_{11} & x_{12} & x_{21} & x_{23} & x_{24} & x_{31} & x_{32} & x_{33} & x_{34} & x_{44} & \\ -1.9 & -1.1 & 0 & 0 & 0 & 0 & 0 & 0 & 0 & 0 & 0 \\ -14.0 & -3.0 & 0 & 0 & 0 & 0 & 0 & 0 & 0 & 0 & 0 \\ 0 & 0 & -1.9 & -2.9 & 13.6 & 0 & 0 & 0 & 0 & 0 & 0 \\ 0 & 0 & -16.5 & 4.0 & -17.0 & 0 & 0 & 0 & 0 & 0 & 0 \\ 0 & 0 & 0 & 0 & 0 & -1.9 & 1.1 & -2.9 & 13.6 & 0 & 0 \\ 0 & 0 & 0 & 0 & 0 & -7.5 & 7.0 & 13.0 & -8.0 & 0 & 0 \\ 1 & 0 & 1 & 0 & 0 & 1 & 0 & 0 & 0 & 0 & 0 \\ 0 & 1 & 0 & 0 & 0 & 0 & 1 & 0 & 0 & 0 & 0 \\ 0 & 0 & 0 & 1 & 0 & 0 & 0 & 1 & 0 & 0 & 0 \\ 0 & 0 & 0 & 0 & 1 & 0 & 0 & 0 & 1 & 1 & 0 \\ -.36 & -.36 & -.889 & -.889 & -.889 & -1.494 & -1.494 & -1.494 & -1.494 & 0 & 1 \end{bmatrix}.$$

Table 7.5b

Inverse of the Final Optimal Basis

$$\hat{B}^{-1} = \begin{bmatrix} -14218 & -5213 & 0 & 0 & 0 & 0 & 0 & 0 & 0 & 0 & 0 \\ 66351 & -9005 & 0 & 0 & 0 & 0 & 0 & 0 & 0 & 0 & 0 \\ -1125 & 43 & -6724 & -5380 & -105 & -179 & -1538 & 1366 & 2017 & 0 & 0 \\ -56625 & 2154 & -5129 & -4102 & -5287 & -8991 & -77428 & 68751 & 101547 & 0 & 0 \\ -12232 & 465 & 5320 & -1626 & -1142 & -1942 & -16727 & 14851 & 21935 & 0 & 0 \\ 15343 & 5171 & 6724 & 5380 & 105 & 179 & 101538 & -1366 & -2017 & 0 & 0 \\ -66351 & 9005 & 0 & 0 & 0 & 0 & 0 & 10^5 & 0 & 0 & 0 \\ 56625 & -2154 & 5129 & 4102 & 5287 & 8991 & 77428 & -68751 & -1547 & 0 & 0 \\ 19586 & -465 & 2034 & 1627 & 8495 & 1942 & 30706 & -22939 & -610 & 0 & 0 \\ -7354 & 0 & -7353 & -1 & -7353 & 1 & -13979 & 8088 & -21325 & 10^5 & 0 \\ -5793 & 14513 & 14939 & 6721 & 14939 & 6722 & 228179 & 85912 & 105333 & 0 & 10^5 \end{bmatrix} \times 10^{-5}.$$

utilized although its PN rating is the highest. Since, according to (7.3.3), the last row of $\hat{B}^{-1}$ is $\tilde{C}B^{-1}$, its entries represent the implicit values of the inputs. Thus, additional supply of alkylate would be worth \$2.28179/bbl, catalytic gasoline \$0.85912/bbl, straight-run gasoline \$1.05333/bbl, and isopentane nothing at all, since there is already a surplus being dumped into motor gasoline.

Although the optimal program calls for blending two grades of aviation gasoline, it is of interest to ascertain how much better such a two-product program is as compared with a simple one-product program. Charnes and Cooper showed that if only the high-priced 100C grade is blended with the residue going into motor gasoline, the total net revenue would be \$65,930 per day, which is somewhat lower as expected but only by less than 5%—a rather unexpectedly small margin. In the remainder of this section, we will attempt to show how we may "ease" our way from the established optimal solution of the original problem into the above-mentioned one-product solution of the reduced problem through transformation and modification of the basis inverse. The computational advantage of this approach as compared with solving the reduced problem from the very beginning should be significant when the basis is large relative to the extent of its reduction in dimensions.

In general, our approach is to introduce appropriate unit vectors whenever necessary into the basis, transform its inverse accordingly, and by deleting the proper rows and columns of the transformed inverse to obtain a reduced inverse which corresponds to a basic feasible solution of the desired smaller problem. The theoretical basis of this approach has already been discussed in Section 7.6. Our starting point here is the basis $\hat{B}$ shown in Table 7.5a which corresponds to the optimal solution to the original three-product problem. Since in the proposed study, 80L and 91C aviation gasoline will not be produced, the first four constraints in (7.7.3) are no longer relevant. Our objective, therefore, is a 7×7 basis inverse (reflecting 6 constraints and 1 row of price coefficients) to be obtained from the full basis mentioned above.

First of all, we notice that in the given optimal solution of the full problem, x_{11} and x_{12} are the only two *basis* variables involving the first two constraints and they are both zero. Hence, we may delete the first two rows and the first two columns of $\hat{B}$ (Table 7.5a) and obtain a 9×9 feasible basis for the reduced two-product problem in

Table 7.6a

x_i	$\hat{B}'^{-1}(10^{-5})$									P_3	Y_3 (10^{-5})	θ_i	θ_i/y_{i3}
x_{21}	−6724	−5380	−105	−179	−1538	1366	2017	0	0	1	−6724	60	—
x_{23}	−5129	−4102	−5287	−8991	−77428	68751	101547	0	0	0	−5129	3023	—
x_{24}	5320	−1626	1142	−1942	−16727	14851	21935	0	0	0	[5320]	653	12275
x_{31}	6724	5380	105	179	101538	−1366	−2017	0	0	0	6724	3740	55610
x_{32}	0	0	0	0	0	10^5	0	0	0	0	0	2652	∞
x_{33}	5129	4102	5287	8991	77428	−68751	−1547	0	0	0	5129	1058	20628
x_{34}	2034	1627	8495	1942	30706	−22939	−610	0	0	0	2034	534	26260
x_{44}	−7353	−1	−7353	1	−13979	8088	−21325	10^5	0	0	−7353	113	—
	14939	6721	14939	6722	228180	85912	105333	0	10^5	0	14939		

Table 7.6b

x_i	$\hat{B}'^{-1}(10^{-5})$									P_4	Y_4 (10^{-5})	θ_i	θ_i/y_{i4}
x_{21}	0	−7435	−1549	−2634	−22681	20138	29744	0	0	0	−7435	886	—
x_{23}	0	−5670	−6388	−10863	−93555	83069	122696	0	0	1	−5670	3655	—
x_3	10^5	−30571	−21469	−36508	−314426	279170	412339	0	0	0	−30571	11283	—
x_{31}	0	7435	1549	2634	122681	−20138	−29744	0	0	0	7435	2914	39192
x_{32}	0	0	0	0	0	10^5	0	0	0	0	0	2652	∞
x_{33}	0	5670	6388	10863	93555	−83069	−22696	0	0	0	[5670]	426	7513
x_{34}	0	2249	8932	2684	37100	−28616	−8995	0	0	0	2249	284	12631
x_{44}	0	−2249	−8932	−2684	−37100	28616	8995	10^5	0	0	−2249	1016	—
	0	11288	18146	12176	275150	44208	43736	0	10^5	0	11288		

which 80L aviation gasoline (the variables x_{11}, x_{12}, x_{13} and x_{14}) and its specifications (the first two constraints) are not considered at all. To determine the inverse of this reduced basis, we observe that $\hat{B}$ can be partitioned into four submatrices and quickly inverted as follows:

$$\hat{B} = \begin{bmatrix} A & O \\ C & \hat{B}' \end{bmatrix} \quad \text{where } A \text{ is } 2 \times 2 \text{ and } \hat{B}' \text{ (the reduced basis) is } 9 \times 9.$$

$$\hat{B}^{-1} = \begin{bmatrix} A^{-1} & O \\ -\hat{B}'^{-1}CA^{-1} & \hat{B}'^{-1} \end{bmatrix}$$

Thus, the inverse of the reduced basis ($\hat{B}'^{-1}$) is simply a submatrix of the inverse of the full basis ($\hat{B}^{-1}$), and is readily available from Table 7.5a without further computation.

As the second step, we shall remove the third and fourth constraints from the problem by introducing x_3 and x_4 into $\hat{B}'$. The revised simplex method is used in the several transformations of the inverse and the detailed computations are displayed in Tables 7.6a, b, and c. After completing Table 7.6c, we are ready to discard the third

Table 7.6c

x_i	$\hat{B}'^{-1}(10^{-5})$										θ_i
x_{21}	0	0	6828	11612	10^5	−88793	−0	0	0		1445
x_{23}	0	0	0	0	0	0	10^5	0	0		4081
x_3	10^5	0	12975	22065	190001	−168720	289969	0	0		14579
x_{31}	0	0	−6828	−11612	0	88793	0	0	0		2355
x_{32}	0	0	0	0	0	10^5	0	0	0		2652
x_4	0	10^5	112669	191593	1650002	−1465067	−400279	0	0		7511
x_{34}	0	0	6398	−1624	0	4327	0	0	0		116
x_{44}	0	0	−6398	1624	0	−4327	0	10^5	0		1184
	0	0	5428	−9451	88899	209583	88919	0	10^5		

and fourth constraints together with the basis variables x_3 and x_4 from $\hat{B}'$ (not shown). Since these two constraints are represented by the first and second rows of $\hat{B}'$, and since x_3 and x_4 are associated with the third and sixth columns (formerly representing x_{24} and x_{33} which have been displaced) of $\hat{B}'$, the first and second columns as well as the third and sixth rows of $\hat{B}'^{-1}$ are deleted as indicated in Table 7.6c. The remainder is then a feasible basis inverse for the 6-constraint reduced problem.

This basis, however, is not the final answer. In the first place, it contains x_{21} and x_{23} which are the output volume of 91C grade aviation gasoline—contrary to our assumption that it will not be produced. Secondly, it is not known whether the basis is optimal. Therefore, in our third and final step, we shall replace x_{21} and x_{23} by variables permitted in the one-product program, and then determine the optimal solution. Fortunately the first part can be accomplished very easily. By reference to (7.7.3), we note that the coefficient vectors of x_{21} and x_{23} are identical with those of x_{41} and x_{43} after the first four constraints are eliminated. Therefore the latter may be substituted for the former in the inverse without disturbing anything except the last row.* Thus we obtain the basis inverse ($\hat{B}''^{-1}$) shown in Table 7.7, which in turn

Table 7.7

x_i	$\hat{B}''^{-1}(10^{-5})$							P_{33}	Y_{33} (10^{-5})	θ_i	$\theta_i/y_{i,33}$
x_{41}	6828	11612	10^5	−88793	0	0	0	−2.9	**131149**	1145	1102
x_{43}	0	0	0	0	10^5	0	0	13.0	100000	4081	4081
x_{31}	−6828	−11612	0	88793	0	0	0	0	−131149	2355	—
x_{32}	0	0	0	10^5	0	0	0	0	0	2652	∞
x_{34}	6398	−1624	0	4327	0	0	0	1.0	−39662	116	—
x_{44}	−6398	1624	0	−4327	0	10^5	0	0	39662	1184	2985
	−642	−19774	0	288520	0	0	10^5	−1.494	−404592	61472	

establishes the relative advantage of x_{33}. The substitution of x_{33} for x_{41} in the basis finally yields the optimal solution of Table 7.8. The

Table 7.8

x_i	$\hat{B}''^{-1}(10^{-5})$							P_0	θ_i
x_{33}	5207	8854	76249	−67704	0	0	0	0	1102
x_{43}	−5207	−8854	−76249	67704	10^5	0	0	0	2979
x_{31}	0	0	10^5	0	0	0	0	3800	3800
x_{32}	0	0	0	10^5	0	0	0	2652	2652
x_{34}	8464	1888	30242	−22526	0	0	0	4081	553
x_{44}	−8464	−1888	−30242	22526	0	10^5	0	1300	747
	20423	16049	308497	14596	0	0	10^5	53818	65930

total net revenue is $65,930 per day which, as mentioned above, is about 5% less than the maximal value realizable under the full program. The total outout of 100C aviation gasoline would be 8,107 barrels per day with 2,979 barrels of straight-run gasoline and 747 barrels of isopentane left over for blending premium motor gasoline.

---

* The price coefficient of x_{21} and x_{23} is 0.889 whereas that of x_{41} and x_{43} is 0.

7.8. Production Smoothing Problem.

A familiar application of linear programming is the scheduling of production to achieve maximum possible economy consistent with the objective of satisfying demand at all times. The problem of fluctuating demand is common to all businesses in varying degrees. Some are due to seasonal factors, some to cyclical ones. Some are more predictable, some are less predictable. But predictability notwithstanding, changing demand, unless offset by sales efforts, is going to impinge upon the production operation of a firm and evoke responses either in the form of inventory accumulation or unstable production. The former leads to higher carrying charges, interest cost on capital and risk of obsolescence while the latter leads to overtime pay, recurrent hiring expenses, uncertainty of labor supply and the intangible disadvantages of poor employee and public relations. The problem then is to find a compromise between the two opposing alternatives so as to meet the inevitable fluctuations in demand at minimum cost. Such a compromise will be stated in terms of the production rates in the time periods affected. Once these rates are determined, the behavior of inventory can readily be ascertained from the interaction of production and demand.*

For the sake of clarity let us define the following:

s_0 = initial inventory,

s_i = inventory at the end of the ith period,

x_j = production in the jth period,

$X_i = \sum_{j=1}^{i} x_j$ = cumulative production up to the end of the ith period,

r_j = demand in the jth period,

$R_i = \sum_{j=1}^{i} r_j$ = cumulative demand up to the end of the ith period,

a = cost of carrying inventory per unit per period,

b = cost of unit increase in production,

---

* As will be seen later, the problem is going to be solved in terms of the inventory levels s_i because of their direct effects on cost. The production schedule is then constructed from s_i with the aid of r_j, the known demands.

$t \quad = \dfrac{b}{a} =$ ratio of inventory carrying cost to production increase cost.

It is clear that the one constraint which has to be satisfied repeatedly is simply

$$s_0 + X_i - R_i \geq 0, \qquad i = 1, \ldots, n,$$

or equivalently

$$s_0 + X_i - R_i - s_i = 0, \qquad i = 1, \ldots, n. \tag{7.8.1}$$

This constraint insures that demand will be met in all of the n periods under planning. An interesting observation about this set of constraints is that s_i is both a decision variable having a role in the objective function and a slack variable in the ith constraint. This coincidence makes possible the use of decision variables to obtain an initial feasible solution, which will reduce the time required for convergence towards the optimal solution.

Turning our attention to the objective function, let us assume that it consists predominantly of two parts: (1) inventory carrying cost, and (2) cost of increasing production from period to period. The part that represents inventory carrying costs is quite simple to construct. Assuming that the rate of demand is fairly constant within any time period, the average inventory in the ith period is obviously $\frac{1}{2}(s_{i-1} + s_i)$ and the cost, $\frac{1}{2}a(s_{i-1} + s_i)$. Hence, the total inventory cost is $\frac{1}{2}a \sum_{i=1}^{n} (s_{i-1} + s_i) = \frac{1}{2}as_0 + as_1 + as_2 + \ldots + as_{n-1} + \frac{1}{2}as_n.$

The other part dealing with production-increase costs is, however, not nearly so simple. An explicit statement of these costs presupposes a production schedule and yet our problem is precisely to determine such a schedule which depends partly upon these costs. Thus, we have a dilemma. The resolution of this dilemma, which we shall discuss presently, is a striking testimony both to the intriguing charm of the simplex method and the importance of its creative use based upon a thorough understanding of its mathematical foundation.

To resolve the dilemma, we define the difference between the production of two successive periods in terms of two separate variables as follows:

(7.8.2) $$x_j - x_{j-1} = u_j - v_j \quad \text{where } u_j, v_j \geq 0, \text{ and } u_j = 0 \text{ if } v_j > 0, \text{ and } v_j = 0 \text{ if } u_j > 0.$$

We shall soon see that the auxiliary conditions imposed on u_j and v_j can be enforced within the framework of the regular simplex method. Meanwhile, assuming that u_j and v_j do behave as expected, then $u_j > 0$ implies that v_j must be zero and x_j must be greater than x_{j-1}. Hence, u_j represents an increase in production from period $j - 1$ to period j. On the other hand, if v_j is positive, then u_j must be zero, and x_j must be smaller than x_{j-1}. Hence, v_j stands for a decrease in production from the $(j - 1)$th to the jth period. Finally if both u_j and v_j are zero, then $x_j = x_{j-1}$, indicating that production has remained constant. Thus we have a unique and workable scheme to represent increases and decreases in production by separable variables as desired, provided if u_j and v_j can be prevented from assuming positive values simultaneously. Now that we have u_j representing only the increases in production, the second part of the objective function can be written simply as $b \sum_{j=1}^{n} u_j$.

Returning to the auxiliary conditions imposed on u_j and v_j in (7.8.2), we note that non-negativity requirements are a regular feature of linear programming, and so present no problem at all. The condition which requires one variable to be zero when the other is positive in any solution is also automatically satisfied by the simplex procedure because u_j and v_j always exist in pairs in the constraints. Therefore, for each j, their coefficient vectors are always identical in the problem, which means that both cannot be in the same basis without destroying the independence of the basis vectors. By definition, a basis is a collection of linearly independent vectors; therefore, in any basic feasible solution, no more than one of the two can be in the basis. This implies that while the one in the basis (either u or v) is assuming a positive value, the other must be outside the basis carrying a value of zero. Furthermore, since linear independence is maintained after each change of basis,* no u_j (v_j) can be introduced into a basis if v_j (u_j) is already in it, unless the latter is the basis vector replaced.

---

* See Chapter IV Section 4.2.

Summarizing the above discussion, we can now formulate the production smoothing problem in the following linear programming form:

(7.8.3)

$$\text{Minimize } f(U, S) = \tfrac{1}{2}a \sum_{i=1}^{n} (s_i + s_{i-1}) + b \sum_{j=1}^{n} u_j,$$

$$\text{subject to:} \quad s_0 + \sum_{j=1}^{i} (x_j - r_j) - s_i = 0, \qquad i = 1, \ldots, n, \quad (7.8.1)$$

$$x_j - x_{j-1} - u_j + v_j = 0, \qquad j = 1, \ldots, n, \quad (7.8.2)$$

$$\text{and} \qquad x_j, s_i, u_j, v_j \geq 0, \qquad i, j = 1, \ldots, n.$$

Assuming that, in addition to the r_j, s_0 (the initial inventory) and x_0 (the production level immediately preceding the first period) are known, this formulation of the production smoothing problem will have $4n$ variables and $2n$ constraints where n is the number of production periods under consideration. Although not prohibitively large, the problem by virtue of its logic does provide room for condensation. In the balance of this section then, we shall develop a condensed version of (7.8.3) (after Gass*) and proceed to solve a numerical problem based on it. The latter will be undertaken for the purpose of illustrating the use of parametric programming in optimality analysis.

First, by retaining the first of the n equations in (7.8.1) and by subtracting from the 1st, 2nd, . . . , $(n - 1)$th equations in (7.8.1), the 2nd, 3rd, . . . , nth equations in the same set correspondingly, we obtain the following n equations:

$$(7.8.4) \qquad s_{j-1} + x_j - r_j = s_j, \qquad j = 1, 2, \ldots, n.$$

Each of these states that inventory at the end of the $(j - 1)$th period plus production in the jth period, minus sales in the jth period is equal to inventory at the end of the jth period. This relationship is of course self-evident and needs no elaboration. Next, starting with the second equation in (7.8.4), subtract each equation from the preceding one and obtain

$$(7.8.5) \qquad s_{j-1} - s_j + x_j - x_{j+1} - r_j + r_{j+1} = s_j - s_{j+1}, \qquad j = 1, \ldots, n - 1.$$

---

* S. I. Gass, *op. cit.*, pp. 158–164.

Substituting (7.8.2) for $x_j - x_{j+1}$ in (7.8.5), we finally arrive at the set of $(n - 1)$ constraints in (7.8.6).

$$s_{j-1} - 2s_j + s_{j+1} - u_{j+1} + v_{j+1} = r_j - r_{j+1}, \quad j = 1, \ldots, n - 1. \tag{7.8.6}$$

In addition, by substituting the first equation in (7.8.2), viz., $x_1 - x_0 + u_1 - v_1$ into the first equation in (7.8.4), we have

$$s_0 + x_0 + u_1 - v_1 - r_1 = s_1. \tag{7.8.7}$$

Now, we collect together (7.8.6), (7.8.7) and after more housekeeping details such as systematizing the subscripts and transposing all constants to the right-hand side of the equations, we present the production smoothing problem in the following condensed form:

(7.8.8)

$$\text{Minimize } f(U, S) = \tfrac{1}{2}a \sum_{j=1}^{n} (s_{j-1} + s_j) + b \sum_{j=1}^{n} u_j,$$

subject to:

$$\begin{aligned} &s_1 - u_1 + v_1 = s_0 + x_0 - r_1, \\ &-2s_1 + s_2 - u_2 + v_2 = -s_0 + r_1 - r_2, \\ &s_{j-2} - 2s_{j-1} + s_j - u_j + v_j = r_{j-1} - r_j, \quad j = 3, \ldots, n, \end{aligned}$$

and

$$s_j, u_j, v_j \geq 0, \quad j = 1, \ldots, n.$$

This formulation not only has fewer constraints (n instead of $2n$) and fewer variables ($3n$ instead of $4n$), but it also possesses the important advantage of having two built-in identity matrices. These two identity matrices, composed of the negative and positive coefficients of u_j and v_j respectively, insure that an initial basic feasible solution can always be obtained from the appropriate u_j and v_j, depending on whether the constants on the right are negative or positive.

We notice that the above reduction in the size of the problem was achieved by substituting (7.8.2) for $x_j - x_{j+1}$ in (7.8.5). Each of these substitutions resulted in the elimination of one constraint and one variable (x_j) from the original system. However, it must be strongly emphasized that this procedure is valid only under special circumstances. Otherwise, every problem with m constraints and n ($> m$) variables would be reducible to an equivalent problem of 1 constraint and $(n - m)$ variables, which would be extremely easy to solve. This is unfortunately too good to be true. It will not work because each variable so eliminated from the system of constraints also escapes the control of the non-negativity constraint. Thus, a feasible solution to the reduced problem might actually imply negative values of the eliminated variables which, of course, would be unacceptable as a solution to the original problem. This difficulty does not exist in the present problem, however, because of the problem's economic logic. For if any x_j is to be negative, s_{j-1} must be greater than r_j, or s_j would be negative which is barred by non-negativity constraints. In other words, the inventory at the end of the $(j - 1)$th period must be in excess of the demand in the jth period. Now, this is obviously uneconomical because production in the $(j - 1)$th period could be decreased with savings in inventory costs, unless of course it is also negative already. In the former case, i.e., $x_{j-1} > 0$ while $x_j < 0$, we argue that the solution cannot be an optimal solution. This in effect eliminates the possibility of negative x_j in any optimal solution that we do get. In the latter case, i.e., $x_{j-1} \leq 0$ while $x_j < 0$, we simply carry the same argument one period back and note at the same time, that with both x_j and x_{j-1} negative, the implication grows into $s_{j-2} > r_{j-1} + r_j$. In this manner we press the argument to one of its two logical conclusions: either in some preceding period, say $(j - k)$th, $x_{j-k} > 0$, in which case the solution is not optimal, or in all preceding periods, $x_{j-k} \leq 0$, which implies that the initial inventory must be greater than the total cumulative demand in the j periods. If the latter is true, there is no need for production in the first j periods, and production scheduling should begin with the $(j + 1)$th. Thus, so long as $s_0 \leq r_1$ in the problem, the condensed formulation given in (7.8.8) will effectively suppress *from its optimal solutions* all solutions which imply negative production.

Now we are ready for a numerical illustration of the foregoing theoretical discussion. But as mentioned earlier, we will present it from the standpoint of parametric programming in order to show how the technique may be used in practice to conduct optimality analysis.

In connection with the present problem of production smoothing, optimality analysis is of interest because the scheduling of production depends only on two costs: inventory-holding cost a and production-increase cost b. As a matter of fact, if we divide the objective function $f(U, S)$ by a (which we may reasonably assume to be non-negative), we will have a new objective function $f'(U, S)$ which has the same minima as $f(U, S)$ but only one parameter, viz., $t = \frac{b}{a}\cdot$ That is, the problem now becomes

$$\text{(7.8.9)} \qquad \text{Minimize } f'(U, S) = \tfrac{1}{2}\sum_{j=1}^{n} (s_{j-1} + s_j) + t\sum_{j=1}^{n} u_j,$$

subject to the same constraints in (7.8.8). From this we can see how the minimal cost varies in response to different values of t, and infer from that the practical value of reducing a (or b) to achieve greater economy.

In order to keep the computational work within limits, let us assume a planning period of 6 months only. During this period the requirements are

Month (j)	1	2	3	4	5	6
Requirement	20	40	50	40	60	10

It is further assumed that the initial inventory is $s_0 = 10$ (less than the first month's requirement) and that the previous production is $x_0 = 30$. Substituting these values into the constraints in (7.8.8) and using (7.8.9) as the objective function, we obtain the following problem:

$$\begin{aligned}\text{Minimize } f'(U, S) = \tfrac{1}{2}s_0 + s_1 + s_2 + s_3 + s_4 + s_5 + \tfrac{1}{2}s_6 + tu_1 \\ + tu_2 + tu_3 + tu_4 + tu_5 + tu_6,\end{aligned}$$

$$\begin{aligned}
\text{subject to:} \qquad s_1 - u_1 + v_1 &= 20,\\
-2s_1 + s_2 - u_2 + v_2 &= -30,\\
s_1 - 2s_2 + s_3 - u_3 + v_3 &= -10,\\
s_2 - 2s_3 + s_4 - u_4 + v_4 &= 10,\\
s_3 - 2s_4 + s_5 - u_5 + v_5 &= -20,\\
s_4 - 2s_5 + s_6 - u_6 + v_6 &= 50,\\
\text{and} \qquad s_j, u_j, v_j = 0, \qquad & j = 1, \ldots, 6.
\end{aligned}$$

It is immediately obvious that an initial feasible basis could be made up of the coefficient vectors of these variables: v_1, u_2, u_3, v_4, u_5, v_6, which meanwhile assume the following values respectively: 20, 30, 10, 10, 20, 50. On the basis of this initial solution, we set up the initial simplex tableau shown in Table 7.9a and proceed to iterate for various optimal solutions for different ranges of t starting with $t = 0$. Since the price coefficients, viz., 1 for s_j and t for u_j cannot be combined into one number, we use two rows for $z_j - c_j$: the upper row marked "h_j" is reserved for the coefficient of t, and the lower marked "g_j", for the numerical part of the evaluation (see the last paragraph of Section 7.2). Similarly, the value of $f'(U, S)$ is also divided into two parts. For instance, the value of the objective function in Table 7.9a reads $60t + 5$. Now, within the context of the present problem negative t is highly improbable because it implies either negative inventory holding cost or negative production increase cost. Hence, we begin

Table 7.9a

c_i	$P_i \backslash P_j$	s_1	s_2	s_3	s_4	s_5	s_6	u_1	u_2	u_3	u_4	u_5	u_6	v_1	v_2	v_3	v_4	v_5	v_6	θ_i
0	v_1	1						−1						1						20
t	u_2	2	−1						1						−1					30
t	u_3	−1	2	−1						1						−1				10
0	v_4		1	−2	1						−1						1			10
t	u_5			−1	②	−1						1						−1		20
0	v_6				1	−2	1						−1						1	50
$z_j - c_j$	h_j	1	1	−2	2	−1	0	−1	0	0	−1	0	−1	0	−1	−1	0	−1	0	60
	g_j	−1	−1	−1	−1	−1	−1	0	0	0	0	0	0	0	0	0	0	0	0	5

with $t = 0$ as already suggested. But for this value of t, only the numerical part of the evaluation remains. Referring to the "g_j" row of Table 7.9a, we observe that the initial solution is already the optimal solution and the total cost is 5. The implied production schedule is:

$$\begin{aligned}
x_1 &= x_0 + u_1 - v_1 = 30 + 0 - 20 = 10,\\
x_2 &= x_1 + u_2 - v_2 = 10 + 30 - 0 = 40,\\
x_3 &= x_2 + u_3 - v_3 = 40 + 10 - 0 = 50,\\
x_4 &= x_3 + u_4 - v_4 = 50 + 0 - 10 = 40,\\
x_5 &= x_4 + u_5 - v_5 = 40 + 20 - 0 = 60,\\
x_6 &= x_5 + u_6 - v_6 = 60 + 0 - 50 = 10,
\end{aligned}$$

which calls for production according to the monthly demand with only the exception of the first month. This schedule makes sense in view of the assumption of $t = 0$ which means that there is no additional cost for increase in production.

Next, we raise the question: For what range of t does the above solution continue to be optimal? As discussed in Section 7.2, this depends on how far t can be increased before the first favorable evaluation of some non-basis variable turns up. Thus we make use of the formula in (7.2.5), but since we are minimizing instead of maximizing, we concentrate only on those variables which have $h_j > 0$. These are s_1, s_2 and s_4. Forming the ratios according to (7.2.5) we find that s_4 has the smallest ratio ($t = \frac{1}{2}$). Hence, s_4 is put into the basis to replace

Table 7.9b

c_i	P_j / P_i	s_1	s_2	s_3	s_4	s_5	s_6	u_1	u_2	u_3	u_4	u_5	u_6	v_1	v_2	v_3	v_4	v_5	v_6	θ_i
0	v_1	1						-1						1						20
t	u_2	②	-1						1						-1					30
t	u_3	-1	2	-1						1						-1				10
0	v_4		1	$\frac{3}{2}$		$\frac{1}{2}$					-1	$-\frac{1}{2}$					1	$\frac{1}{2}$		0
1	s_4			$-\frac{1}{2}$	1	$-\frac{1}{2}$						$\frac{1}{2}$						$-\frac{1}{2}$		10
0	v_6			$\frac{1}{2}$		$-\frac{3}{2}$	1					$-\frac{1}{2}$	-1					$\frac{1}{2}$	1	40
$z_j - c_j$	h_j	1	1	-1	0	0	0	-1	0	0	-1	-1	-1	0	-1	-1	0	0	0	40
	g_j	-1	-1	$-\frac{3}{2}$	0	$-\frac{3}{2}$	-1	0	0	0	0	$\frac{1}{2}$	0	0	0	0	0	$-\frac{1}{2}$	0	15

u_5,* and after the routine simplex transformation, Table 7.9b is obtained giving the following optimal production schedule for $t = \frac{1}{2}$:

$$\begin{aligned}
x_1 &= x_0 + u_1 - v_1 = 30 - 20 = 10, \\
x_2 &= x_1 + u_2 - v_2 = 10 + 30 = 40, \\
x_3 &= x_2 + u_3 - v_3 = 40 + 10 = 50, \\
x_4 &= x_3 + u_4 - v_4 = 50, \\
x_5 &= x_4 + u_5 - v_5 = 50, \\
x_6 &= x_5 + u_6 - v_6 = 50 - 40 = 10.
\end{aligned}$$

Now only s_1 and s_2 have positive h_j and both yield the same ratio which is $t = 1$. Therefore we select one of them, say s_1, to go into the basis, and by the usual simplex decision rule decide that u_2 is to

Table 7.9c

c_i	P_j / P_i	s_1	s_2	s_3	s_4	s_5	s_6	u_1	u_2	u_3	u_4	u_5	u_6	v_1	v_2	v_3	v_4	v_5	v_6	θ_i
0	v_1		$\frac{1}{2}$					-1	$-\frac{1}{2}$					1	$\frac{1}{2}$					5
1	s_1	1	$-\frac{1}{2}$						$\frac{1}{2}$						$-\frac{1}{2}$					15
t	u_3		$\frac{3}{2}$	-1					$\frac{1}{2}$	1					$-\frac{1}{2}$	-1				25
0	v_4		①			$\frac{1}{2}$					-1	$-\frac{1}{2}$					1	$\frac{1}{2}$		0
1	s_4			$-\frac{1}{2}$	1	$-\frac{1}{2}$						$\frac{1}{2}$						$-\frac{1}{2}$		10
0	v_6			$\frac{1}{2}$		$-\frac{3}{2}$	1					$-\frac{1}{2}$	-1					$\frac{1}{2}$	1	40
$z_j - c_j$	h_j	0	$\frac{3}{2}$	-1	0	0	0	-1	$-\frac{1}{2}$	0	-1	-1	-1	0	$-\frac{1}{2}$	-1	0	0	0	25
	g_j	0	$-\frac{3}{2}$	$-\frac{3}{2}$	0	$-\frac{3}{2}$	-1	0	$\frac{1}{2}$	0	0	$\frac{1}{2}$	0	0	$-\frac{1}{2}$	0	0	$-\frac{1}{2}$	0	30

be replaced. The result is Table 7.9c which implies the following production schedule:

---

* There is a tie between v_4 and u_5, so the solution is degenerate. However, since either one is an optimal solution for $t = \frac{1}{2}$, degeneracy is not a problem here. Incidentally, both solutions yield identical production schedules.

$$\begin{aligned}
x_1 &= 30 - 5 = 25,\\
x_2 &= 25,\\
x_3 &= 25 + 25 = 50,\\
x_4 &= 50 - 0 = 50,\\
x_5 &= 50,\\
x_6 &= 50 - 40 = 10.
\end{aligned}$$

As we look for the next critical value of t, we discover an interesting phenomenon—the optimal solution given in Table 7.9c is valid for $t = 1$ only. As soon as t exceeds 1, s_2 becomes favorable and ought to be put into the basis in place of v_4 yielding a new optimal schedule. But since v_4 is originally zero, its displacement does not have any effect upon the existing production schedule. The entry of s_2 into the basis merely opens the way for further increase in the value of t. We will simply state without showing detailed computations that the next critical value of t is 3. The interested reader may verify this value himself and perhaps even carry this parametric analysis further.

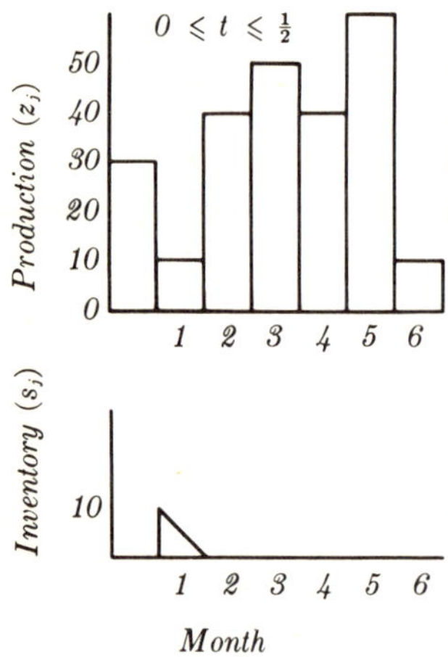

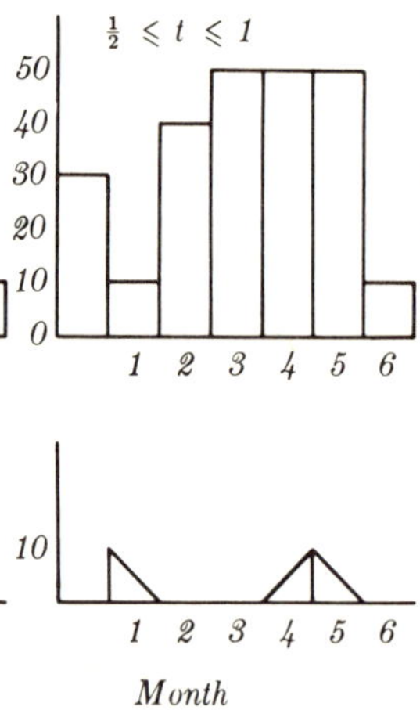

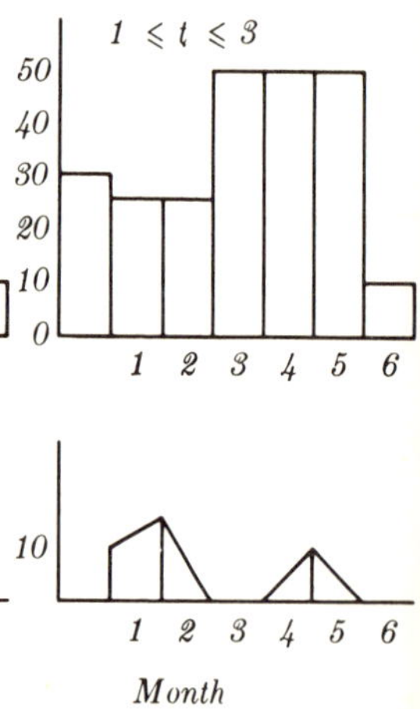

FIGURE 7.1a FIGURE 7.1b FIGURE 7.1c

To summarize the results of the foregoing partial parametric study of t, we present the production schedules and inventory fluctuations implied by the three optimal solutions for three different ranges of t in Figures 7.1a, 7.1b and 7.1c. The total costs that correspond to the various ranges of t are:

Range of t:	$0 \leq t \leq \frac{1}{2}$	$\frac{1}{2} \leq t \leq 1$	$1 \leq t \leq 3$
Total cost:	$60t + 5$	$40t + 15$	$25t + 30$

It is obvious from the diagrams that as t increases in value, i.e., as production-increase cost becomes higher relative to inventory-holding cost, both the number of production increases and the amounts of such increases become progressively smaller. On the other hand, the average inventory over the 6-month period becomes increasingly larger. Such is the nature of production smoothing.

EXERCISES

1. Given the following linear programming problem:

$$\begin{aligned} \text{Maximize } f(X) &= -x_1 + 2x_2 - x_3, \\ \text{subject to: } \quad & 3x_1 + x_2 - x_3 \leq 10, \\ & -x_1 + 4x_2 + x_3 \geq 6, \\ & x_2 + x_3 \leq 4, \\ \text{and} \quad & x_j \geq 0, \quad j = 1, 2, 3. \end{aligned}$$

   (a) Determine the optimal solution.
   (b) Find the separate ranges of b_1 and b_2 consistent with the above solution.
   (c) Use the dual simplex method to study the changes in the composition of the optimal basis as b_2 is varied beyond its range as determined in (b).

2. Given the following linear programming problem:

$$\begin{aligned} \text{Minimize } f(X) &= x_1 + x_2 + x_3, \\ \text{subject to: } \quad & 2x_1 - x_2 + 2x_3 \geq 2, \\ & 3x_1 + x_2 - x_3 \geq 6, \\ & -x_1 - 2x_2 + x_3 \geq 4, \\ \text{and} \quad & x_j \geq 0, \quad j = 1, 2, 3. \end{aligned}$$

   (a) Determine the optimal solution.
   (b) Find the range of b_1 consistent with the above optimal solution.
   (c) Find the ranges of b_2 and b_3 separately which are consistent with the above optimal solution.

3. The following mathematical formulation describes a problem of allocating three resources to the annual production of three com-

modities by a hypothetical manufacturing firm. The amounts of the three products to be produced are represented by x_1, x_2 and x_3 respectively. The objective function reflects the dollar contribution to overhead of these products.

$$\begin{aligned} \text{Maximize } f(X) &= 10x_1 + 15x_2 + 5x_3, \\ \text{subject to:} \quad & 2x_1 + x_2 \leq 5{,}000, \\ & 3x_1 + 3x_2 + x_3 \leq 9{,}000, \\ & x_1 + 2x_2 + 2x_3 \leq 4{,}000, \\ \text{and} \quad & x_j \geq 0, \quad j = 1, 2, 3. \end{aligned}$$

(a) Use the revised simplex method to find the optimal production program.

(b) What are the implicit costs of the resources under the program determined in (a).

(c) If the Research and Development Department proposes a new product whose production coefficients are represented by the vector $P_7 = \begin{bmatrix} 2 \\ 4 \\ 1 \end{bmatrix}$, and whose contribution to overhead is calculated to be \$12 per unit, should the product be produced? If so, what would be the new optimal program?

(d) What is the minimal contribution to overhead that should be expected before production of this new product would actually increase profit?

(e) If additional research holds great promise of increasing its contribution from \$12 to \$16, what would be the maximal budget that should be authorized for this research?

4. In connection with Exercise 3, if instead of a new product, the R & D Department discovers a new process of manufacturing the second product, which is represented by $P_2' = \begin{bmatrix} 1 \\ 2 \\ 3 \end{bmatrix}$ and is expected to increase the contribution to overhead from \$15 to \$18, would it be profitable to put the new process into effect? Assume that no capital outlay is required for this changeover. Start with the optimal solution in Exercise 3 (a) in evaluating the new process and explain your conclusion.

5. In connection with Exercise 3, if instead of a new product, the R & D Department discovers a new process of manufacturing the second product, which is represented by $P_2' = \begin{bmatrix} 2 \\ 2 \\ 2 \end{bmatrix}$ and is expected to in-

crease the contribution to overhead from \$15 to \$18, would it be profitable to put the new process into effect? If the new process has an estimated economic life of 5 years and requires \$5,000 initial outlay, would it still be profitable? Start evaluating the new process with the optimal solution in Exercise 3 (a) and explain your conclusions.

6. (a) Carry the parametric analysis of the production-smoothing example in Section 7.7 through three more intervals of t.
   (b) Determine the optimal schedule in the limiting case of $t = \infty$.

7. In connection with Exercise 2 of Chapter I, assume a fixed cost of \$20 and \$15 respectively for each 15-ton and 10-ton trailer-truck dispatched. Let the distance be a parameter t instead of the 1000 miles originally given. Perform a parametric analysis of the effects of t on the optimal assignment of trucks to move 3000 tons of materials.

8. In parametric programming as discussed in Section 7.2:
   (a) What is the implication if for $\mathrm{t} = \mathrm{t}_L$, all $\mathrm{h}_j \geq 0$?
   (b) What is the implication if for $t = t_L$, and $\min\limits_{h_j<0} \left\{-\frac{g_j}{h_j}\right\} = -\frac{g_k}{h_k}$, all $y_{ik} < 0$?
   (c) Is it possible that for some value of t, no optimal feasible solution exists?

9. In the gasoline blending problem discussed in Section 7.7, the use of isopentane, although desirable for its high PN rating, is hampered by its high RVP index. Compute the increase in net revenue in the two-product program if its RVP index is reduced by 1. Would the optimal basis obtained in the text still be optimal? What would be the effect on net revenue if the PN rating of alkylate is decreased by 1? If the PN rating of catalytic gasoline is decreased by 1?

10. (a) Solve the problem described in Exercise 3 of Chapter I.
    (b) What would happen to the optimal solution determined in (a) if Eskimo dressing is sold for \$3 per quart? For \$3.25 per quart?
    (c) What would happen to the solution in (a) if the cost of A is \$2 per quart? \$2.50 per quart?

THE DUAL PROBLEM

8.1. Introduction—The Discovery of the Dual.

Suppose that we have a maximizing problem with objective function $f(X) = CX$ and constraints $AX \leq P_0$ and $X \geq O$, where C, X, A, P_0 and O are all matrix notations as defined in Chapter III Section 3.1. If a finite maximum exists, then the solution must be such that all the evaluations are non-negative, i.e., $Z - C \geq O$. Referring specifically to (6.2.6) in Chapter VI, we notice that $Z - C = \tilde{C}B^{-1}A - C$. Therefore, the condition for optimality in a maximizing problem is:

$$\tilde{C}B^{-1}A \geq C. \tag{8.1.1}$$

Here B stands for the basis associated with a particular basic feasible solution to $AX \leq P_0$. As the basis changes in the iterative process of the simplex method, $\tilde{C}B^{-1}$ changes accordingly, failing at first (8.1.1) but eventually satisfying it and thereby indicating that the optimal solution in X has been obtained. This variability of $\tilde{C}B^{-1}$ suggests that we may regard it as a variable (row vector), in which case (8.1.1) may be viewed as a system of linear inequalities. Thus, we may define a new variable row vector $W = [w_1\ w_2 \ldots w_m] = \tilde{C}B^{-1}$ and rewrite (8.1.1) as follows:

$$WA \geq C, \tag{8.1.2}$$

or equivalently in full form:

$$
\begin{aligned}
w_1a_{11} + w_2a_{21} + \ldots + w_ma_{m1} &\geq c_1, \\
w_1a_{12} + w_2a_{22} + \ldots + w_ma_{m2} &\geq c_2, \\
\vdots \qquad \vdots \qquad\quad\; \vdots \quad &\quad \vdots \\
w_1a_{1n} + w_2a_{2n} + \ldots + w_ma_{mn} &\geq c_n.
\end{aligned}
\tag{8.1.3}
$$

Viewed in this light, we may think of the linear programming problem as one of finding a *coupled* X-W vector pair that satisfies $AX \leq P_0$, $WA \geq C$ at the same time. The word "coupled" is emphasized because of the determined relationship between X and W through the latter's definition in terms of B. The simplex methods we have described so far all are similar in one respect, viz., the use of this relationship defined in terms of the basis B that is associated with X, not W. For instance, the regular simplex starts with an X that satisfies $AX \leq P_0$ and $X \geq O$ but is coupled with a W ($= \tilde{C}B^{-1}$ where B is the basis associated with X) that does not satisfy $WA \geq C$. In the end, however, the simplex decision rules lead to the selection of an X that satisfies $AX \leq P_0$, $X \geq O$ and is coupled with a W that also satisfies $WA \geq C$, thus signifying optimality. The dual simplex method on the other hand starts with an X that satisfies $AX \leq P_0$ but not $X \geq O$. However, this initial X solution is coupled (still through its basis B) with a W that satisfies $WA \geq C$. By changing the basis of X, we arrive eventually at a coupled X-W solution that satisfies all three conditions, $AX \leq P_0$, $X \geq O$ and $WA \geq C$. This then is our optimal solution.

There is another interesting and significant relationship between X and W that must be mentioned. We note that the constraint $AX \leq P_0$ can be converted to $AX + IX_s = P_0$, where X_s is a column vector of slack variables $x_{n+1}, x_{n+2}, \ldots, x_{n+m}$. We know from previous chapters that at optimal solution the evaluations of these slack variables must also be non-negative. Hence, $\tilde{C}B^{-1}I \geq O$ or, equivalently, $W \geq O$ since the price coefficients of these slack variables are all zero. This means that the optimal solution is a coupled X-W solution such that not only $AX \leq P_0$ and $WA \geq C$ but also $X \geq O$ and $W \geq O$.

With the above information at hand, an inquisitive reader would almost naturally raise the question: Cannot the problem be solved through W instead of X? That is, if the maximal X-solution to AX

$\leq P_0$ and $X \geq O$ implies that $WA \geq C$ and $W \geq O$, why couldn't a search among the W-solutions to $WA \geq C$ and $W \geq O$ turn up the same maximal X-solution? This would be a good question and would get an affirmative answer. In the next section we shall demonstrate that this approach is indeed feasible and develop a "compass" with which to guide the search. For the sake of clarity, we shall hereafter call X the primal variables, $AX \leq P_0$ the primal constraints, W the dual variable and $WA \geq C$ the dual constraints.

8.2. The Duality Theorem.

The main conclusion of the preceding section, to repeat, is that the optimal solution of a linear programming problem is a coupled X-W solution which satisfies not only $AX \leq P_0$ and $X \geq O$ but also $WA \geq C$ and $W \geq O$. Let this optimal solution be X^*-W^* with $W^* = \tilde{C}^*B^{*-1}$ where $\tilde{C}^*$ is the reduced price coefficient vector and B^* the basis associated with the optimal solution X^* respectively. Multiplying both sides by P_0, we have,

$$W^*P_0 = \tilde{C}^*B^{*-1}P_0.$$

But $B^{*-1}P_0 = \tilde{X}^*$. Therefore

$$W^*P_0 = \tilde{C}^*\tilde{X}^* = CX^* = \text{Max} f(X), \tag{8.2.1}$$

which means that the W-solution to $WA \geq C$ and $W \geq O$ contains a particular solution W^* which when multiplied by the $m \times 1$ column vector P_0 yields the maximal value of the objective function $f(X)$.

Furthermore, since $AX \leq P_0$, we can multiply both sides by a non-negative $1 \times m$ row vector $W = [w_1 \, w_2 \, . \, . \, . \, w_m]$ and obtain

$$WAX \leq WP_0. \tag{8.2.2}$$

If we then restrict W to those vectors that satisfy $WA \geq C$, since X is required to be non-negative, then

$$WAX \geq CX. \tag{8.2.3}$$

Combining (8.2.2) and (8.2.3), we have the following important conclusion:

$$CX \leq WP_0, \tag{8.2.4}$$

which states that any W-solution to $WA \geq C$ and $W \geq O$ yields a value of WP_0 that is never smaller than (i.e., always greater than or equal to) the value of CX yielded by any X that is a solution to $AX \leq P_0$ and $X \geq O$. Coupling this with (8.2.1) which states that $W^*P_0 = CX^* = \max f(X)$, we conclude that W^*P_0 is the minimal value of all WP_0, and W^* is the minimizing solution to $WA \geq C$ and $W \geq O$. In other words, if we denote WP_0 by $g(W)$, i.e., a linear function of W, we can say that $g(W^*) = \min g(W)$ is equal to $f(X^*) = \max f(X)$. Consequently the problem of maximizing $f(X)$ is equivalent to the problem of minimizing $g(W)$, with X and W subjected of course to the appropriate sets of linear constraints as discussed before.

It is now possible to summarize the above by a somewhat formal statement of the Duality Theorem.

Given a (primal) linear programming problem as follows:

(I) Maximize $f(X) = CX$,

$$\text{subject to:} \quad AX \leq P_0 \quad \text{and} \quad X \geq O,$$

where C is a $1 \times n$ row vector, X an $n \times 1$ column vector, A an $m \times n$ matrix and P_0 an $m \times 1$ column vector, there is always a related (dual) linear programming problem defined as follows:

(II) Minimize $g(W) = WP_0$,

$$\text{subject to:} \quad WA \geq C \quad \text{and} \quad W \geq O,$$

where W is a $1 \times m$ row vector, such that if there exists a finite maximal solution in X to (I), there is always a finite minimal solution in W to (II), and furthermore,

$$\text{maximal } f(X) = \text{minimal } g(W).$$

This theorem clearly establishes a new approach to the solution of linear programming problems. It possesses important practical values not only as an additional computational device but also as an interpretive tool with which to obtain a clearer analysis of the problem and its solutions. We shall discuss the former in Section 8.5 and the latter in Section 8.6 of this chapter.

The above theorem was stated with the condition that a finite optimam has to exist. If it does not, i.e., if the maximal value of $f(X)$ is infinity, then it must be impossible to solve the dual as defined in (II) since all values of $g(W)$ must be greater than or equal to any

$f(X)$ (8.2.4), and it is impossible to have any value greater than infinity.*

It has been found helpful to represent the structure of the primal (I) and the dual (II) by a tabular arrangement of the relevant coefficients as shown in Figure 8.1. The core of the table is the matrix A.

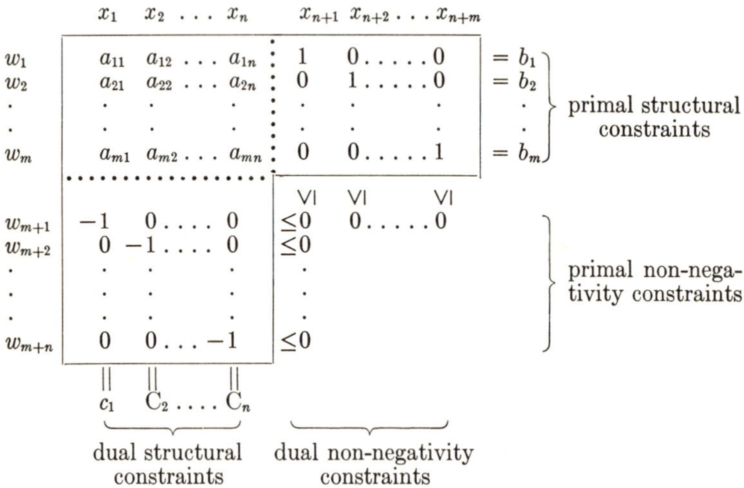

FIGURE 8.1

To its right is added the identity matrix representing the coefficients of the slack variables in the primal and to its bottom is added the negative identity matrix representing the coefficients of the slack variables in the dual. The columns and rows are identified by the primal and dual variables as shown. Finally, the constants (b_i) and price coefficients (c_j) are recorded on the right and lower margins respectively. With this arrangement the primal constraints can be quickly obtained by reading across the table from left to right and the dual constraints from top to bottom. This arrangement is also

---

* The absence of a solution to the dual may manifest itself in the form of an optimal basic feasible solution containing one or more artificial vectors as discussed in Chapter V.

very helpful in identifying the relationships between the primal and dual variables, as will be made clear in Section 8.4.

Example.

$$\begin{aligned} \text{Maximize } f(X) &= 2x_1 - x_2, \\ \text{subject to:} \quad & x_1 + x_2 \leq 10, \\ & -2x_1 + x_2 \leq 2, \\ & 4x_1 + 3x_2 \geq 12, \\ \text{and} \quad & x_1, x_2 \geq 0. \end{aligned}$$

We first rewrite the third constraint to conform to (I) in the Duality Theorem, which gives us $-4x_1 - 3x_2 \leq -12$. Forming a mental picture of the arrangement illustrated in Figure 8.1 with the coefficients and constants of the present problem and reading them from top to bottom column by column, we arrive at the following dual:

$$\begin{aligned} \text{Minimize } g(W) &= 10w_1 + 2w_2 - 12w_3, \\ \text{subject to:} \quad & w_1 - 2w_2 - 4w_3 \geq 2, \\ & w_1 + w_2 - 3w_3 \geq -1, \\ \text{and} \quad & w_1, w_2, w_3 \geq 0. \end{aligned}$$

In order to make use of the regular simplex to solve the dual, we multiply the second constraint through by -1 and add the necessary slack and artificial variables to obtain the following system:

$$\begin{aligned} \text{Minimize } g(W) &= 10w_1 + 2w_2 - 12w_3 + Mw_6, \\ \text{subject to:} \quad & w_1 - 2w_2 - 4w_3 - w_4 + w_6 = 2, \\ & -w_1 - w_2 + 3w_3 + w_5 = 1, \\ \text{and} \quad & w_1, w_2, w_3, w_4, w_5, w_6 \geq 0. \end{aligned}$$

The solution of the above linear programming problem is shown in Tables 8.1a and 8.1b. In these Tables, it will be noticed that $P_{(i)}$, P_j

Table 8.1a

$c_{(i)}$	$P_{(i)}$ \ P_j		P_1	P_2	P_3	P_4	P_5	P_6	P_0
M	$P_{(1)}$	P_6	(1)	-2	-4	-1	0	1	2
0	$P_{(2)}$	P_5	-1	-1	3	0	1	0	1
$z_j - c_j$		M	1	-2	-4	-1	0	0	2
		N	-10	-2	$+12$	0	0	0	0

Table 8.1b

10	$P_{(1)}$	P_1	1	−2	−4	−1	0	2
0	$P_{(2)}$	P_5	0	−3	−1	−1	1	3
$z_j - c_j$			0	−22	−28	−10	0	20

and $z_j - c_j$ all refer to the dual problem rather than the original primal. After the first iteration, the artificial vector is ejected from the basis and the minimal value of $g(W)$ is found to be 20. The same result is also obtained as the maximal value of $f(X)$ by solving the primal directly in the following sequence of tables.

Table 8.2a

$c_{(i)}$	$P_{(i)}$ \ P_j		P_1	P_2	P_3	P_4	P_5	P_6	P_0
0	$P_{(1)}$	P_3	1	1	1	0	0	0	10
0	$P_{(2)}$	P_4	−2	1	0	1	0	0	2
−M	$P_{(3)}$	P_6	(4)	3	0	0	−1	1	12
$z_j - c_j$		M	−4	−3	0	0	1	0	−12
		N	−2	1	0	0	0	0	0

Table 8.2b

0	$P_{(1)}$	P_3	0	$\frac{1}{4}$	1	0	($\frac{1}{4}$)		7
0	$P_{(2)}$	P_4	0	$\frac{5}{2}$	0	1	$-\frac{1}{2}$		8
2	$P_{(3)}$	P_1	1	$\frac{3}{4}$	0	0	$-\frac{1}{4}$		3
$z_j - c_j$			0	$\frac{5}{2}$	0	0	$-\frac{1}{2}$		6

Table 8.2c

0	$P_{(1)}$	P_5	0	1	4	0	1		28
0	$P_{(2)}$	P_4	0	3	2	1	0		22
2	$P_{(3)}$	P_1	1	1	1	0	0		10
$z_j - c_j$			0	3	2	0	0		20

8.3. Correspondence of the Primal and the Dual Variables.

The equivalence of the primal and the dual optimal solutions does not end with their having the same value for their respective objective functions. If we look a little closer for a moment at Tables 8.1b and 8.2c we notice that disregarding zeroes, the numbers appearing under P_0, viz., 2 and 3, in Table 8.1b are the same as those that show up in the $z_j - c_j$ row in Table 8.2c. Similarly, the numbers in P_0 column in Table 8.2c also appear in the $z_j - c_j$ row in Table 8.1b but with the signs reversed. In other words, the optimal X-solutions show up as the evaluations of the W-variables in the optimal W-solution and vice versa. This phenomenon is no mere coincidence. It is actually rooted in the logic of the primal-dual relationships.

To understand this, let us once again refer to Figure 8.1. Let X^* be the optimal primal solution and B^* be its associated basis. Then

$$z_j - c_j = \tilde{C}^* B^{*-1} P_j - c_j.$$

But we have already seen in Section 8.2 that $\tilde{C}^* B^{*-1}$ is $W^* = [w_1^* \, w_2^* \ldots w_m^*]$ which is the optimal dual solution. Therefore

$$z_j - c_j = W^* P_j - c_j.$$

If P_j is a primal slack vector, i.e., $j = n + i$, $i = 1, 2, \ldots, m$, then P_j is a unit vector and c_j is zero. Hence

$$\text{(8.3.1)} \qquad \begin{aligned} z_{n+i} - c_{n+i} = W^* P_{n+i} - 0 &= w_1^*(0) + w_2^*(0) + \ldots \\ &+ w_i^*(1) + \ldots + w_m^*(0) = w_i^*, \qquad i = 1, 2, \ldots, m. \end{aligned}$$

If on the other hand P_j is a structural vector, i.e., $j = 1, 2, \ldots, n$, then, bearing in mind that W^* must satisfy the dual constraint $WP_j \geq c_j$ (or equivalently) $WP_j - w_{m+j} = c_j$, where w_{m+j} is the jth slack variable in the dual (cf. Figure 8.1), then

$$\text{(8.3.2)} \qquad z_j - c_j = W^* P_j - c_j = w_{m+j}^*, \qquad j = 1, 2, \ldots, n.$$

In conclusion, we may assert firmly that the evaluations of the primal variables in the optimal program are in fact the optimal dual solution and vice versa. If we have n structural and m slack variables in the primal, we will have m structural and n slack in the dual, and in the final solution, the evaluations of the structural and slack variables of the primal are the same as the optimal slack and structural variables respectively of the dual.

Applying the above conclusion to the Example of Section 8.2, we find in Table 8.2c:

$$z_1 - c_1 = 0 = w_4,$$
$$z_2 - c_2 = 3 = w_5,$$
$$z_3 - c_3 = 2 = w_1,$$
$$z_4 - c_4 = 0 = w_2,$$
$$z_5 - c_5 = 0 = w_3,$$

which check with results either displayed or implied in Table 8.1b. The careful reader will also observe that if a primal variable is in the basis in the optimal solution, its counterpart in the dual must be out of the basis and the same is true *mutatis mutandis*. This particular relationship has an economic interpretation which will be discussed in Section 8.7 in connection with the warehousing problem.

There is but one question now unanswered: Why are the signs of the primal variables in Table 8.2c reversed when they show up as evaluations of the dual variables in Table 8.1b? To answer this, we only have to think of the w_i as the primal variables and the x_j as their duals. In the light of this interchange, the evaluator of w_{m+1} can be written according to (8.3.1) and Figure 8.1 as $a_{m+1,1}x_1 + a_{m+2,2}x_2 + \ldots + a_{m+1,n}x_n$ which is $(-1)x_1 + (0)x_2 + \ldots + (0)x_n = -x_1$. Similarly, the evaluator of w_1, viz., $(a_{11}x_1 + a_{12}x_2 + \ldots + a_{1n}x_n) - b_1$, which must be smaller than or equal to zero in a minimal solution is actually $-x_{n+1}$. The same thing is true of all other evaluators. In general, the evaluators of $w_1, w_2, \ldots, w_{m+n}$ in the optimal dual solution represent the negative of the x values in the optimal primal solution with the individual variables in the two sets paired as described in the preceding paragraph.

8.4. The General Case of the Dual.

In the foregoing sections we have studied the dual of the linear programming problem whose structural constraints are $AX \leq P_0$. We are in luck with this type of constraints because the dual constraints associated with the primal slack variables turn out to be $WI \geq O$ or simply $W \geq O$, which are the non-negativity constraints of the dual variables. The identity matrix involved here is the coefficient-matrix of the slack variables to be added to $AX \leq P_0$. In other words, the case of $AX \leq P_0$ is really a special case of $AX = P_0$, where a part of A is an identity matrix; furthermore, the variables associated with the identity matrix all have a price coefficient of zero. It behooves us therefore to give some consideration to the general case where no especially accommodating conditions prevail.

If our primal problem now is:

$$\text{(III)} \qquad \begin{aligned} &\text{Maximize } f(X) = CX, \\ &\text{subject to:} \quad AX = P_0, \\ &\text{and} \quad X \geq O, \end{aligned}$$

our dual will be simply:

$$\text{(IV)} \qquad \begin{aligned} &\text{Minimize } g(W) = WP_0, \\ &\text{subject to:} \quad WA \geq C. \end{aligned}$$

We will leave the proof to the reader as a test of his understanding of the Duality Theorem. We must point out that we cannot in general assume the presence of an identity submatrix in A—therefore we cannot always obtain the dual constraints $WI \geq O$. This means that we cannot impose the non-negativity constraints on the dual structural variables, although we may do so on the dual slack variables as they are introduced to convert the constraints in (IV) into equations. When such a dual problem arises, the simplex method has to be modified somewhat before it can be used. Specifically, both Decision Rules I and II have to be relaxed in so far as they are applied to those variables which are not constrained to non-negativity. These exceptional considerations tend so to complicate the procedure that the solution of (III) by way of its dual (IV) is not usually advised. It is nevertheless possible as the Example in this section will demonstrate.

One final thought before we leave this topic concerns the case where the A matrix in (III) does contain an identity submatrix which is, however, associated with a set of variables whose price coefficients are not zeroes. The relevant dual constraints will be $W \geq C'$ where C' is a row vector consisting of the price coefficients of these variables, and not $W \geq O$. In using the simplex method to solve the dual, the basis will have to be expanded to include these additional constraints, which makes it undesirable.

In short, the general case of the dual presents no new theoretical complications. Its formulation is just as straightforward as the special dual discussed earlier provided that the underlying rationale of the dual explained in Sections 8.1 and 8.2 is firmly grasped. The usefulness of the general dual as a method of attacking linear programming problems is, however, quite a different matter.

Example. In the interest of variation, we present the following problem calling for the minimization of $f(X)$ subject to mixed constraints, to be solved via its dual. In so far as its solution makes use of the dual simplex method and some *ad hoc* variations of the simplex decision rules, it is considered a good teaching device for review purposes and for testing the reader's fundamental grasp of the simplex theory. It is not intended to demonstrate an efficient way of solving the original problem.

$$\begin{array}{lrcl}
\text{Minimize } f(X) = & 3x_1 - x_2 - 2x_3, & & \\
\text{subject to:} & x_1 + x_2 + x_3 & = & 12, \\
& 2x_1 \qquad - x_3 & \leq & 6, \\
& x_1 + x_2 \qquad & \geq & 3, \\
\text{and} & x_1, x_2, x_3 & \geq & 0.
\end{array}$$

Since it is a minimizing problem, we will rewrite the second constraint in the "$\geq$" form and then convert it as well as the third constraint into equations. We have thus:

$$\begin{array}{lrcl}
\text{Minimize } f(X) = & 3x_1 - x_2 - 2x_3, & & \\
\text{subject to:} & x_1 + x_2 + x_3 \qquad\qquad & = & 12, \\
& -2x_1 \qquad + x_3 - x_4 \qquad & = & -6, \\
& x_1 + x_2 \qquad\qquad - x_5 & = & 3, \\
\text{and} & x_1, x_2, x_3, x_4, x_5 & \geq & 0.
\end{array}$$

This primal problem can be very easily solved by the dual simplex method using P_3, P_4, and P_5 to form the initial basis. After one iteration, as the reader can quickly verify the optimal solution is found to be $X^T = [3\ 0\ 9\ 9\ 0]$ and the minimal $f(X) = -9$. The optimal dual solution can also be obtained by noticing the evaluators: $z_1 - c_1 = -w_4 = 0$, $z_2 - c_2 = -w_5 = -6$, $z_3 - c_3 = -w_6 = 0$, $z_4 - c_4 = -w_2 = 0$, $z_5 - c_5 = -w_3 = -5$. w_1 is not represented directly by any primal evaluator because the first constraint does not have any slack variable. However, it can be readily determined from the third dual constraint which is $w_1 + w_2 + w_6 = -2$. Since $w_2 = 0$ and $w_6 = 0$, w_1 must be equal to -2.

Now, to solve this problem via its dual, we first formulate the dual in accordance with (IV) above:

$$\begin{aligned}
\text{Maximize } g(W) &= 12w_1 - 6w_2 + 3w_3,\\
\text{subject to:} &\quad w_1 - 2w_2 + w_3 \le 3,\\
&\quad w_1 + w_3 \le -1,\\
&\quad w_1 + w_2 \le -2,\\
&\quad -w_2 \le 0,\\
\text{and} &\quad -w_3 \le 0.
\end{aligned}$$

The fourth and fifth constraints above mean that w_2, w_3 must be non-negative, while w_1 can be either positive, negative or zero. Keeping this in mind and adding proper slack variables to the inequalities, we arrive at the following dual problem:

$$\begin{aligned}
\text{Maximize } g(W) &= 12w_1 - 6w_2 + 3w_3,\\
\text{subject to:} &\quad w_1 - 2w_2 + w_3 + w_4 = 3,\\
&\quad w_1 - w_3 + w_5 = -1,\\
&\quad w_1 + w_2 + w_6 = -2,\\
\text{and} &\quad w_2, w_3, w_4, w_5, w_6 \ge 0.
\end{aligned}$$

Designating the coefficient vectors P_1, P_2 etc., we see that P_4, P_5 and P_6 form a convenient initial basis, which, however, yields an infeasible solution. Hence, we will use the dual simplex method. Upon checking the evaluations of the vectors we discover that the evaluations of w_1 and w_3 (which are -12 and -3 respectively) are of the wrong sign. Following the discussion in Chapter VI, Section 6.6, Example 2, we introduce a new variable w_a defined by $w_a + w_1 + w_3 = \text{M}$. Since w_1 has the most negative evaluation, we substitute $w_1 = \text{M} - w_a - w_2$ into $g(W)$ and the constraints and obtain:

$$\begin{aligned}
\text{Minimize } g'(W) &= -12w_a - 6w_2 - 9w_3 + 12\text{M}\\
\text{subject to:} &\quad -w_a - 2w_2 + w_4 = -\text{M} + 3\\
&\quad -w_a - 2w_3 + w_5 = -\text{M} - 1\\
&\quad -w_a + w_2 + w_6 = -\text{M} - 2\\
&\quad w_1 + w_a + w_3 = \text{M}\\
\text{and} &\quad w_2, w_3, w_4, w_5, w_6 \ge 0
\end{aligned}$$

The problem is now ready for iterative solution by the dual simplex algorithm as carried out in the sequence of tableaus, Tables 8.3a, b and c. The solution in Table 8.3c shows w_a, w_3 and w_5 are all non-negative as required. That w_1 is negative is quite acceptable since it is not constrained to be non-negative. However, we must be certain that $g(W)$ cannot be further decreased by replacing P_1 in the basis with some other non-basis vector. This is found to be the case as all coefficients

Table 8.3a

$b_{(i)}$	$P_{(i)}$ \ P_j		P_1	P_a	P_2	P_3	P_4	P_5	P_6	P_0
0	$P_{(1)}$	P_4	0	−1	−2	0	1	0	0	−M + 3
0	$P_{(2)}$	P_5	0	−1	0	−2	0	1	0	−M − 1
0	$P_{(3)}$	P_6	0	−1	1	[−1]	0	0	1	−M − 2
0	$P_{(4)}$	P_1	1	1	0	1	0	0	0	M
Evaluators			0	12	6	9	0	0	0	12M

Table 8.3b

$b_{(i)}$	$P_{(i)}$		P_1	P_a	P_2	P_3	P_4	P_5	P_6	P_0
0	$P_{(1)}$	P_4	0	[−1]	−2	0	1	0	0	−M + 3
0	$P_{(2)}$	P_5	0	1	−2	0	0	1	−2	M + 3
−9	$P_{(3)}$	P_3	0	1	−1	1	0	0	−1	M + 2
0	$P_{(4)}$	P_1	1	0	1	0	0	0	1	− 2
Evaluators			0	3	15	0	0	0	9	3M − 18

Table 8.3c

$b_{(i)}$	$P_{(i)}$		P_1	P_a	P_2	P_3	P_4	P_5	P_6	P_0
−12	$P_{(1)}$	P_a	0	1	2	0	−1	0	0	M − 3
0	$P_{(2)}$	P_5	0	0	−4	0	1	1	−2	6
−9	$P_{(3)}$	P_3	0	0	−3	1	1	0	−1	5
0	$P_{(4)}$	P_1	1	0	1	0	0	0	1	− 2
Evaluators			0	0	9	0	3	0	9	− 9

in the P_1 row are non-negative. Hence the solution contained in Table 8.3c, viz. $w_1 = -2$, $w_2 = 0$, $w_3 = 5$, $w_4 = 0$, $w_5 = 6$, $w_6 = 0$, is the optimal solution and the minimal value of $g'(W)$ [or the maximal value of $g(W)$] is -9 as mentioned earlier.

The optimal values of the primal variables may also readily be identified in this table. The evaluation of w_1 is the value of the first primal slack variable. It is zero, which agrees with the fact that the first primal constraint is an equation and does not have a slack variable. The evaluations of w_2, w_3, w_4, w_5 and w_6 should represent respectively the optimal values of x_4, x_5, x_1, x_2 and x_3. Hence, $x_4 = 9$, $x_5 = 0$, $x_1 = 3$, $x_2 = 0$ and $x_3 = 9$ which, of course, agree with the results given earlier by a direct solution of the primal.

8.5. Appraisal of the Dual Approach.

After having treated extensively the theory of the dual, we may question its practical value, especially in view of the fact that the optimal dual solution appears inevitably as a "by-product" of the optimal primal solution. The answer is partly that some linear programming problems can be solved more expeditiously by means of it and partly that it is more suitable for studying the effect of certain structural changes in a problem on the character of its optimal solution.

According to published reports on computational experiences, the number of iterations required to reach the optimum generally ranges from 1.5 to 3 times the number of structural constraints in the problem.* Thus, if the number of structural inequality constraints (m) is larger than the number of structural variables (n) in the primal (not counting the slack variables), then there is a strong presumption that the number of iterations will be reduced by solving the dual since the dual has only n structural constraints. Considering the high cost of modern digital computers, the savings in time and money made possible by the judicious use of the dual approach should not be lightly dismissed. Furthermore, computers have only limited high-speed internal memory (whether drum or core) which is generally used to store the inverse of the basis employed in the revised simplex method. If m is greater than n in the primal, then the basis in the dual, which is $n \times n$, will be smaller. Consequently, larger problems may be solved on a given computer using the dual formulation. The reader is cautioned, however, that in certain circumstances, e.g., when $m < n$, or when the lack of slack variables causes some dual variables to be freed from non-negativity constraints, complications arise in the use of the simplex method to solve the dual, rendering it neither practicable nor efficient (cf. Example of Section 8.3). In short, the computational value of the dual approach is subject to some severe limitations.

The second advantage of the dual approach is the ease with which the effects of a change in the structural constraints of the primal on the optimality of its solutions can be studied. Suppose we wish to add

---

* Cf. William Orchard-Hays, "Evolution of Linear Programming Computing Techniques," *Management Science*, IV (1958), No. 2, pp. 183–190.

a constraint to the original system and observe its effects on the optimal solution, or delete a constraint, or replace an existing constraint with a different one (which is the same as deleting the old and adding the new in succession): How can this be accomplished? If one were to do this through the primal, changes would be extensive if not exhausting since all P_j are affected in terms either of their contents or their dimensions. On the other hand, if one chooses the dual approach while realizing the direct correspondence between the number of constraints (m) in the primal and the number of structural variables (m) in the dual, the task becomes much simpler as the addition or deletion of primal constraints means merely the addition or deletion of dual variables, leaving the existing dual problem unaltered. If new dual variables are added, they should be evaluated to determine whether they should be ignored or introduced into the basis to improve the solution. This is very simple if the revised simplex method is employed as it involves only the multiplication of the new P_j by the inverse of the present basis as described in Chapter VI, Section 6.4. If existing dual variables are dropped, nothing has to be done unless they happen to be basis variables. In that event the task becomes essentially one of finding the best sub-optimal solution through a transformation of the existing basis, which is assumed to be optimal. This again is a simple task the details of which are left for the reader himself to discover as an exercise (cf. Chapter V Exercise 5).

One should not leave the topic of an appraisal of the dual approach without taking especial notice of the fact that the dual simplex method discussed in Chapter VI is really a method of solving the dual with the dual solution disguised in the primal form. We have already explained the connection briefly in Section 8.1 of this chapter. Bearing in mind the correspondence of the primal variables with the dual evaluations and that of the primal evaluations with the dual variables, we can see that by insisting on the "right" sign for all primal evaluations we are in effect seeking and maintaining a solution to the dual. By allowing the primal variables to be negative in the dual simplex, we are actually noting the fact that some dual evaluations are favorable, which means that the present dual solution is not yet the optimal one. Through the dual simplex algorithm the negative primal variables are removed from the basis without violation of the sign requirement of the primal evaluations. This is equivalent to substituting non-basis dual variables with favorable evaluations into the basis until no such variables can be found any more, which then signifies the optimality of the solution.

In addition to the above merits, the dual approach sometimes makes possible great simplification of the system of constraints, permitting the use of special and efficient methods of solution. One example is the Warehousing Problem, which is the topic of the next section. Another is the Transportation Problem, which will be taken up in the next chapter.

8.6. The Warehousing Problem.*

"Warehousing Problem" is a generic term representing a class of linear programming problems whose coefficient matrices possess a special structure. It is so called only because Cahn who first formulated the problem in 1948 chose the operation of a warehouse as a vehicle to describe the abstract model. Our interest obviously is in the general structure of the problem and the character of its solution rather than in warehousing as such.

We define a warehouse as a decision-making profit-seeking organization engaged in buying, storing and selling a single commodity.** It has a fixed storage capacity but the storage cost is negligible and the market is unlimited. We assume that decisions on how much to buy and how much to sell are made regularly in what we may refer to as time-periods. Prices and costs are assumed to be constant within each time-period but are free to fluctuate from period to period. Furthermore, for a reasonable number of periods in the future, they may be reliably estimated. We also assume that the lead-time for the delivery of purchase-orders is one time-period, i.e., supplies ordered by the warehouse at the beginning of any period will not be available for sale until the beginning of the next period. Now, if the warehouse has some initial inventory to begin with, how should it program its purchases and sales in the foreseeable future in order to maximize overall profits?

Let us first of all write down the various aspects of this problem in precise mathematical language. To do this, we have to decide what

---

* This section draws heavily on Charnes and Cooper, "Generalization of the Warehousing Model," *Operations Research Quarterly*, VI (1955), No. 4, pp. 131–172.

** For an extension to multiple commodities, see *ibid.*, pp. 149–152.

our decision variables should be. This is often the hardest part of the problem to beginners. Fortunately, the present problem is an exception in this respect. Since the problem asks for a program of the amounts to be purchased and sold in different periods, we will simply let x_i be the amount to be purchased in the ith period and y_i be the amount to be sold in the ith period with i ranging from 1 through n, if n is the number of periods under consideration. Let us further designate the capacity of the warehouse as C, the initial inventory as I, and the price and cost of the commodity in the ith period as p_i and c_i respectively. Since the overall profit in the n periods is $(p_1y_1 + p_2y_2 + \ldots + p_ny_n) - (c_1x_1 + c_2x_2 + \ldots + c_nx_n)$, the objective of the problem is to

$$\text{Maximize } f(X, Y) = \sum_{i=1}^{n} (p_iy_i - c_ix_i).$$

What are the factors that restrict this overall profit? First we realize that in any period, there must not be more goods in storage than the warehouse can hold. Thus, for periods 1, 2, etc., we have

$$\begin{gathered} \mathrm{I} + x_1 - y_1 \leq \mathrm{C}, \\ \mathrm{I} + x_1 - y_1 + x_2 - y_2 \leq \mathrm{C}, \\ \text{etc.} \end{gathered}$$

In general,

$$\sum_{t=1}^{i} (x_t - y_t) \leq \mathrm{C} - \mathrm{I}, \qquad i = 1, 2, \ldots, n.$$

Secondly, it is clear that the sales in any period cannot be more than the amount available in that period. The amount available in any period is the sum of the initial inventory and the cumulative purchases delivered up to that period minus the cumulative sales up to the preceding period. Bearing in mind the one-period lead-time required for delivery, we have for periods 1, 2, etc.

$$\begin{gathered} y_1 \leq \mathrm{I}, \\ y_2 \leq \mathrm{I} + x_1 - y_1, \\ \text{etc.} \end{gathered}$$

In general,

$$y_1 \leq \mathrm{I},$$

$$y_i - \sum_{t=1}^{i-1} (x_t - y_t) \leq \mathrm{I}, \qquad i = 2, 3, \ldots n.$$

Finally, for obvious reasons x_i and y_i cannot be negative. Therefore, we add to the above the usual non-negativity constraints and arrive at the following linear programming problem:

$$\text{(8.6.1)} \qquad \text{Maximize } f(X, Y) = \sum_{i=1}^{n} (p_i y_i - c_i x_i),$$

$$\text{(8.6.2)} \qquad \text{subject to:} \qquad \sum_{t=1}^{i} (x_t - y_t) \leq \mathrm{C} - \mathrm{I}, \qquad i = 1, 2, \ldots, n,$$

$$\text{(8.6.3)} \qquad y_1 \leq \mathrm{I},$$

$$\text{(8.6.4)} \qquad y_i - \sum_{t=1}^{i-1} (x_t - y_t) \leq \mathrm{I}, \qquad i = 2, 3, \ldots, n,$$

$$\text{(8.6.5)} \qquad \text{and} \qquad x_i, y_i \geq 0, \qquad i = 1, 2, \ldots, n.$$

While this problem can readily be solved by the simplex method, the labor involved may be reduced substantially if we formulate and solve its dual instead, as will be presently made clear.

In order to write the dual, let us first take a look at the coefficient matrix A of the warehousing problem, shown in Figure 8.2. There are

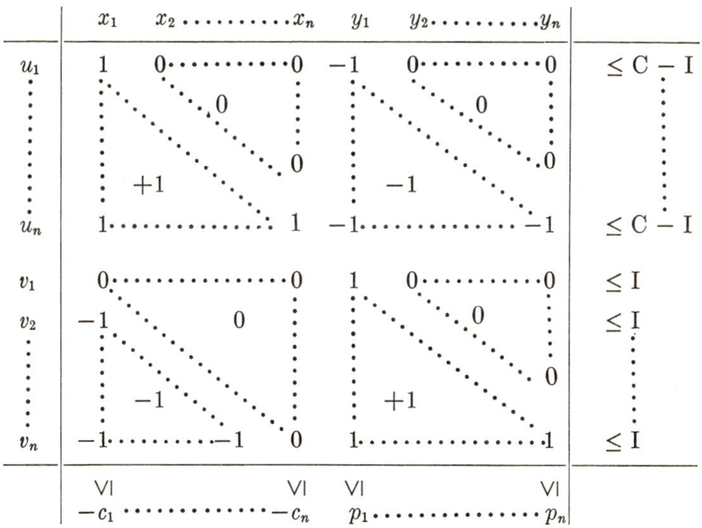

FIGURE 8.2

three points of interest about the structure of A. One is that the coefficients are either $+1$, -1 or 0. Another is that the $+1$'s and -1's arrange themselves neatly in four separate groups, all triangular in shape. Lastly, due to the assumed lead-time of one period for delivery, the dimensions of the lower left triangle are one element smaller than the other triangles. Let us call the dual variables associated with the first set of n constraints (8.6.2), $u_1, u_2, \ldots, u_n$ and those associated with the second set (8.6.3) (8.6.4), $v_1, v_2, \ldots, v_n$. In accordance with Section 8.2 and Figure 8.1 the dual problem is:

Minimize $g(U, V)$

$$
\begin{aligned}
&= (u_1 + u_2 + \ldots + u_n)(\mathrm{C} - \mathrm{I}) + (v_1 + v_2 + \ldots + v_n)\mathrm{I}, \\
\text{subject to:}\quad & (u_1 + u_2 + \ldots + u_n) - (0 + v_2 + \ldots + v_n) \geq -c_1, \\
& (0 + u_2 + \ldots + u_n) - (0 + 0 + v_3 + \ldots + v_n) \geq -c_2, \\
& \qquad \vdots \qquad\qquad\qquad\qquad \vdots \qquad\qquad \vdots \\
& (0 + 0 + \ldots + 0 + u_n) - (0 + 0 + 0 + \ldots + 0) \geq -c_n, \\
& -(u_1 + u_2 + \ldots + u_n) + (v_1 + v_2 + \ldots + v_n) \geq p_1, \\
& -(0 + u_2 + \ldots + u_n) + (0 + v_2 + \ldots + v_n) \geq p_2, \\
& \qquad \vdots \qquad\qquad\qquad\qquad \vdots \qquad\qquad \vdots \\
& -(0 + \ldots + 0 + u_n) + (0 + \ldots + 0 + v_n) \geq p_n, \\
\text{and}\quad & u_1, u_2, \ldots, u_n, v_1, v_2, \ldots, v_n \geq 0.
\end{aligned}
$$

Comparing the dual with the primal, we notice that the major change is in the objective functions rather than in the constraints. Whereas in the primal objective function, there are $2n$ variables and $2n$ price coefficients, in the dual there are $2n$ variables but only 2 different price coefficients. The dual can be simplified further by the following transformation of variables:

$$U_i = \sum_{t=i}^{n} u_t, \qquad i = 1, 2, \ldots, n, \tag{8.6.6}$$

$$V_i = \sum_{t=i}^{n} v_t, \qquad i = 1, 2, \ldots, n. \tag{8.6.7}$$

Substituting these new variables into the original dual problem, the objective function becomes simply $U_1(C - I) + V_1I$, and the structural constraints transform straightforwardly into:

$$\left.\begin{array}{r} U_1 - V_2 \geq -c_1 \\ U_2 - V_3 \geq -c_2 \\ \vdots \\ U_{n-1} - V_n \geq -c_{n-1} \end{array}\right\} \text{ or in general, } U_i - V_{i+1} \geq -c_i, \quad i = 1, \ldots, n-1,$$

$$U_n \geq -c_n$$

$$\left.\begin{array}{r} -U_1 + V_1 \geq p_1 \\ -U_2 + V_2 \geq p_2 \\ \vdots \\ -U_n + V_n \geq p_n \end{array}\right\} \text{ or in general, } -U_i + V_i \geq p_i, \quad i = 1, \ldots, n.$$

The non-negativity constraints of u_i, v_i, however, have two different interpretations in terms of U_i and V_i. First, it is obvious that they mean U_i and V_i must be also non-negative. Second, they imply that given the definitions of U_i and V_i, U_i must be greater than or equal to U_{i+1}, and V_i must likewise be greater than V_{i+1}. Thus the original dual in u_i and v_i is fully equivalent to the following problem in U_i and V_i:

(8.6.8) Minimize $g(U, V) = U_1(\mathrm{C} - \mathrm{I}) + V_1\mathrm{I}$,

(8.6.9) subject to: $U_i - V_{i+1} \geq -c_i$, $i = 1, 2, \ldots, n-1$,

(8.6.10) $U_n \geq -c_n$,

(8.6.11) $-U_i + V_i \geq p_i$, $i = 1, 2, \ldots, n$,

(8.6.12) $U_i \geq U_{i+1}$, $i = 1, 2, \ldots, n-1$,

(8.6.13) $V_i \geq V_{i+1}$, $i = 1, 2, \ldots, n-1$,

(8.6.14) $U_i, V_i \geq 0$, $i = 1, 2, \ldots, n$.

At first glance the merit of this transformation seems open to question. The same number of variables ($2n$) are present in the new formulation and, what is worse, the number of structural constraints has increased from $2n$ to $2n + 2(n - 1)$. But a second look reveals the true advantage of the modified formulation, which is that no single constraint involves more than two variables. Furthermore, the objective function involves also only two variables, viz., U_1 and V_1. Since both are constrained to be non-negative (8.6.14) and (C − I) and I are also non-negative, the minimization of the objective function clearly calls for the minimization of U_1 and V_1 to the extent consistent with the constraints. From (8.6.11) and (8.6.13) we see that V_1 can be decreased only to the extent permitted by $U_1 + p_1$ or V_2, whichever is larger. Next we notice from (8.6.9) and (8.6.12) that U_1 can be lowered only to the extent allowed by $V_2 - c_1$ or U_2, whichever is larger. In general, we can trace the determinants of V_i to U_i and V_{i+1}, and the determinants of U_i to V_{i+1} and U_{i+1}, until finally we have V_n determined by U_n alone, and U_n by c_n in accordance either with (8.6.10), or with the non-negativity requirement of (8.6.14). Schematically we can represent the "genealogical" relationships between the variables by the following diagram:

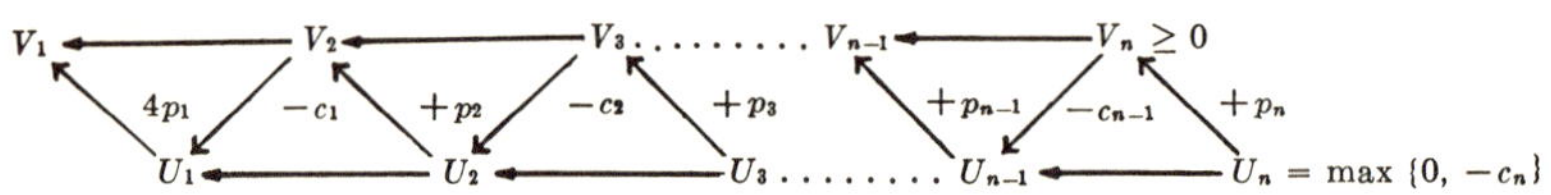

FIGURE 8.3

The discussion has shown that the problem of minimizing the objective function (8.6.8) is reducible to the simple problem of minimizing U_n subject to $U_n \geq -c_n$ and $U_n \geq 0$. Hence the optimal solution is:

(8.6.15) $\quad U_n = \text{Max}\,\{-c_n, 0\}$ (either $-c_n$ or 0 whichever is greater),

(8.6.16) $\quad V_n = \text{Max}\,\{U_n + p_n, 0\}$.

Once these two variables are determined, the other U_i's and V_i's can be solved alternately and successively by simple arithmatic. For instance, since U_i is to have the minimal possible value, it must be equal to either $V_{i+1} - c_i$ due to (8.6.9) or to U_{i+1} due to (8.6.12), whichever is greater. Hence,

(8.6.17) $\quad U_i = \text{Max}\,\{V_{i+1} - c_i, U_{i+1}\}, \qquad i = 1, 2, \ldots, n - 1.$

Similarly, due to (8.6.11) and (8.6.13),

(8.6.18) $\quad V_i = \text{Max}\,\{U_i + p_i, V_{i+1}\}, \qquad i = 1, 2, \ldots, n - 1.$

The optimal value of $g(U, V)$ is readily obtained once U_1 and V_1 have been solved from (8.6.17) and (8.6.18) above. Thus, we have succeeded in reducing a complex linear programming problem by the use of the Duality Theorem to a chain of simple two-variable decision problems, all of which can be solved by simple arithmetic. The analytical advantage of the dual approach, in addition to the computational merits discussed before, is obviously not to be dismissed lightly.

However, the solution is not yet complete since the original problem requires a set of optimal values of x_i and y_i, i.e., the optimal amount of purchases and sales in the n periods. To find these, we first observe that (8.6.15) through (8.6.18) indicate that U_1 and V_1 are both linear functions of p_i and c_i, $i = 1, 2, \ldots, n$. Let us write:

$$\left.\begin{aligned} U_1^* &= \sum_{i=1}^{n} d_i p_i - \sum_{i=1}^{n} e_i c_i \\ V_1^* &= \sum_{i=1}^{n} r_i p_i - \sum_{i=1}^{n} s_i c_i \end{aligned}\right\} \text{where } d_i,\ e_i,\ r_i \text{ and } s_i \text{ are all coefficients, and } U_1^*,\ V_1^* \text{ are the optimal solutions.}$$

The optimal value of the objective function (8.6.8) is then:

$$\begin{aligned} g(U^*, V^*) &= U_1^*(\mathrm{C} - \mathrm{I}) + V_1^*\mathrm{I} \\ &= \left(\sum_{i=1}^{n} d_i p_i - \sum_{i=1}^{n} e_i c_i\right)(\mathrm{C} - \mathrm{I}) + \left(\sum_{i=1}^{n} r_i p_i - \sum_{i=1}^{n} s_i c_i\right)\mathrm{I} \\ &= \sum_{i=1}^{n} [(\mathrm{C} - \mathrm{I})d_i + \mathrm{I}r_i]p_i - \sum_{i=1}^{n} [(\mathrm{C} - \mathrm{I})e_i + \mathrm{I}s_i]c_i. \end{aligned}$$

According to the Duality Theorem, $\text{Min}\, g(U, V) = \text{Max}\, f(X, Y)$. Hence, $g(U^*, V^*) = f(X^*, Y^*)$ or

$$\sum_{i=1}^{n} [(\mathrm{C} - \mathrm{I})d_i + \mathrm{I}r_i]p_i - \sum_{i=1}^{n} [(\mathrm{C} - \mathrm{I})e_i + \mathrm{I}s_i]c_i = \sum_{i=1}^{n} y_i^* p_i - \sum_{i=1}^{n} x_i^* c_i,$$

where y_i^* and x_i^* represent the optimal values maximizing (8.6.1). Since both sides contain like terms of p_i and c_i, one solution of x^* and y^* is:

(8.6.19) $$x_i^* = (\mathrm{C} - \mathrm{I})e_i + \mathrm{I}s_i \qquad i = 1, 2, \ldots, n.$$

(8.6.20) $$y_i^* = (\mathrm{C} - \mathrm{I})d_i + \mathrm{I}r_i \qquad i = 1, 2, \ldots, n.$$

The method by which we arrived at the optimal primal solution above makes use of the Regrouping Principle: it was obtained by regrouping the components of the optimal dual solution with respect to p_i and c_i.† This was possible because the special structure of the constraints (8.6.9) through (8.6.14) makes the optimal solutions of U_1 and V_1 linear functions of p_i and c_i.

Example. Assume that a warehouse is committed to a policy of planning its weekly sales and purchases six weeks ahead of time. During one such planning period the prices and costs of the commodity are estimated to be as follows:

Week	1	2	3	4	5	6
Price (\$/unit) p_i	6	5	7	6	10	6
Cost (\$/unit) c_i	4	5	5	7	7	7

The warehouse has an initial stock of 100 units and a storage capacity of 500 units. Assuming that one week is required for a replacement order to be delivered at the warehouse, what should be its purchases and sales during each of the ensuing six weeks in order to maximize its profits?

The formulation of the primal and the dual of this problem is exactly as discussed in the text and will be omitted here. We proceed immediately with the solution of U_i and V_i. According to (8.6.15),

$$U_n = U_6 = \text{Max}\,\{-c_6, 0\} = \text{Max}\,\{-7, 0\} = 0.$$

Employing (8.6.16), we obtain

$$V_n = V_6 = \text{Max}\,\{U_6 + p_6, 0\} = \text{Max}\,\{0 + 6, 0\} = 6 = p_6.$$

In succession (8.6.17) and (8.6.18) yield the following:

$$\begin{aligned}
U_5 &= \max\,\{V_6 - c_5, U_6\} = \max\,\{6 - 7, 0\} = 0,\\
V_5 &= \max\,\{U_5 + p_5, V_6\} = \max\,\{0 + 10, 6\} = 10 = p_5,\\
U_4 &= \max\,\{V_5 - c_4, U_5\} = \max\,\{3, 0\} = 3 = p_5 - c_4,\\
V_4 &= \max\,\{U_4 + p_4, V_5\} = \max\,\{9, 10\} = 10 = p_5,\\
U_3 &= \max\,\{V_4 - c_3, U_4\} = \max\,\{5, 3\} = 5 = p_5 - c_3,\\
V_3 &= \max\,\{U_3 + p_3, V_4\} = \max\,\{12, 10\} = 12 = p_5 - c_3 + p_3,\\
U_2 &= \max\,\{V_3 - c_2, U_3\} = \max\,\{7, 5\} = 7 = P_5 - c_3 + p_3 - c_2,
\end{aligned}$$

† A more efficient way of determining x_1^*, y_1^* from U_1^*, V_1^* is suggested in Exercise 8.

$$V_2 = \max\{U_2 + p_2, V_3\} = \text{Max}\{12, 12\}$$

$$= 12 = \begin{cases} p_5 - c_3 + p_3 - c_2 + p_2, & \text{(i)} \\ p_5 - c_3 + p_3, & \text{(ii)} \end{cases}$$

$$U_1 = \max\{V_2 - c_1, U_2\} = 8 = \begin{cases} p_5 - c_3 + p_3 - c_2 + p_2 - c_1, & \text{(i)} \\ p_5 - c_3 + p_3 - c_1, & \text{(ii)} \end{cases}$$

$$V_1 = \max\{U_1 + p_1, V_2\}$$

$$= 14 = \begin{cases} p_5 - c_3 + p_3 - c_2 + p_2 - c_1 + p_1, & \text{(i)} \\ p_5 - c_3 + p_3 - c_1 + p_1. & \text{(ii)} \end{cases}$$

Attention is called to the solution of V_2 which has two expressions in terms of p_i and c_i for its minimal value. Evidently this is the result of a tie between $U_2 + p_2$ and V_3. These alternative expressions of V_2 followed by similarly diverse expressions of U_1 and V_1 lead ultimately to different optimal programs of purchases and sales, but with no difference in profits resulting.

The amount of profit implied by the above solution of U_1 and V_1 is readily found from (8.6.8):

$$g(U, V) = U_1(\text{C} - \text{I}) + V_1\text{I} = 8(500 - 100) + 14(100) = \$4{,}600.$$

Applying the Regrouping Principle to obtain the optimal primal solution, we have from the (i) expressions of U_1, V_1 above

$$g(U, V) = (p_5 - c_3 + p_3 - c_2 + p_2 - c_1)(\text{C} - \text{I}) + (p_5 - c_3 + p_3 - c_2 + p_2 - c_1 + p_1)\text{I}$$

$$= \text{C}p_5 + \text{C}p_3 + \text{C}p_2 + \text{I}p_1 - \text{C}c_3 - \text{C}c_2 - \text{C}c_1. \quad (1)$$

Or, using the (ii) expressions of U_1 and V_1,

$$g(U, V) = (p_5 - c_3 + p_3 - c_1)(\text{C} - \text{I}) + (p_5 - c_3 + p_3 - c_1 + p_1)\text{I}$$

$$= \text{C}p_5 + \text{C}p_3 + \text{I}p_1 - \text{C}c_3 - \text{C}c_1. \quad (2)$$

Hence, according to (i), y_5 is the coefficient of p_5, x_3 is the coefficient of c_3, etc. and the entire program emerges as follows:

Month	1	2	3	4	5	6
Sales (y_i)	100	500	500	0	500	0
Purchases (x_i)	500	500	500	0	0	0

(i)

Alternatively,

Sales (y_i)	100	0	500	0	500	0
Purchases (x_i)	500	0	500	0	0	0

(ii)

To complete the solution, let us also determine from (8.6.6) and (8.6.7) the optimal values of the original dual variables:

$$\begin{aligned} u_6 &= U_6 = 0, & v_6 &= V_6 = 6, \\ u_5 &= U_5 - U_6 = 0, & v_5 &= V_5 - V_6 = 4, \\ u_4 &= U_4 - U_5 = 3, & v_4 &= V_4 - V_5 = 0, \\ u_3 &= U_3 - U_4 = 2, & v_3 &= V_3 - V_4 = 2, \\ u_2 &= U_2 - U_3 = 2, & v_2 &= V_2 - V_3 = 0, \\ u_1 &= U_1 - U_2 = 1, & v_1 &= V_1 - V_2 = 2, \end{aligned}$$

The interpretation of these dual variables will be discussed in the next section.

8.7. Economic Interpretation of the Dual Variables.

As often as not, a scientific inquiry into a particular problem leads unexpectedly to a host of other related problems, the understanding of which significantly reinforces and enriches the correct interpretation of the original one. The dual problem and its solution are a kind of analytical by-product born of the attempt to solve the primal. And like the common experience in industrial production of having by-products available before markets are found, the dual problem presents an intriguing question as to what use could be made of it and its solution. The impeccable mathematical logic that is the basis of the correspondence between the primal and the dual encourages the speculation that the relationship is substantive as well as formal. If that be the case, then it is reasonable to expect that our comprehension of the original problem would be made more complete by a correct interpretation of the dual. Since we are primarily concerned here with the economic application of linear programming, we shall concentrate only on the economic interpretation of the dual.

The basic ingredients of all economic problems are inputs, outputs and profit. Despite a wide range of variations, the typical linear programming problem in economic context is: How much of each of n different outputs (x_j) should be produced from given amounts of m different inputs (b_i) in order to maximize revenue and profit (since costs and inputs are fixed by assumption) if the prices of these products (c_j) and the amounts of different inputs required to make one unit of the various outputs (a_{ij}) are known. The mathematical formulation of this economic problem is the standard one:

$$\text{Maximize } f(X) = \sum_{j=1}^{n} c_j x_j,$$

$$\text{subject to:} \quad \sum_{j=1}^{n} a_{ij} x_j \leq b_i, \quad i = 1, 2, \ldots, m,$$

$$\text{and} \quad x_j \geq 0. \quad j = 1, 2, \ldots, n.$$

The dual is then:

$$\text{Minimize } g(W) = \sum_{i=1}^{m} w_i b_i,$$

$$\text{subject to:} \quad \sum_{i=1}^{m} w_i a_{ij} \geq c_j, \quad j = 1, 2, \ldots, n,$$

$$\text{and} \quad w_i \geq 0, \quad i = 1, 2, \ldots, m.$$

Since according to the Duality Theorem the maximum of $f(X)$ must be equal to the minimum of $g(W)$, the two linear functions must have the same unit of measurement, in this case, dollars. If the unit of $w_i b_i$ is dollars and b_i is the amount of input, then it stands to reason that w_i must be a kind of cost of the input factor measured in dollars per unit of that input. The dual variables (w_i), however, are not the accounting costs paid by the business firm for the services of the input factors. They are rather what the economists call *implicit* or *opportunity* costs, and are a measure of the value of alternative uses of the input. Optimal utilization of resources dictates that the total opportunity cost of all inputs contained in one unit of output must be equal to the price of that output. This means that if the total opportunity cost of some output, i.e., z_j, is greater than the price, viz., c_j, then these resources are not allocated optimally and better results could be obtained by assigning them to alternative uses. Recalling the equivalence of the dual variables (w_i) and the primal evaluators $(z_j - c_j)$ discussed in Section 8.3, we have, therefore, the inequalities $WA \geq C$ as the dual structural constraints in a dual problem. Any $W = [w_1 \ldots w_m]$ that satisfies $WA \geq C$, must imply a specific production program in which all resources would be efficiently allocated in the above sense. But it must immediately be pointed out that such a program may or may not be feasible. That is to say, it may call for the use of more of some resources than is available. In such events, some primal (slack) variables will become negative, and due to the equivalence discussed in Section 8.3, the evaluators of the corresponding dual variables will become positive, thereby indicating the need for further improvement. Accordingly, we employ the simplex method to arrive at a different W, or a different set of assumed implicit costs of the resources that reduces $g(W)$. In this manner we complete one cycle of the adjustment of W, which is then repeated with sufficient frequency to yield an optimal W that is both efficient (i.e., $Z - C \geq O$ in the primal and $W \geq O$ in the dual) and feasible (i.e., $X \geq O$ in the primal and all evaluators are non-positive in the dual).

In the light of the foregoing it should be easy to understand that if a particular resource is not fully utilized (i.e., if the corresponding slack variable is in the basis of the primal, or equivalently, if the corresponding dual evaluator is negative) then its opportunity cost is zero since no opportunity is lost. This will be indicated by the fact that the evaluator of the slack variable and its corresponding dual are both zero. On the other hand, if a particular resource is fully utilized (i.e., the corresponding slack variable is not in basis), then its opportunity cost is positive (i.e., the evaluation of the slack is positive and the corresponding dual variable is in basis and hence assumes a positive value). Consequently, the dual variables measure the economic gain (or loss) of a unit increase (or decrease) of a particular resource.

In the warehousing problem discussed in Section 8.6, the objective function of the original dual problem is $g(U, V) = u_1(\mathrm{C} - \mathrm{I}) + \ldots + u_n(\mathrm{C} - \mathrm{I}) + v_1\mathrm{I} + \ldots + v_n\mathrm{I}$. In view of the foregoing discussion therefore, u_1 may be interpreted as the value of an additional unit of warehouse capacity in period one after the space requirement of the initial inventory is met, and v_1 the value of an additional unit of initial inventory in period one. But since an increase in I decreases the remaining capacity (C − I) which as stated before has a value of u_1, such an increase in period one has only a net value of $v_1 - u_1$. If we consolidate the C and I terms in $g(U, V)$, we have $g(U, V) = \sum_{i=1}^{n} u_i\mathrm{C} + \sum_{i=1}^{n} (v_i - u_i)\mathrm{I}$. This indicates that if the warehouse capacity is increased by one unit throughout the n periods, other things being equal, total profit will be increased by $\sum_{i=1}^{n} u_i$ dollars. Similar interpretation can be given to a permanent increase in I although the exact meaning of a permanent increase is rather difficult to explain.

These dual variables can obviously have considerable value in the policy-making decisions of business firms. They may be used, for instance, as a basis for assessing the admissible costs of acquiring additional resources for temporary or permanent expansion. Or they may be used to price an order which would demand the use of existing resources already optimally employed. Increasingly today decision-makers in business and industry realize that sound decisions must be based on opportunity costs rather than on accounting or historical costs. Since the determination of opportunity costs depends on one's knowledge of the optimal opportunity that exists, it is evident that linear programming can contribute importantly to the making of sound business policies.

EXERCISES

1. Solve the duals of the problems in Exercise 1 (a), (b), (c), (e) and (g) of Chapter V. In each case, obtain the optimal primal solution from the final tableau of the dual.

2. Given the linear programming problem: Minimize $f(X) = CX$ subject to $AX \geq P_0$ and $X \geq O$, prove that if a finite minimum exists, then there is an equivalent problem: Maximize $g(W) = WP_0$ subject to $WA \leq C$ and $W \geq O$ such that minimal $f(X)$ = maximal $g(W)$.

3. The following simplex tableau contains the optimal solution to a linear programming problem. It is known that P_5, P_6 and P_7 are the slack vectors of the 1st, 2nd and 3rd constraints respectively.

c_i	P_1	P_2	P_3	P_4	P_5	P_6	P_7	P_0
10	0	$\frac{1}{2}$	1	0	$\frac{1}{2}$	0	0	$\frac{5}{2}$
6	1	$-\frac{1}{2}$	0	$\frac{1}{3}$	$-\frac{1}{6}$	$\frac{1}{3}$	0	$\frac{5}{2}$
0	0	0	0	$\frac{2}{3}$	$-\frac{1}{3}$	$-\frac{1}{3}$	1	3
$z_j - c_j$	0	4	0	1	4	2	0	40

(a) Construct the dual of the original problem.
(b) Obtain the optimal solution to the dual from the above tableau.

4. Write the dual of the 6-month production-smoothing example discussed in Chapter VII Section 7.7. Assuming that $t = 1$, solve the dual and give an economic interpretation of the dual variables.

5. Write the dual of the transportation problem as presented in Chapter I Section 1.3. Point out the significant structural characteristics of this dual, if any.

6. Write the dual of the rail-freight movement problem as presented in Chapter I Section 1.3. Give an interpretation of the dual variables used.

7. Referring to the numerical example of the warehousing problem in Section 8.8, what would be the optimal purchase-and-sale program if additional storage space of 200 units could be rented for \$2.50 per unit per month? What would be the maximal profit?

8. A more practical and efficient method of solving the warehousing problem is one that takes full advantage of the genealogical relationship between U_i and V_i as shown in Figure 8.3. To explain this method, let us solve the numerical example in Section 8.6 again. We first draw two rows of circles as shown below. These are marked U_1,

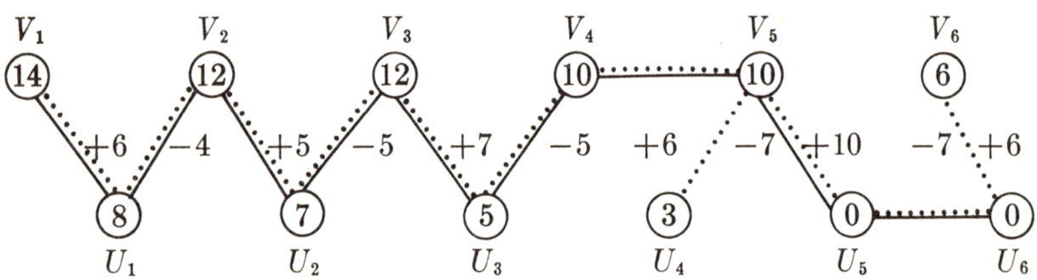

FIGURE 8.4

V_1, etc. and their values will be filled in later as they are determined. In addition, data on prices (positive) and costs (negative) are placed between each U-V pair to facilitate computation. As before, we begin with $U_6 = \max\{-c_6, 0\} = 0$ which is now entered in the circle for U_6. Since $V_6 = \max\{U_6 + p_6, 0\}$, the value of U_6 enables us quickly to determine the value of V_6 from the figure. This optimal value is 6, and is so recorded in the circle for V_6. In the meantime, a dotted line is drawn between U_6 and V_6 to show that V_6 is obtained from U_6. Next, we determine the optimal value of $U_5 = \max\{V_5 - c_5, U_6\}$. This can easily be accomplished in the figure with all the necessary information already available there. Hence, we enter the value of 0 in the circle for U_5 and link U_5 with U_6 by a dotted line. In like manner, we determine all U_i, V_i directly from the figure by using alternately (8.6.17) and (8.6.18), and complete simultaneously the charting of their genealogy with dotted lines. Now, a unique and connected *path* is traced from U_6 all the way to V_1 guided by the dotted lines. This is indicated by the solid line in the diagram. The undulations in this path reflect the fluctuations of inventory, which in turn reflect a certain purchase-and-sale program that represents the optimal solution of the problem. For instance, the path falls from V_1 to U_1 indicating sale. Since only the initial inventory is available for sale, therefore $y_1 = 100$. From U_1 to V_2 the path rises, indicating rise in inventory due to purchases. As the warehouse is now empty after the sale of y_1, purchases should amount to the full capacity of the warehouse. Hence, $x_1 = 500$. This goes on until V_4 is reached. Between V_4 and V_5, the path remains at a "high" level implying that the full inventory is carried through period 4. Hence in this period there is neither selling nor buying. On the other hand, the path remains flat at a "low" level between U_5 and U_6 indicating no inventory as well as no purchase and no sale in period 5.

(a) Explain the validity of this method by means of a proper interpretation of the dual variables u_i, v_i as discussed in Section 8.7.

(b) Use the above method to solve the same warehousing problem with the following different price-cost expectations:

Week	1	2	3	4	5	6
Price (\$/unit)	8	9	12	11	12	12
Cost (\$/unit)	10	9	10	12	11	10

THE TRANSPORTATION PROBLEM

9.1. Introduction.

The "Transportation Problem" refers to a special class of linear programming problems dealing with the distribution of a single commodity from various sources of supply to various points of demand in such a manner that the total transportation costs are minimized. It was first studied by F. L. Hitchcock in 1941, then separately by T. C. Koopmans in 1947, and finally placed in the framework of linear programming and solved by the simplex method by G. B. Dantzig in 1951. Since then, improved methods of solution have been developed and the range of application has been steadily widened to include the caterer problem, the tank-routing problem, etc. It is now accepted as one of the most important analytical and planning tools in business and industry.

There are two ways of approaching this problem: the general and the specific. In the former, one views the transportation problem as a special case of the general linear programming problem and explains how the general simplex method—its rules and its algorithm—applies to this special case. This approach tends to be mechanical and would probably add little to the reader's knowledge except a firm conviction that one *is* the special case of the other. In the latter approach, the procedure for solving the transportation problem is developed independently with each step supported by close economic and mathematical reasoning so that the reader may gain further insight into the problem. There seems little doubt that this approach is more beneficial to the reader; therefore it will be followed in this chapter.

9.2. Properties of the Transportation Problem.

Let us assume in general that there are m sources $S_1, \ldots, S_m$ with capacities $a_1, \ldots, a_m$ and n destinations $D_1, \ldots, D_n$ with requirements $b_1, \ldots, b_n$ respectively. The shipping cost from the ith source to the jth destination is c_{ij} and the amount shipped is x_{ij}. If the total capacity of all sources is equal to the total requirement of all destinations, what must be the values of x_{ij} with $i = 1, \ldots, m$ and $j = 1, \ldots, n$ for the total shipping cost to be a minimum?

Upon examining the above statement of the problem, we realize that it has an objective function which is $f(X) = c_{11}x_{11} + c_{12}x_{12} + \ldots + c_{1n}x_{1n} + c_{21}x_{21} + \ldots + c_{2n}x_{2n} + \ldots + c_{m1}x_{m1} + \ldots + c_{mn}x_{mn} = \sum_{i=1}^{m}\sum_{j=1}^{n} c_{ij}x_{ij}$. Secondly, in view of the condition that the total capacity is equal to the total requirement, i.e., $\sum_{i=1}^{m} a_i = \sum_{j=1}^{n} b_j$, the individual capacity of each source must be fully utilized and the individual requirement of each destination must likewise be fully satisfied. Hence, we have m number of the so-called capacity constraints and n number of requirement constraints. The capacity constraints impose on the solution the condition that the total shipments to all destinations from any source must be equal to the capacity of that source. Thus, $x_{i1} + x_{i2} + \ldots + x_{in} = a_i$ for $i = 1, 2, \ldots, m$. On the other hand, the requirement constraints require that the demand of every destination be fully satisfied by the total shipments from all sources. Hence, $x_{1j} + x_{2j} + \ldots + x_{mj} = b_j$ for $j = 1, 2, \ldots, n$. Thirdly, there are the usual non-negativity constraints, i.e., $x_{ij} \geq 0$ for all i and j. To sum up, we have the following mathematical formulation of the transportation problem:

$$\text{(9.2.1)} \qquad \text{Minimize } f(X) = \sum_{i=1}^{m}\sum_{j=1}^{n} c_{ij}x_{ij},$$

$$\text{(9.2.2)} \qquad \text{subject to:} \qquad \sum_{j=1}^{n} x_{ij} = a_i, \qquad i = 1, \ldots, m,$$

$$\text{(9.2.3)} \qquad \sum_{i=1}^{m} x_{ij} = b_j, \qquad j = 1, \ldots, n,$$

$$\text{(9.2.4)} \qquad \text{and} \qquad x_{ij} \geq 0, \qquad i = 1, \ldots, m, \text{ and } j = 1, \ldots, n.$$

Alternatively the problem may be described in long-hand form as shown in (9.2.5). This mode of expression has the advantage of revealing more clearly the structure of the coefficient matrix of the

$$\begin{array}{llllr}
\multicolumn{5}{l}{\text{Minimize } f(X) = c_{11}x_{11} + \ldots + c_{1n}x_{1n} + \ldots + c_{m1}x_{m1} + \ldots + c_{mn}x_{mn},} \\
\text{subject to: } x_{11} + \ldots + x_{1n} & & & & = a_1, \\
 & x_{21} + \ldots + x_{2n} & & & = a_2, \\
 & \qquad \vdots & & & \vdots \\
\text{(9.2.5)} & & & x_{m1} + \ldots + x_{mn} & = a_m, \\
\quad x_{11} & + x_{21} & \ldots + x_{m1} & & = b_1, \\
\qquad x_{12} & \quad + x_{22} & \quad \ldots + x_{m2} & & = b_2, \\
\qquad \vdots & \quad \vdots & \quad \vdots & & \vdots \\
\qquad\quad x_{1n} & \qquad + x_{2n} & \qquad \ldots + x_{mn} & & = b_n, \\
\multicolumn{5}{l}{\quad \text{and} \qquad x_{ij} \geq 0, \qquad i = 1, \ldots, m \ \text{ and } \ j = 1, \ldots, n.}
\end{array}$$

transportation problem. As shown in Figure 9.1, the matrix is characterized by staggered rows of 1's with slanted columns of 1's directly beneath, and 0's elsewhere. Many students of linear programming believe that similarly structured problems lend themselves to similar methods of attack, therefore careful study in the taxonomy of linear programming problems would seem to hold great promise for the future. However, such study is beyond the scope and purpose of this text.

There should be no doubt that (9.2.1) through (9.2.4) show that the transportation problem is a linear programming problem. It is important, however, that we point out the following special properties of a transportation problem.

(a) The dimensions of the basis of a transportation problem are $(m + n - 1) \times (m + n - 1)$. This means that a transportation problem has only $m + n - 1$ independent structural constraints and that its basic feasible solution has only $m + n - 1$ positive components (fewer if degenerate). The reason for this can readily be

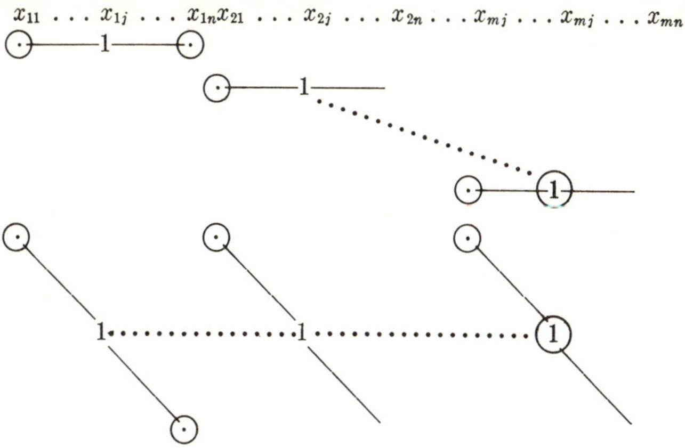

FIGURE 9.1

traced to the assumed condition that $\sum_{i=1}^{m} a_i = \sum_{j=1}^{n} b_j$. For instance, suppose that we add together the m capacity constraints in (9.2.2):

$$\sum_{i=1}^{m} \sum_{j=1}^{n} x_{ij} = \sum_{i=1}^{m} a_i = \sum_{j=1}^{n} b_j. \tag{9.2.6}$$

Now let us add together the first $(n - 1)$ requirement constraints in (9.2.3):

$$\sum_{j=1}^{n-1} \sum_{i=1}^{m} x_{ij} = \sum_{j=1}^{n-1} b_j. \tag{9.2.7}$$

Now subtract (9.2.7) from (9.2.6). The difference is clearly

$$\sum_{i=1}^{m} x_{in} = b_n$$

which is the last requirement constraint. Hence, one of the $m + n$ constraints can always be derived from the remaining $m + n - 1$ constraints and the problem in effect has only $m + n - 1$ independent constraints. This shows the truth of what was said earlier about the basis and the basic feasible solution.

(b) A transportation problem always has a basic feasible solution. Commonsense almost tells us as much. For as long as the total supply at the sources is equal to the total demand at the destinations, there must be *a* way of distributing the supply to satisfy the demand, (barring strikes, acts of God, and such special cases). We learn from Chapter III that the solutions of a linear programming problem form a convex set and that a feasible solution in the set can always be expressed as a convex combination of the extreme points (solutions) of the set. Hence, a transportation problem always has a basic feasible solution (cf. Chapter III Section 3.6). For instance, we may decide as a working rule to distribute the supply at the ith source in strict proportion to the requirements of all destinations, i.e., let $x_{ij} = k_i b_j$ where k_i is the proportionality factor for the ith source. Since this supply must be completely distributed, $\sum_{j=1}^{n} x_{ij} = k_i \sum_{j=1}^{n} b_j = a_i$. Or, $k_i = \dfrac{a_i}{\sum_{j=1}^{n} b_j} = \dfrac{a_i}{\sum_{i=1}^{m} a_i}$. Consequently,

$$x_{ij} = k_i b_j = \left(\frac{a_i}{\sum_{j=1}^{n} b_j}\right) b_j, \quad i = 1, \ldots, m \text{ and } j = 1, \ldots, n. \tag{9.2.8}$$

Does this represent a feasible solution? We can find out by substituting x_{ij} into the capacity and requirement constraints and determining whether the latter are satisfied. Thus,

$$\sum_{j=1}^{n} x_{ij} = \sum_{j=1}^{n} \left(\frac{a_i}{\sum_{j=1}^{n} b_j}\right) b_j = \frac{a_i}{\sum_{j=1}^{n} b_j} \sum_{j=1}^{n} b_j = a_i, \qquad i = 1, \ldots, m,$$

and

$$\sum_{i=1}^{m} x_{ij} = \sum_{i=1}^{m} \left(\frac{a_i}{\sum_{j=1}^{n} b_j}\right) b_j = \frac{b_j}{\sum_{i=1}^{m} a_i} \sum_{i=1}^{m} a_i = b_j, \qquad j = 1, \ldots, n,$$

which show that the constraints are indeed satisfied. Furthermore, since all a_i and b_j are positive, x_{ij} determined from (9.2.8) must be all positive. Therefore (9.2.8) yields a feasible solution and, as explained earlier, implies the existence of a basic feasible solution.

(c) The values of the basis variables in a basic feasible solution are all differences between partial sum of a_i and partial sum of b_j. In other words,

$$(9.2.9) \qquad x_{ij} = \pm \sum_{i=1}^{m} r_i a_i \mp \sum_{j=1}^{n} s_j b_j, \qquad \text{where } r_i, s_j \text{ are either 1 or 0.}$$

For explanation let us refer to Figure 9.1 which shows schematically the constraint system of (9.2.5). There are two important observations to be made. First, a basis variable (represented by circled dots) appear in two and only two of the constraints, i.e., one capacity constraint and one requirement constraint (circled dots show up in vertical pairs). Secondly, every constraint must be represented by at least one basis variable (one circled dot), for otherwise that constraint will not be satisfied. We recall from (a) above that there are only $m + n - 1$ basis variables. Therefore, there are only $2(m + n - 1)$ circled dots in the system. However, they have to embrace $m + n$ constraints. Consequently there must be at least two constraints which contain only one basis variable each. These two basis variables can be equated to the constants of the relevant constraints and determined immediately, i.e., $x_{ij} = a_i$ or b_j. If $x_{ij} = a_i$, then the jth requirement contraint will have one less unknown and b_j must be adjusted to $b'_j = b_j - a_i$ to signify the reduced amount of unsatisfied demand. On the other hand, if $x_{ij} = b_j$, then the ith capacity constraint will have one less unknown and a_i must be adjusted to $a'_i = a_i - b_j$ to signify the reduced supply still available. In the meantime, the total number of undetermined basis variables is reduced to $m + n - 3$ and the number of unsolved constraints to $m + n - 2$. Thus, by the same argument employed before, we reason that there must be two more constraints which contain one basis variable each. Furthermore, these two basis variables must be equal to either some other a_i, b_j or the adjusted a'_i, b'_j obtained above. By repeated substitution of determined basis variables into unsolved constraints in this manner, all basis variables will be eventually determined and found equal to differences between partial sums of a_i on the one hand and partial sums of b_j on the other as stated in (9.2.9).* This fact explains both the feasibility of the Northwest Corner Rule as a method of obtaining an initial solution to the transportation problem and the

* This property of the basic feasible solution of a transportation problem was first brought to light by G. B. Dantzig in "Simplex Method: Application of the Simplex Method to a Transportation Problem," *Activity Analysis of Production and Allocation*, ed. Koopman (New York: John Wiley & Sons, Inc., 1951), p. 365.

nature of degeneracy in it. These will engage our close attention in the sections immediately following.

9.3. The Initial Solution—The Northwest Corner Rule and Vogel's Approximation Method.

Let us explain by a specific example how the two methods work. Suppose that a steel firm has 4 plants which purchase coal for their coke ovens from 3 mines. The costs of shipping one ton of coal from each source (mine) to each destination (plant) are given in Table 9.1.

Table 9.1

Plant		Shipping cost ($/ton)				Capacity (000 tons)
		1	2	3	4	
Mine	1	3	1	4	5	50
	2	7	3	8	6	50
	3	2	3	9	2	75
Requirement		40	55	60	20	175

The capacity of each source and the requirement of each destination are also shown in the table. How much coal should the firm purchase from each source in order to satisfy the demand of each destination at minimal shipping expense?

We will first take up the Northwest Corner Rule. To obtain an initial basic solution by this rule, we arbitrarily choose to begin assigning supply to demand from the northwest corner (the upper left corner) of the cost table. Disregarding shipping cost altogether, we consider first the requirement of Plant 1 and satisfy it as much as possible from Mine 1. Since Mine 1 in this instance has more coal than Plant 1 needs, we decide to purchase the entire needs of Plant 1, i.e., 40,000 tons from Mine 1. Thus, $x_{11} = 40$ (000). The supply at Mine 1 presently is reduced by the same amount and the remainder is assigned to Plant 2. Hence, $x_{12} = 10$ (000). This reduces the needs of Plant 2 to 45,000 tons which is entirely supplied by Mine 2, i.e., $x_{22} = 45$. Repeating the procedure for the remaining source and destinations, we finally obtain the initial basic solution (always feasible) shown in the upper left corners of the appropriate

Table 9.2
(x_{ij} and c_{ij} are entered in the upper left and lower right corners of each cell respectively)

S_i \ D_j	1	2	3	4	Capacity
1	40 / 3	10 / 1	4	5	50
2	7	45 / 3	5 / 8	6	50
3	2	3	55 / 9	20 / 2	75
Requirement	40	55	60	20	175

cells of Table 9.2. To summarize, the rule is to *assign an amount to Route ij* (R_{ij}), *i.e., from Source i* (S_i) *to Destination j* (D_j), *which either exhausts the available supply at* S_i *or completely fulfills the remaining demand of* D_j, *whichever is smaller, and then proceed to* $R_{i+1,j}$ *in the case of the former or to* $R_{i,j+1}$ *in the case of the latter.* This rule quite obviously produces a basic solution wherein x_{ij} are all differences between partial sums of a_i and partial sums of b_j as required by (9.2.9). For instance, in this specific solution just determined,

$$x_{11} = \min\{a_1, b_1\} = \min\{50, 40\} = 40,$$
$$x_{12} = \min\{a_1 - b_1, b_2\} = \min\{10, 55\} = 10,$$
$$x_{22} = \min\{a_2, b_2 - a_1 + b_1\} = \min\{50, 45\} = 45,$$
$$x_{23} = \min\{a_2 - b_2 + a_1 - b_1, b_3\} = \min\{5, 60\} = 5,$$
$$x_{33} = \min\{a_3, b_3 - a_2 + b_2 - a_1 + b_1\} = \min\{75, 55\} = 55,$$
$$x_{34} = \min\{a_3 - b_3 + a_2 - b_2 + a_1 - b_1, b_4\} = \min\{20, 20\} = 20.$$

Notice that according to the Northwest Corner Rule, the value of each basis variable is determined so as to satisfy one constraint. Due to the assumption of $\sum_{i=1}^{m} a_i = \sum_{j=1}^{n} b_j$, however, the last basis variable assumes a value which satisfies sumultaneously both the last capacity constraint and the last requirement constraint. This method, therefore, always yields $m + n - 1$ basis variables, i.e., one less than the $m + n$ constraints in the problem, and consequently always guarantees a basic feasible solution as defined in Section 9.2 above. Nonetheless, it has the serious drawback of complete divorcement from considerations of cost, a defect which usually leads to prolonged iterative process and slow convergence to the optimal solution. There are many methods that attempt to correct this. Of these perhaps the most elaborate is Vogel's Approximation Method which will be discussed in the remainder of this section.

The core of Vogel's method is the idea that a penalty will be incurred if the best (lowest-cost) route is not used for a source or a destination. Thus, for each source and destination we compute a 'penalty' rating which is the difference in cost of the two cheapest routes for that source or destination. For instance, for S_1 in Table 9.1 the two cheapest routes are R_{11} and R_{12} with a difference of 2 between them. This "penalty" rating of S_1 is then recorded. After all ratings have been computed and recorded, we choose that source or destination with the highest rating to attend to first on the theory that it is more important to avoid the high penalty associated with a wrong assignment there. For the sake of clarity let this be S_i. We then assign the minimum of a_i and b_j to R_{ij} whose c_{ij} is the lowest for S_i. In other words, we let $x_{ij} = \min \{a_i, b_j\}$. This eliminates either one source or one destination from further consideration. If S_i is eliminated, then b_j of D_j will be reduced by a_i to reflect the diminished unfulfilled requirement of D_j. On the other hand, if D_j is eliminated, then a_i of S_i will be reduced by b_j for similar reasons. This adjustment of the row or column total concludes the determination of one basis variable. We are now ready for the determination of the next basis variable by repeating the same steps. This process is continued, with each cycle bringing about the elimination of one row or one column until the last row and the last column are eliminated simultaneously as a result of the determination of the last [or the $(m + n - 1)$th] basis variable. The initial basic feasible solution is then complete.

For the sake of illustration, we reproduce in Tables 9.3a through 9.3e the steps leading to an initial basic feasible solution of the coal-shipment problem (Table 9.1) by Vogel's Approximation Method. In Table 9.3a, for instance, the highest "penalty" rating is 4 for Plant 3 (D_3) and R_{13} is the cheapest route of supply. Hence we let $x_{13} = \min \{a_1, b_3\} = \min \{50, 60\} = 50$. This assignment accordingly eliminates S_1 (shaded in Table 9.3b) and reduces the total requirement of D_3 to $b_3 - a_1$ or 10 as shown in Table 9.3b. Repeating the same pro-

Table 9.3a

S_i \ D_j	D_1	D_2	D_3	D_4	Penalty rating
S_i	3	1	(4)	5	2
S_2	7	3	8	6	3
S_3	2	3	9	2	0
P.R.	1	2	4	3	

S_i \ D_j	D_1	D_2	D_3	D_4	a_i
S_i			50		~~50~~
S_2					50
S_3					75
b_j	40	55	~~60~~	20	
			10		

Table 9.3b

S_i \ D_j	D_1	D_2	D_3	D_4	Penalty rating
S_1					
S_2	7	3	8	6	3
S_3	(2)	3	9	2	0
P.R.	5	0	1	4	

S_i \ D_j	D_1	D_2	D_3	D_4	a_i'	
S_1			50		0	
S_2					50	
S_3	40				~~75~~	35
b_j'	~~40~~	55	10	20		

Table 9.3c

S_1					
S_2		3	8	6	3
S_3		3	9	(2)	1
P.R.		0	1	4	

S_1			50		0	
S_2					50	
S_3	40			20	~~35~~	15
b_j'	0	55	10	~~20~~		

Table 9.3d

S_1					
S_2		3	8		5
S_3		(3)	9		6
P.R.		0	1		

S_1			50		0
S_2					50
S_3	40	15		20	~~15~~
b_j'	0	~~55~~	10	0	
		40			

Table 9.3e

S_i \ D_j	D_1	D_2	D_3	D_4	Penalty rating
S_1					
S_2		(3)	(8)		5
S_3					
P.R.					

S_i \ D_j	D_1	D_2	D_3	D_4	a_i'
S_1			50		0
S_2		40	10		~~50~~
S_3	40	15		20	0
b_j'	0	~~40~~	~~10~~	0	

cedure for the remaining sources and destinations, we find that D_1 now has the highest penalty rating and accordingly assign its entire requirement (b_1) to its cheapest route R_{31}. This eliminates D_1 from further consideration and reduces the available supply of S_3 to 75 − 40 = 35. In this manner the plants and mines are successively assigned shipments of coal until as in Table 9.3e the last available supply at S_2 is exactly divided between the two remaining requirement deficits at D_2 and D_3, thus completing the initial basic feasible solution. In Vogel's Approximation Method, again each assignment coincides with the elimination of one source or one destination with the exception of the last. Hence, normally we have $m + n - 1$ basis variables in the initial solution, thus constituting a basic feasible solution. It is of interest to note that the Northwest Corner Rule yields an initial solution carrying a total cost of \$840,000 whereas Vogel's initial solution costs only \$565,000. The contrast clearly shows the merit of Vogel's method although its somewhat more lengthy process must also be borne in mind.

In either method if a particular assignment other than the final one causes both a source and a destination to be simultaneously eliminated, then the initial solution will eventually contain less than $m + n - 1$ basis variables. This signifies that the solution is degenerate. In such cases it is important that either the source be retained with zero supply or the destination be retained with zero demand so as to maintain continuity. The effect of this measure is that subsequently a basis variable of zero value will be determined to make up the full basis. After the initial basic feasible solution is established we may improve upon it by methods discussed in Sections 9.4 and 9.5.

9.4. The Optimal Solution via the Long Method.

As explained in Section 9.1 we will attempt to develop the optimizing method of this section by means of stressing its fundamental economic rationale rather than by reliance on its mathematical affinity to the simplex method of solving general linear programming problems. However, from time to time the reader's attention will be called to the corresponding concepts, rules and operations in the simplex method with which he is assumed to be familiar by now. The exact equivalence of the two methods may advantageously be left to the reader to work out as a test of his understanding.

The basic approach in solving the transportation problem is in fact the same as that employed by the simplex method. First, an initial basic feasible solution is obtained. Then a new and better basic feasible solution is determined by means of the substitution of a basis variable by a non-basis variable, which tends to improve the value of the objective function. In the transportation problem a basis variable identifies a route (from specific source to specific destination) that is in use while a non-basis variable identifies a route that is not in use. The problem is a composite of three parts: (1) how can non-basis routes be evaluated for their effects on the transportation costs; (2) how can a favorable non-basis route be inserted into the basis to obtain a new basic feasible solution; and (3) how can an optimal solution be recognized so that iteration may be ended? Obviously the second question is answered by Decision Rule I and the various transformation rules if the simplex method is employed. The first and third questions also have their answers in Decision Rule II as discussed in Chapter IV. However, it must be realized that the use of the simplex method in a transportation problem would require a tableau with $m + n - 1$ rows and mn columns. It would be enormous even for a small problem, say 10 sources and 20 destinations. Fortunately, due to the special properties discussed in Section 9.2, Dantzig developed a much more efficient procedure which can be explained as follows.

Let us return to the coal-shipment problem whose initial solution is exhibited in Table 9.2 and reproduced in Table 9.4a. Scanning from left to right and from top to bottom we notice that the first non-basis route is R_{13} (from Mine 1 to Plant 3). It is clear that if this route is

Table 9.4a
(numbers in the upper left and lower right corners of each cell represent respectively x_i and c_{ij})

S_i \ D_j	D_1	D_2	D_3	D_4	Cap.
S_1	40 / 3	$10-\theta$ / 1	$+\theta$ / 4	5	50
S_2	7	$45+\theta$ / 3	$5-\theta$ / 8	6	50
S_3	2	3	55 / 9	20 / 2	75
Req.	40	55	60	20	175

used, there will be a direct shipping cost of \$4 per ton. On the other hand, the use of this route will make possible a corresponding decrease in shipment via R_{23} which is now in use, thus saving \$8 per ton. This will prevent overshipment to D_3 but at the same time create a surplus at S_2. Fortunately, however, this surplus can be re-assigned to R_{22} (also in the basis) and be made to offset exactly the decrease in shipment via R_{12} (also in the basis) brought about by the use of the non-basis route R_{13} as originally assumed. Cost-wise the increase in shipment via R_{22} means an additional cost of \$3 per ton and the decrease in shipment via R_{12} means a saving of \$1 per ton. In summary then, the shipment of 1 ton by R_{13} calls for a closed-loop adjustment of -1 ton by R_{23}, $+1$ ton by R_{22} and finally -1 ton by R_{12} thereby leaving all row and column totals unchanged. The net cost involved in this rearrangement is $c_{13} - c_{23} + c_{22} - c_{12} = 4 - 8 + 3 - 1 = -2$ or a saving of \$2 per ton.* We assert then that the evaluation of R_{13} is favorable and that its inclusion in the basis after all

* This method of evaluating a non-basis route is nothing more than determining

$$z_j - c_j = \sum_{(i)=(1)}^{(m)} c_{(i)}y_{(i)j} - c_j$$

in disguise. It will be noted that $y_{(i)j}$ are always ± 1 or 0 in a transportation problem.

necessary adjustments are made will bring about a saving of \$2 per ton shipped by the new route. In linear programming language the routes involved in the reshuffling comprise what is known as a *loop* (dotted line in Table 9.4a). Without going into details, we know that each non-basis route is associated with a unique loop composed of basis routes only (discounting the non-basis route itself). The reader may satisfy his own curiosity by proving it in terms of the concept of a basis. In simplex language the loop is composed of those basis vectors (routes) represented by ± 1 in the Y vector (of linear-combination coefficients) of the non-basis vector (route) being evaluated.

Now that R_{13} is picked to go into the basis, the next question is how much can be and should be shipped through it. Since each ton promises a saving of \$2, naturally as many tons as possible should be shipped through R_{13} in order to minimize costs. However, each ton shipped through R_{13} also calls for a concurrent decrease of one ton shipped through R_{23} and R_{12}. Hence, the maximal amount that can be re-routed is the minimum of x_{23} and x_{12}, or one of them will turn negative which is of course taboo in linear programming. Designating this maximal amount by θ, we can thus write: $\theta = \min \{x_{23}, x_{12}\} = \min \{5, 10\} = 5$, and add or subtract this amount from basis routes in the loop as indicated in Table 9.4a. These adjustments will bring x_{23} to zero, thereby removing it from the basis and generating a new basic feasible solution as desired. To an observant reader, this part of the transportation routine is clearly the counterpart of Decision Rule I (4.2.6) and (4.2.7), (4.2.8) which determine the new values of the basis variables.

We are now ready for the third question: Is the new solution optimal? To answer this, we once again evaluate the non-basis routes in the new solution systematically (Table 9.4b). The first one considered is R_{14} which is associated with a loop that includes R_{13}, R_{33} and R_{34} (not indicated in Table 9.4b. Its direct cost is $c_{14} = 5$ and its indirect saving is $c'_{14} = c_{13} - c_{33} + c_{34} = 4 - 9 + 2 = -3$. Hence, the net cost is $c_{14} - c'_{14} = 5 - (-3) = 8$, i.e., the use of R_{14} will cost an additional \$8 per ton as compared with the present shipping program. Obviously this route should not be used and we pass on to evaluate R_{21}, R_{23} (which is unfavorable since it has just been removed from the basis) and R_{24}. On similar grounds we reject all of them. But in R_{31} we find a favorable non-basis route, as its evaluation is $c_{31} - c'_{31} = c_{31}$

Table 9.4b

S_i \ D_j	D_1	D_2	D_3	D_4	Cap.
S_1	$40 - \theta$ / 3	5 / 1	$5 + \theta$ / 4	5	50
S_2	7	50 / 3	8	6	50
S_3	$+\theta$ / 2	3	$55 - \theta$ / 9	20 / 2	75
Req.	40	55	60	20	175

$-(c_{11} - c_{13} + c_{33}) = 2 - (3 - 4 + 9) = -6$ (loop shown in Table 9.4b). Next we ascertain that the maximal amount that can be shifted around this loop is $\theta = \min\{x_{11}, x_{33}\} = \min\{40, 55\} = 40$. Consequently we make the necessary adjustments for the routes in this loop and arrive at the new (and better) basic feasible solution shown in Table 9.4c. Although not essential it is of interest to point out that

Table 9.4c

S_i \ D_j	D_1	D_2	D_3	D_4	Cap.
S_1	3	$5 - \theta$ / 1	$45 + \theta$ / 4	5	50
S_2	7	50 / 3	8	6	50
S_3	40 / 2	$+\theta$ / 3	$15 - \theta$ / 9	20 / 2	75
Req.	40	55	60	20	175

the degree of improvement in each new solution can readily be calculated. In the present case, since the saving is $6 per ton and the amount re-routed is 40 (40,000 tons), the total savings measure up to \$6 × 40,000 = \$240,000.

Repeating the evaluation of non-basis routes in Table 9.4c, we discover that R_{32} may be used to advantage and that the loop consists of R_{33}, R_{13} and R_{12}. Of these three routes in the basis, R_{12} and R_{33} are going to be affected negatively by any re-routing through R_{32}. Hence $\theta = \min \{x_{12}, x_{33}\} = \min \{5, 15\} = 5$. We are thus taken to the new solution in Table 9.4d and finally to the optimal solution in Table 9.4e. In Table 9.4e, the evaluations of all non-basis routes turn out to be unfavorable. It is this criterion that determines the optimality of the solution.

The reader may have noticed that in Table 9.4d, R_{23}, which has been eliminated from the basis in Table 9.4a, is selected to return to the basis again. This is reasonable because the desirability of a specific route is strictly a matter relative to the basis routes currently in use. Whereas R_{23} is relatively unfavorable with respect to the solutions in Tables 9.4b and c, it is relatively favorable with respect to the solution in Table 9.4d. Hence it is chosen to enter into the basis which turns out to be the optimal one.

Table 9.4d

D_j / S_i	D_1	D_2	D_3	D_4	Cap.
S_1	3	1	50 / 4	5	50
S_2	7	$50 - \theta$ / 3	$+\theta$ / 8	6	50
S_3	40 / 2	$5 + \theta$ / 3	$10 - \theta$ / 9	20 / 2	75
Reg.	40	55	60	20	175

The total shipping cost implied by the optimal solution is: $c_{13}x_{13} + c_{22}x_{22} + c_{23}x_{23} + c_{31}x_{31} + c_{32}x_{32} + c_{34}x_{34} = 4 \cdot 50 + 3 \cdot 40 + 8 \cdot 10 + 2 \cdot 40 + 3 \cdot 15 + 2 \cdot 20 = 565$ thousand dollars. Two points are worth noting. The first is that the cheapest route in the whole problem, R_{12}, is not used at all. This may be taken as a strong argument for the merits of the analytical approach over the intuitive to the transportation problem. The other observation is that the optimal

Table 9.4e

S_i \ D_j	D_1	D_2	D_3	D_4	Cap.
S_1	3	1	50 / 4	5	50
S_2	7	40 / 3	10 / 8	6	50
S_3	40 / 2	15 / 3	9	20 / 2	75
Req.	40	55	60	20	175

solution is identical with the initial solution obtained by Vogel's method (Table 9.3e), which seems to speak strongly for that method despite its complexity and relative analytic obscurity.

9.5. The Optimal Solution via the Short Method.

One of the serious drawbacks of the long method discussed above is the need for identifying the loop before a non-basis route can be evaluated. Since any number of these non-basis routes may turn out to be unfavorable, the amount of time and effort invested in their evaluation ought to be kept at a minimum. The short method discussed in this section achieves this objective by postponing loop identification until *after* a favorable route has been found and selected to go into the basis. Thus, instead of numerous loop identifications for many non-basis routes, there is only one loop identification for one non-basis route, viz., the incoming route. The saving in time and effort is considerable as we shall see presently.

Let us denote the coefficient vector of x_{ij} by P_{ij}. From (9.2.5) it is clear that,

$$P_{ij} = \begin{bmatrix} 0 \\ \cdot \\ \cdot \\ \cdot \\ 1 \\ 0 \\ \cdot \\ \cdot \\ \cdot \\ 1 \\ 0 \\ \cdot \\ \cdot \\ \cdot \end{bmatrix} \begin{matrix} \\ \\ \\ \\ \leftarrow i\text{th row} \\ \\ \\ \\ \\ \leftarrow j\text{th row} \\ \\ \\ \\ \\ \end{matrix}$$

In other words all P_{ij} are column vectors whose components are all zeroes except two 1's—one in the ith row and one in the jth row. Furthermore, from (6.2.6) in Chapter VI we know that the evaluation of the non-basis route can be written (with suitable change from single to double subscripts employed in the transportation problem) as follows:

$$z_{ij} - c_{ij} = \tilde{C}B^{-1}P_{ij} - c_{ij},$$

where $\tilde{C}$ is the row vector of the basis variables' price coefficients and B the basis. But due to the above-mentioned special property of P_{ij}, $\tilde{C}B^{-1}P_{ij}$ must be the scalar sum of the ith and jth components of $\tilde{C}B^{-1}$. We shall designate these two components by two variables u_i and v_j called respectively the *row value* and the *column value*. As a result we can write for all non-basis routes

$$z_{ij} - c_{ij} = u_i + v_j - c_{ij}. \tag{9.5.1}$$

To determine the values of these u_i and v_j, let us recall that the evaluation of a basis vector is always zero. Hence, for a *basis route* we have,

$$z_{(ij)} - c_{(ij)} = u_i + v_j - c_{(ij)} = 0, \quad (ij) = \text{index pertaining to basis,}$$

Or, simply

$$u_i + v_j = c_{(ij)}, \quad (ij) = \text{index pertaining to basis.} \tag{9.5.2}$$

Bearing in mind that there are m row values and n column values to be determined but only $m + n - 1$ routes in the basis to which (9.5.2) applies, we realize that (9.5.2) represents an indeterminate system of linear equations which can be solved by assigning any arbitrary value to any one of the u_i, v_j variables. Customarily u_1 is set equal to zero. Substituting $u_1 = 0$ into those equations of (9.5.2) in which u_1 appears, we then determine by simple subtractions from $c_{(1j)}$ the values of the v_j in them. Each of these v_j is then substituted in turn into other equations of the system (9.5.2) to obtain other u_i. In this manner u_i and v_j are alternately determined from each other and the process fans out to embrace all rows and columns since each row and column must be represented by at least one basis route. This phase of the short method terminates, of course, when all of the m row values (u_i) and n column values (v_j) are determined.

As an illustration, let us take the initial solution shown in Table 9.2, and reproduced in Table 9.5a. A zero is first entered for row 1 in

Table 9.5a

S_i \ D_j	u_i \ v_j	D_1	D_2	D_3	D_4	Cap. a_i
		3	1	6	−1	
S_1	0	40 / 3	$10 - \theta$ / 1	θ / 4	5	50
S_2	2	7	$45 + \theta$ / 3	$5 - \theta$ / 8	6	50
S_3	3	2	3	55 / 9	20 / 2	75
Requirement b_j		40	55	60	20	175

the column of u_i. As R_{11} is in the basis, $u_1 + v_1 = c_{11}$ (9.5.2). Hence, $v_1 = c_{11} - u_1 = 3 - 0 = 3$. Similarly v_2 is found to be $c_{12} - u_1 = 1 - 0 = 1$. Since there is no more basis route in row 1, no other v_j can be determined from $u_1 = 0$ and we proceed to determine other u_i from the v_1 and v_2 just obtained. Since there is no other basis route than R_{11} in column 1, no additional u_i is determined from v_1. But from v_2 we can determine $u_2 = c_{22} - v_2 = 3 - 1 = 2$ because R_{22} is in the basis. u_2 then "determines" v_3, and v_3 in turn "determines" u_3 which finally completes the solution of u_i, v_j by yielding v_4. These row and column values are recorded in the column marked u_i and the row marked v_j in Table 9.5a.

Once the row and column values are determined, we are ready to evaluate the non-basis routes in accordance with (9.5.1). For instance, the evaluation of R_{13} is

$$z_{13} - c_{13} = u_1 + v_3 - c_{13} = 0 + 6 - 4 = 2,$$

which is favorable according to simplex interpretation for a minimizing problem. Note that this result coincides with the finding in the previous section by the "loop" method. As a matter of fact, it is a relatively simple task to prove that the two methods are fully equivalent since the costs of the basis routes used in the "loop" method can all be broken down into row and column values defined by (9.5.2). The details, however, are left as an exercise to the reader.

After the incoming route is selected, the loop can then be identified and the maximal amount of transfer determined with subsequent adjustments of the variables in the loop in the same manner as described in the previous section. For the sake of illustration, the results of one complete iteration on the initial solution in Table 9.5a together with the re-determined row and column values of the new solution are exhibited in Table 9.5b. The reader is advised to continue the iterative process until $u_i + v_j \leq c_{ij}$ for all R_{ij} not in the basis. The prevailing solution is the optimal (or minimal) cost solution.

Table 9.5b

S_i \ D_j	u_i \ v_j	D_1	D_2	D_3	D_4	Cap. a_i
		3	1	4	−3	
S_1	0	40 / 3	5 / 1	5 / 4	/ 5	50
S_2	2	/ 7	50 / 3	/ 8	/ 6	50
S_3	5	/ 2	/ 3	55 / 9	20 / 2	75
Requirement b_j		40	55	60	20	175

To summarize, the short method of solving the transportation problem consists of the following steps:

(a) Obtain an initial basic feasible solution by one of the methods discussed in Section 9.3, or some other suitable method not discussed here.

(b) Determine the row and column values by first assuming $u_1 = 0$ and then alternatively determining the other u_i and v_j by the use of (9.5.2).

(c) Evaluate the non-basis routes in some suitable order:

(1) If a non-basis route is found for which $u_i + v_j > c_{ij}$, it should be introduced into the basis. Proceed to (d).

(2) If no non-basis route can be found for which $u_i + v_j > c_{ij}$, then the optimal solution is the last solution obtained and computation ends.

(d) Identify the loop either by visual inspection or by the use of "lineage chart" explained below. Mark those routes in the loop whose shipments will be reduced by the use of the incoming route.

(e) Determine the maximal amount of allowable transfer by ascertaining the minimal shipment carried by those basis routes marked in (d).

(f) Reduce the shipments of the marked basis routes and increase the shipments of all other basis routes in the loop by the amount of transfer. This completes the determination of a new basic feasible solution and iteration begins anew with (b).

It is not always easy to identify the loop of a non-basis route in a sizable transportation problem by visual inspection. A more systematic method especially useful in programming electronic computers to solve transportation problems makes use of the so-called 'lineage chart' of the row and column values. From previous illustrations in this section, we recall that each u_i is determined from some unique v_j and vice versa by the use of (9.5.2). The only exception is u_1 which is customarily assigned some arbitrary value, say zero. Thus, the successive determination of u_i and v_j may be represented by a lineage chart or family tree describing the origin of each u_i and v_j and terminating at u_1. Figure 9.2 shows such a chart for the initial solution exhibited in Table 9.5a. If as before, R_{13} is selected to go into the basis, then by linking u_1 and v_3 (row value of row 1 and column value of column 3, the row and column indicated by R_{13}) a loop is immediately obtained embracing the lineage relationships between u_1 and v_3. According to Figure 9.2, u_1 leads to v_2, implying that the basis route R_{12} is in the loop. Similarly, as v_2 leads to u_2, the basis route R_{22} is in the loop. Finally, as u_2 leads to v_3, R_{23} is also in the loop. Hence, the

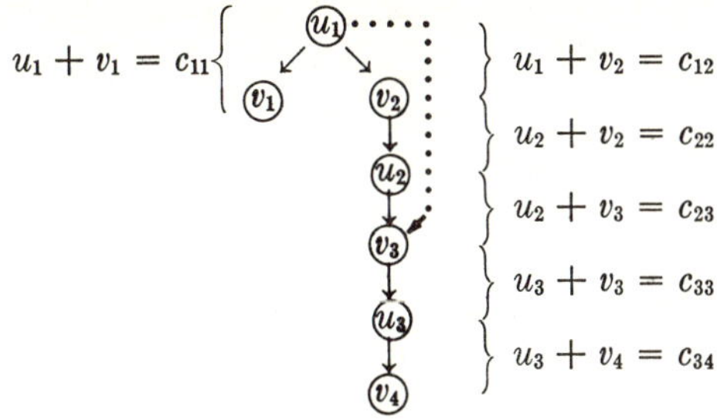

FIGURE 9.2

loop for R_{13} involves R_{12}, R_{22} and R_{23}. No matter how complicated the problem, it is possible to obtain quickly the loop associated with any non-basis route by this method so long as the lineage chart is correctly prepared. When a new solution is found, the new lineage chart can be constructed from the old with equal facility. Just as the introduction of a new basis route, e.g., R_{13}, is represented by the forging of a new link in the lineage chart, e.g., u_1 with v_3, the elimination of an old basis route may be likened to the severance of an old link. Thus, if we assume that R_{23} is eliminated in the new solution, then the link between u_2 and v_3 should be severed and the lineage chart in Figure 9.2 becomes the one shown in Figure 9.3 for the new solution. By this method, the tedious process of re-determining the lineal relationships between all u_i and v_j for each new solution is greatly simplified.

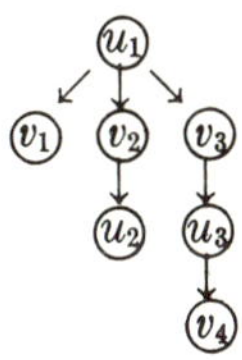

FIGURE 9.3

Finally, it should be noted that the row and column values are not just arbitrary computational devices but bear a close relationship with the dual of the transportation problem. For if we designate the dual variables associated with the first m constraints of (9.2.5), $u_1, u_2, \ldots, u_m$ and those associated with the remaining n constraints, $v_1, v_2, \ldots, v_n$, we may write the dual of the transportation problem as follows:

$$\text{Maximize } g(U, V) = \sum_{i=1}^{m} u_i a_i + \sum_{j=1}^{n} v_j b_j,$$

$$\begin{array}{llllllll}
\text{subject to:} & u_1 & & & + v_1 & & & \leq c_{11}, \\
& u_1 & & & & + v_2 & & \leq c_{12}, \\
& \cdot & & & & \cdot & & \cdot \\
& \cdot & & & & & \cdot & \cdot \\
& \cdot & & & & & \cdot & \cdot \\
& u_1 & & & & & + v_n & \leq c_{1n}, \\
& & u_2 & & + v_1 & & & \leq c_{21}, \\
& & u_2 & & & + v_2 & & \leq c_{22}, \\
& & \cdot & & & \cdot & & \cdot \\
& & \cdot & & & & \cdot & \cdot \\
(9.5.3) & & \cdot & & & & \cdot & \cdot \\
& & u_2 & & & & + v_n & \leq c_{2n}, \\
& & & \cdot & & & & \cdot \\
& & & \cdot & & & & \cdot \\
& & & \cdot & & & & \cdot \\
& & & u_m & + v_1 & & & \leq c_{m1}, \\
& & & u_m & & + v_2 & & \leq c_{m2}, \\
& & & \cdot & & \cdot & & \cdot \\
& & & \cdot & & & \cdot & \cdot \\
& & & \cdot & & & \cdot & \cdot \\
& & & u_m & & & + v_n & \leq c_{mn},
\end{array}$$

The close resemblance between the dual constraints above and (9.5.2) is self-evident. Of course, the row and column values determined from (9.5.2) do not always satisfy all dual constraints in (9.5.3). But the set derived from the optimal basis does. Hence, the final set of the row and column values represents a solution to the dual of the transportation problem. Finally, since according to the Duality Theorem, the optimal value of the primal objective function is equal to the optimal value of the dual objective function, this final set of row and column values represents an optimal solution to the dual and yields a value of the dual objective function (9.5.3) equal to the minimum desired.

9.6. The Personnel Assignment Problem—a Case of Degeneracy.

We recall from Chapter IV Section 4.4 that a degenerate solution is a solution with less than m positive values in the solution vector, where m is the number of structural constraints. Alternatively it means a solution in which one or more basis variables assumes the value of zero. The tendency for degenerate solutions to develop is particularly strong in a transportation problem because the values of basis variables are all differences between partial sums of a_i and partial sums of b_j (9.2.9). Should any partial sum of a_i be equal to some partial sum of b_j, then a degenerate solution could appear even though it may not actually do so. The danger implicit in a degenerate solution is that "cycling" may result as discussed in Chapter IV Section 4.4. To eliminate this danger, Dantzig and Orden have suggested a method involving essentially the addition of a small number ϵ to each a_i and $m \times \epsilon$ to b_n, which mathematically rules out the possibility of equality between partial sums.* However, as we have stated before, the probability of cycling remains small even in a transportation problem where the frequency of degeneracy is high. In most cases the difficulty caused by degeneracy can be successfully overcome by a constant effort to maintain a full basis on the one hand and by the use of a çonsistent (even if arbitrary) rule to break ties on the other. The following Personnel Assignment Problem will serve as an illustration.

The typical Personnel Assignment Problem deals with the question of assigning n individuals to n tasks such that the total efficiency of these men is maximized. We start out by assuming that each man has a certain independently determined efficiency rating for each of the n tasks. The rating of the ith person for the jth task is denoted r_{ij}. We further assume that the *tasks are unrelated* so that the rating of one for a certain task does not depend on who are assigned to the other tasks. Consequently, the total efficiency may logically be regarded as the sum of the separate efficiency ratings of the n individuals for the specific tasks to which they are assigned. With this important reservation, we can formulate the problem as a transportation problem and solve it by essentially the same procedure as described in the earlier sections of this chapter.

Let x_{ij} be the proportion of the ith person's time assigned to the jth task. If he is entirely assigned to the task, x_{ij} will be 1. If he is not assigned to it at all, x_{ij} will be 0. Theoretically x_{ij} is a continuous variable within the range of 0 and 1. But due to the nature of the con-

* *Activity Analysis of Production and Allocation,* ed. Koopmans (New York: John Wiley & Sons, Inc., 1951), pp. 365–367.

straints which we are about to write, x_{ij} will only assume the value of either 0 or 1 in a basic feasible solution. In any event, it is correct to state that the various portions of the ith person's time assigned to the n tasks must add up to the whole. Hence, $x_{i1} + x_{i2} + \ldots + x_{in} = 1$, or simply $\sum_{j=1}^{n} x_{ij} = 1$ for $i = 1, 2, \ldots, n$. This does not preclude over- or under-assignment from the standpoint of the various tasks; therefore we further stipulate that the various portions of the n persons' time assigned to any single task, say the jth, must also add up to the whole. That is to say, $x_{1j} + x_{2j} + \ldots + x_{nj} = 1$, or simply $\sum_{i=1}^{n} x_{ij} = 1$, for $j = 1, 2, \ldots, n$. Since the objective is to maximize the total efficiency rating of n specific assignments, the objective function is $\sum_{i=1}^{n}\sum_{j=1}^{n} r_{ij}x_{ij}$. Combining these considerations with the usual non-negativity constraints, we obtain the following special "transportation" problem:

$$\text{Maximize } f(X) = \sum_{i=1}^{n}\sum_{j=1}^{n} r_{ij}x_{ij},$$

$$\text{(9.6.1)} \qquad \text{subject to:} \quad \sum_{j=1}^{n} x_{ij} = 1, \qquad i = 1, 2, \ldots, n,$$

$$\sum_{i=1}^{n} x_{ij} = 1, \qquad j = 1, 2, \ldots, n,$$

$$\text{and} \qquad x_{ij} \geq 0, \qquad i, j = 1, 2, \ldots, n.$$

This problem differs from the general transportation problem in two respects. One, to repeat a previous statement, is that x_{ij} will be either 0 or 1 in a basic feasible solution. The reason is now clear. It is that x_{ij} must assume values which are differences of partial sums of a_i's and b_j's (9.2.9) and that in this problem a_i and b_j are all ones. This being so, there can be only n positive assignments, i.e., $x_{ij} = 1$, out of a total of $2n - 1$ assignments required for a basic feasible solution. Hence, there must at all times be $n - 1$ zeroes in a basic feasible solution, which means that the personnel assignment problem is inherently degenerate. The second difference is that in a personnel assignment problem, the objective is to maximize total efficiency rather than to minimize total cost. Accordingly, the criterion for selecting a new basis route is $u_i + v_j < r_{ij}$ instead of $> r_{ij}$. With these considerations in mind, let us now turn to a numerical example in which 4 men are to be assigned optimally to 4 tasks for which their separate efficiency ratings are given as follows:

	Tasks (j) 1	2	3	4
Persons (i) 1	200	420	340	400
2	350	450	100	250
3	360	180	200	360
4	300	210	180	300

To obtain an initial solution, we may, of course, follow the Northwest Corner Rule, especially in regard to the question of degeneracy as discussed in Section 9.3. Such a solution could be $x_{11} = 1$, $x_{12} = 0$, $x_{22} = 1$, $x_{23} = 0$, $x_{33} = 1$, $x_{34} = 0$ and $x_{44} = 1$, with total efficiency rating of 1,150. However, we will adopt a rule which makes the first assignment on the basis of the highest efficiency rating in the table. For instance, in the table above the highest efficiency rating is 450, so the first assignment is for the second person to take the second task. After that, we eliminate the row and column already assigned and obtain a reduced efficiency table. The same rule is then applied again to the reduced table in making the second assignment. Further assignments continue in this manner until all personnel are assigned. Using this method, the initial solution for the present example is shown by the 1's in Table 9.6a, which carries a larger total efficiency

Table 9.6a

Total efficiency rating = 1,390

	Tasks (j) 1	2	3	4
Persons (i) 1		0		1
2	0	1		
3	1		0	
4			1	

rating of 1,390. To make up a full basis, we add a zero in the first row and in the column containing a 1 in the second row. Or, in general, a zero is added in the ith row, $(i = 1, 2, \ldots, n - 1)$ and the jth column where $x_{i+1,j} = 1$.* Once the full basic feasible solution is established we may use the short method discussed in Section 9.5 to obtain the optimal solution shown in Table 9.6b. The only modifications of the iterative procedure that have to be introduced in solving the personnel assignment problem are: (1) in selecting the new "route" to be included in the basis, the criterion is $u_i + v_j < r_{ij}$ rather than $> r_{ij}$ and (2) in selecting the old "route" to be removed from the basis: whenever ties exist among the several routes eligible for eviction, it is always the one with the lowest i that is chosen.

Table 9.6b

Total efficiency rating = 1,450

		Tasks (j)			
		1	2	3	4
Persons (i)	1			1	
	2	0	1		
	3				1
	4	1		0	0

If the reader tries to verify the results in Table 9.6b by going through the computation himself, he will probably find that most of the time no improvement in total efficiency is made and that changes are largely confined to shifting zeroes around the basis. In anticipation of his bewilderment as to the efficiency of the method, we hasten to add that better methods, developed by Kuhn and Flood, are available. The interested reader is referred to other published works for the details.**

* This method of introducing zeroes always produces a full basis, for the same reason that the Northwest Corner Rule does as discussed in Section 9.2.

** These methods, based on a study of matrix properties by the Hungarian mathematician E. Egeváry, are known as the "flow technique" or the "Hungarian method of assignment." A competent description may be found in C. M. Churchman, R. L. Ackoff, and E. L. Arnoff, *Introduction to Operations Research* (New York: John Wiley & Sons, Inc., 1957), XII.

9.7. The Production Scheduling Problem—a Case of Imbalance between Total Capacity and Requirement.

So far, our discussion of the transportation problem has been conducted entirely on the basis of an assumed balance between total capacity and total requirement. In practice, however, such a perfect balance is rather unlikely. When an imbalance exists, how can the transportation method be adapted to find the optimal solution? This will be the main concern of the present section which, as an illustration, will discuss the important application of the transportation problem to production scheduling in industry.

On the conceptual level, the question of imbalance is quite easy to answer. If the total capacity of the m sources is greater than the total requirement of the n destinations, i.e., $\sum_{i=1}^{m} a_i > \sum_{j=1}^{n} b_j$, then some excess capacity is inevitable. Denote the idle capacity of the ith source (S_i) by $x_{i,n+1}$. The capacity constraints (9.2.2) then become $\sum_{j=1}^{n} x_{ij} + x_{i,n+1} = a_i$, $i = 1, 2, \ldots, m$. The requirement constraints (9.2.3), remain the same. Also, since $x_{i,n+1}$ represents idle capacity, its cost coefficient must be zero, i.e., $c_{i,n+1} = 0$. Consequently the old objective function (9.2.1) remains unchanged, and we arrive at the following modified transportation problem:

$$\text{Minimize } f(X) = \sum_{i=1}^{m} \sum_{j=1}^{n} c_{ij}x_{ij},$$

$$\text{subject to:} \quad \sum_{j=1}^{n} x_{ij} + x_{i,n+1} = a_i, \qquad i = 1, 2, \ldots, m.,$$

$$\sum_{i=1}^{m} x_{ij} = b_j, \qquad j = 1, 2, \ldots, n., \tag{9.7.1}$$

$$\text{and} \quad x_{ij} \geq 0, \qquad i = 1, \ldots, m, \quad \text{and } j = 1, \ldots, n+1,$$

which can be solved by the standard procedure discussed in either Section 9.4 or 9.5. An interpretation of the above formulation is to treat the excess capacities of the various sources as the requirements of an additional fictitious destination, which can then be handled computationally in the same way as the other m real destinations. This point will be made clear by the production scheduling example to be discussed presently. Meanwhile the reader should understand that an imbalance in the opposite direction, i.e., more requirement than capacity, can be handled in a similar manner by the addition of the variables $x_{m+1,j}$, representing shipments from a fictitious source of supply.

Assume that a manufacturing firm is confronted with the problem of scheduling its regular and overtime production to meet a fixed schedule of sales.* Let the sales requirement be $a_1, a_2, \ldots, a_n$ for the n periods under consideration. Assume that the regular and overtime production capacities are b_i and b_i' respectively for the ith period, and that the regular and overtime production costs are k_i and k_i' dollars per unit respectively in the ith period. Assume further that the cost of holding inventory is d dollars per unit per period. How should production be scheduled in each of the n periods in order to satisfy all sales requirements at minimal cost?

There are two important reasons why this should be a problem. First, there is the physical reason that the sales requirement in some period may exceed the combined regular and overtime production capacity of that period, thereby making planned inventory buildup necessary. Secondly, there is the economic reason that the excess of demand over production can be met by either overtime production or inventory accumulation. A question arises as to what constitutes a logical basis for choosing one or the other. As we shall see, linear programming in the form of a transportation problem may be employed advantageously to solve this problem.

Let us define x_{ij} and y_{ij} as the amounts of regular and overtime production respectively in the ith period, which are sold in the jth period ($j \geq i$). $j - i$ then is the number of periods x_{ij} and y_{ij} have been held in inventory. If c_{ij} represents the unit cost of x_{ij} and c_{ij}' the unit cost of y_{ij} they must each be the sum of the relevant produc-

* The production scheduling problem was first formulated as a transportation problem by E. H. Bowman in "Production Scheduling by the Transportation Method of Linear Programming," *Journal of Operations Research,* IV (1956), No. 1, pp. 100–103, and later extended by A. S. Manne in "A Note on the Modigliana-Hohn Production Smoothing Model," *Management Science,* III (1957), No. 4, pp. 371–379.

tion cost (k_i or k'_i) and the inventory holding cost $[d(j - i)]$. Hence

$$c_{ij} = k_i + d(j - i), \qquad i = 1, 2, \ldots, n; \quad j \geq i, \tag{9.7.2}$$

$$c'_{ij} = k'_i + d(j - i), \qquad i = 1, 2, \ldots, n; \quad j \geq i. \tag{9.7.3}$$

Since no product can be sold before it is produced, we require that x_{ij} and y_{ij} be zero for $i > j$ and insure that it is so by making c_{ij} and c'_{ij} positive infinity for $i > j$. Then there are the capacity constraints for both regular and overtime production in any period: $\sum_{j=1}^{n} x_{ij} \leq b_i$ and $\sum_{j=1}^{n} y_{ij} \leq b'_i$ for $i = 1, \ldots, n$. Finally we assume that the sales requirement in any period must be met: $\sum_{i=1}^{n} (x_{ij} + y_{ij}) = a_j$, $j = 1, 2, \ldots, n$. Consequently we have the following formulation of the production scheduling problem whose structure is that of a transportation problem:

$$\text{Minimize } f(X, Y) = \sum_{i=1}^{m} \sum_{j=1}^{n} (c_{ij}x_{ij} + c'_{ij}y_{ij}),$$

$$\text{subject to:} \quad \sum_{j=1}^{n} x_{ij} \leq b_i, \qquad i = 1, 2, \ldots, n,$$

$$\sum_{j=1}^{n} y_{ij} \leq b'_i, \qquad i = 1, 2, \ldots, n,$$

$$\sum_{i=1}^{m} (x_{ij} + y_{ij}) = a_j, \qquad j = 1, 2, \ldots, n, \tag{9.7.4}$$

$$\begin{aligned} \text{and} \quad & x_{ij}, y_{ij} \geq 0 && \text{for } i, j = 1, 2, \ldots, n, \\ \text{where} \quad & c_{ij}, c'_{ij} = \infty && \text{for } i > j, \\ \text{and} \quad & c_{ij} = k_i + d(j - i) && \text{for } i \leq j, \\ & c'_{ij} = k'_i + d(j - i) && \text{for } i \leq j. \end{aligned}$$

We shall now assume some numerical data (Table 9.7) and solve a simple 4-period scheduling problem for the purpose of illustration. We see that the problem can readily be fitted into the standard transportation format by having two separate rows for the regular and overtime production in each ith production period, and one column for each jth sales period. For the present problem with 4 planning periods, we will begin with 8 rows and 4 columns. A 5th column is then set up to allow for the slack variables representing the idle (or excess) regular and overtime capacities in each of the 4 periods. These together with the usual data on capacity and sales requirement for

Table 9.7

Period	1	2	3	4
b_i	100	80	100	100
b_i'	30	20	20	30
a	80	80	110	120
k_i	5	6	5	6
k_i'	7	8	8	8

each period give us Table 9.7a. As before, the unit cost data, i.e., c_{ij} and c'_{ij} computed from (9.7.2) and (9.7.3), are entered in the lower right corner of each cell. The inventory holding cost d is assumed to be 1 dollar per unit per period. Since x_{ij} is restricted to zero for $i > j$,

Table 9.7a

Production Period (i)		u_i \ v_j	Sales period (j) 1	2	3	4	Excess capacity	Capacity
			5	6	8	8	0	
1	Reg.	0	80 / 5	6	7	8	20 / 0	100
	O.T.	0	7	8	9	10	30 / 0	30
2	Reg.	0		80 / 6	7	8	0 / 0	80
	O.T.	0		8	9	10	20 / 0	20
3	Reg.	−3			100 / 5	6	0	100
	O.T.	0			10 / 8	9	10 / 0	20
4	Reg.	−2				100 / 6	0	100
	O.T.	0				20 / 8	10 / 0	30
Sales requirement			80	80	110	120	90	480

we simply remove those cells from further consideration by shading them. Now we are ready to proceed with the solution.

To obtain an initial solution we follow the rule of thumb that production will be scheduled for the sales requirement in the current period first, using regular time and then overtime if required. Only when the combined capacity of regular and overtime production still falls short of the current sales requirement will inventory (representing production in previous periods) be dipped into. Accordingly, we schedule production from the last period back to the first, always assigning as much of the current sales requirement as possible to the cell representing regular time production, then as much of the excess requirement as possible (if any) to the overtime cell, and finally, if there is still unsatisfied demand, scheduling it for production in the period immediately preceding. This yields the initial production schedule, $x_{44} = 100$, $y_{44} = 20$, $x_{33} = 100$, $y_{33} = 10$, $x_{22} = 80$, $x_{11} = 80$, with the following idle capacities: $y_{45} = 10$, $y_{35} = 10$, $y_{25} = 20$, $y_{15} = 30$ and $x_{15} = 20$. These are entered in the upper left corners of their respective cells in Table 9.7a. This problem has 13 constraints calling for 12 positive components in a basic feasible solution, but only 11 have been identified in the initial solution. Hence, one zero is assigned to x_{25} to complete the basic feasible solution.* The short method is then applied to determine the row and column values shown in Table 9.7a starting with the assumption of $u_1 = 0$. Since we are minimizing cost, we discover that x_{13} could be brought into the basis with advantage as $u_1 + v_3 \geq c_{13}$. The subsequent adjustments around the loop then bring about the solution in Table 9.7b, which proves to be optimal. In contrast with the \$2,220 total cost of the initial solution, the minimal total cost is \$2,210. The \$10 reduction is the result of replacing the 10-unit overtime production in period 3, with the same amount of regular-time production in period 1 for sale in period 3, i.e., x_{13} in place of y_{33}. Apparently it is cheaper to do so even if the output has to be held in inventory for two periods before it is sold.

The above example is a simple one, for according to the assumed data, the cumulative production capacity in any period is always

---

* This zero may be added to any unassigned cell so long as its addition makes possible the determination of a unique set of row and column values starting with, say, $u_1 = 0$. Thus, instead of x_{25}, y_{22} or y_{12} may be added to the basis at zero value; but not y_{11}, x_{35} or x_{45}, for these three will all fail to bring about a unique and complete determination of u_i and v_j.

Table 9.7b

Production Period (i)		u_i \ v_j	Sales Period (j) 1	2	3	4	Excess capacity	Capacity
			5	6	7	8	0	
1	Reg.	0	80; 5	6	10; 7	8	10; 0	100
1	O.T.	0	7	8	9	10	30; 0	30
2	Reg.	0	(shaded)	80; 6	7	8	0; 0	80
2	O.T.	0	(shaded)	8	9	10	20; 0	20
3	Reg.	−2	(shaded)	(shaded)	100; 5	6	0	100
3	O.T.	0	(shaded)	(shaded)	8	9	20; 0	20
4	Reg.	−2	(shaded)	(shaded)	(shaded)	100; 6	0	100
4	O.T.	0	(shaded)	(shaded)	(shaded)	20; 8	10; 0	30
Sales requirement			80	80	110	120	90	480

greater than the cumulative sales requirement in that period. That is to say, $\sum_{i=1}^{k} (b_i + b_i') \geq \sum_{i=1}^{k} a_i$ for $k = 1, \ldots, 4$. It is possible to extend the problem to allow for the opposite condition when cumulative sales requirement exceeds cumulative production capacity in some period. In such instances, backlog of demand will appear as x and y entries in the shaded region of Table 9.7b. If some rational penalty coefficients could be obtained, possibly as a function of the number of periods delivery is delayed, a meaningful scheduling problem could be designed and solved with the objective of minimizing the total of production cost and penalty. If the cumulative sales requirement of the entire planning interval of 4 periods is greater than the corresponding cumulative production capacity, then the problem calls for the addition of a fictitious source instead of a fictitious demand as in the example just completed.

9.8. Transhipment Problem.

Whereas the ordinary transportation problem permits shipment only from source to destination, it is of interest to consider the broader class of transportation problems, called transhipment problem, which allows shipment both from source to source and from destination to destination. From the standpoint of business and industry, this is a more realistic statement of the distribution problem confronting them. Indeed, a firm may not infrequently find it necessary to ship its products from one warehouse to another in order to meet an abnormal rise in demand in the second supply area. In such cases the second warehouse behaves both as a source and as a destination, and the distinction between source and destination is blurred. As a matter of fact, an ordinary m-source and n-destination transportation problem, when converted into a transhipment problem by removing the restrictions on the sources and destinations to receive and send respectively, becomes one of $m + n$ sources and $m + n$ destinations. Thus, the problem is enlarged. But, with minor modifications, it can still be solved by the standard transportation method as we shall see presently.*

Let us designate these sources and destinations in a transhipment problem as terminals (T). The amount shipped from T_i to T_j is designated x_{ij}. For obvious reasons, the amount of shipment from one terminal to itself should be nothing. Hence, $x_{ii} = 0$. Assume that at m terminals ($T_1, T_2, \ldots, T_m$) total out-shipment exceeds total in-shipment by the amounts $a_1, a_2, \ldots, a_m$, respectively, and at the remaining n terminals ($T_{m+1}, T_{m+2}, \ldots, T_{m+n}$), total in-shipment exceeds total out-shipment by the amounts $b_{m+1}, b_{m+2}, \ldots, b_{m+n}$ respectively. Let the total in-shipment at $T_1, T_2, \ldots, T_m$ be $t_1, t_2, \ldots, t_m$ respectively and the total out-shipment at $T_{m+1}, \ldots, T_{m+n}$ be $t_{m+1}, \ldots, t_{m+n}$ respectively. We can then write the following constraints for the transhipment problem where $\sum_{i=1}^{m} a_i = \sum_{j=m+1}^{m+n} b_j$ and whose objective function is $\sum_{i}^{m+n} \sum_{j}^{m+n} c_{ij}x_{ij}$.

* The transhipment problem was first considered by A. Orden in "The Transhipment Problem," *Management Science*, II (1956), No. 3, pp. 276–285.

(9.8.1) $$x_{i1} + \dots + x_{i,i-1} + x_{i,i+1} + \dots + x_{i,m+n} - t_i = a_i, \quad i = 1, \dots, m,$$

(9.8.2) $$t_i = x_{1i} + \dots + x_{i-1,i} + x_{i+1,i} + \dots + x_{m+n,i}, \quad i = 1, \dots, m,$$

(9.8.3) $$x_{1j} + \dots + x_{j-1,j} + x_{j+1,j} + \dots + x_{m+n,j} - t_j = b_j, \quad j = m + 1, \dots, m + n,$$

(9.8.4) $$t_j = x_{j1} + \dots + x_{j,j-1} + x_{j,j+1} + \dots + x_{j,m+n}, \quad j = m + 1, \dots, m + n.$$

These constraints are very similar to those of a transportation problem with $m + n$ sources and $m + n$ destinations. The only differences are: (1) in the transhipment constraints there are no x_{ii} or x_{jj} terms and (2) it can be inferred from (9.8.2) that $b_j = 0$ for $j = 1, 2, \dots, m$, and similarly from (9.8.4) that $a_i = 0$ for $i = m + 1, m + 2, \dots, m + n$. However, the t_i and t_j in these constraints may be looked upon as algebraic substitutes for x_{ii} and x_{jj}. With this one change, the transhipment constraints can be equated completely to an enlarged set of transportation constraints and solved within the framework of an enlarged transportation problem.

Such a recasting of the transhipment problem is shown in Table 9.8. The m constraints in (9.8.1) are represented by the first m rows

Table 9.8

Terminals		T_j				Capacity
		$1 \dots$	m	$m+1 \dots$	$m+n$	
T_i	1	$-t_1 \dots$	x_{1m}	$x_{1,m+1} \dots$	$x_{1,m+n}$	a_1
	⋮	⋮ ⋱	⋮	⋮	⋮	⋮
	m	$x_{m1} \dots$	$-t_m$	$x_{m,m+1} \dots$	$x_{m,m+n}$	a_m
	$m+1$	$x_{m+1,1} \dots$	$x_{m+1,m}$	$-t_{m+1} \dots$	$x_{m+1,m+n}$	0
	⋮	⋮	⋮	⋮ ⋱	⋮	⋮
	$m+n$	$x_{m+n,1} \dots$	$x_{m+n,m}$	$x_{m+n,m+1} \dots$	$-t_{m+n}$	0
Requirement		$0 \dots$	0	$b_{m+1} \dots$	b_{m+n}	$\Sigma a_i = \Sigma b_j$

of the table; (9.8.4) by the remaining n rows; (9.8.2) by the first m columns and finally (9.8.3) by the remaining n columns. All t's are placed on the main diagonal with a negative sign so that the t's as well as the x_{ij}'s may be constrained to non-negative values. The procedure for solving this problem is identical with that used for solving a transportation problem. Due attention must be given to the sign of t, however, when it is involved in readjustment as follows:

Example. Let us solve the coal shipment example in Section 9.3 as a transhipment problem with the following additional cost data:

	D_1	D_2	D_3	D_4
D_1	0	2	3	2
D_2	2	0	1	3
D_3	3	1	0	2
D_4	2	3	2	0

	S_1	S_2	S_3
S_1	0	2	3
S_2	2	0	1
S_3	3	1	0

We will use the final solution in Table 9.4e as the initial solution to the transhipment problem and obtain Table 9.9a. The row and column values determined according to the short method (Section 9.5) then reveal that "route" R_{56} (D_2 to D_3) ought to be substituted for route R_{26} (S_2 to D_3) in the loop indicated. Since t has a minus sign in a transhipment problem, the maximal amount of transfer is limited only by x_{26} in this case. After adjusting the relevant shipments for

Table 9.9a

Terminals	u_i \ v_j	$T_1(S_1)$	$T_2(S_2)$	$T_3(S_3)$	$T_4(D_1)$	$T_5(D_2)$	$T_6(D_3)$	$T_7(D_4)$	Capacity
		0	−4	−4	−2	−1	4	−2	
$T_1(S_1)$	0	−0 / 0	2	3	3	1	50 / 4	5	50
$T_2(S_2)$	4	2	−0 / 0	1	7	$40+\theta$ / 3	$10-\theta$ / 8	6	50
$T_3(S_3)$	4	3	1	−0 / 0	40 / 2	15 / 3	9	20 / 2	75
$T_4(D_1)$	2	3	7	2	−0 / 0	2	3	2	0
$T_5(D_2)$	1	1	3	3	2	$-(0+\theta)$ / 0	$+\theta$ / 1	3	0
$T_6(D_3)$	−4	4	8	9	3	1	−0 / 0	2	0
$T_7(D_4)$	2	5	6	2	2	3	2	−0 / 0	0
Requirement		0	0	0	40	55	60	20	175

Table 9.9b

Terminals	u_i \ v_j	$T_1(S_1)$	$T_2(S_2)$	$T_3(S_3)$	$T_4(D_1)$	$T_5(D_2)$	$T_6(D_3)$	$T_7(D_4)$	Capacity
		0	0	0	2	3	4	2	
$T_1(S_1)$	0	−0 / 0	2	3	3	$+\theta$ / 1	$50-\theta$ / 4	5	50
$T_2(S_2)$	0	2	−0 / 0	1	7	50 / 3	8	6	50
$T_3(S_3)$	0	3	1	−0 / 0	40 / 2	15 / 3	9	20 / 2	75
$T_4(D_1)$	−2	3	7	2	−0 / 0	2	3	2	0
$T_5(D_2)$	−3	1	3	3	2	$-(10+\theta)$ / 0	$10+\theta$ / 1	3	0
$T_6(D_3)$	−4	4	8	9	3	1	−0 / 0	2	0
$T_7(D_4)$	−2	5	6	2	2	3	2	−0 / 0	0
Requirement		0	0	0	40	55	60	20	175

this amount, we arrive at the solution in Table 9.9b. 10(000 tons) of coal are now shipped from S_2 to D_3 not directly but by way of D_2. The total cost is thereby reduced from \$565,000 to \$525,000.

Table 9.9c

Terminals	u_i \ v_j	$T_1(S_1)$	$T_2(S_2)$	$T_3(S_3)$	$T_4(D_1)$	$T_5(D_2)$	$T_6(D_3)$	$T_7(D_4)$	Capacity
		0	−2	−2	0	1	2	0	
$T_1(S_1)$	0	−0 / 0	2	3	3	50 / 1	4	5	50
$T_2(S_2)$	2	2	−0 / 0	1	7	50 / 3	8	6	50
$T_3(S_3)$	2	3	1	−0 / 0	40 / 2	15 / 3	9	20 / 2	75
$T_4(D_1)$	0	3	7	2	−0 / 0	2	3	2	0
$T_5(D_2)$	−1	1	3	3	2	−60 / 0	60 / 1	3	0
$T_6(D_3)$	−2	4	8	9	3	1	−0 / 0	2	0
$T_7(D_4)$	0	5	6	2	2	3	2	−0 / 0	0
Requirement		0	0	0	40	55	60	20	175

Repeating the same procedure for the new solution, we find that further improvement can be made by using the route R_{15} (S_1 to D_2) for a maximal transfer of 50 (000 tons). Finally, the optimal solution is displayed in Table 9.9c (also Figure 9.4), which shows that S_1 also supplies D_3 with coal by way of D_2. The total cost is as a result sharply reduced from \$525,000 to \$425,000.

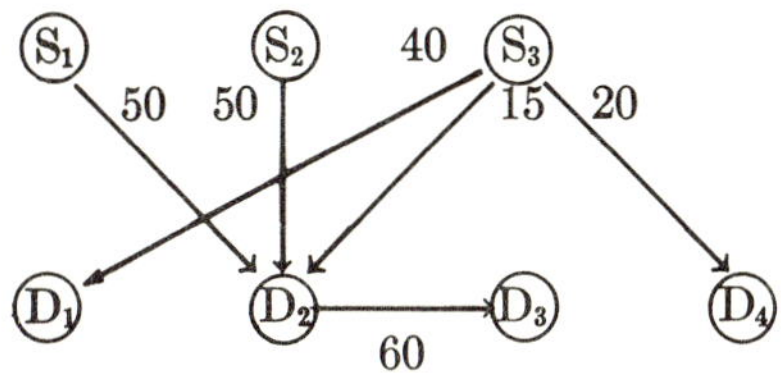

FIGURE 9.4

EXERCISES

1. Given the following cost matrices, and row and column totals, solve the transportation problems by means of the short method, using the Northwest Corner Rule to obtain the necessary intitial solutions:

(a)

S_i \ D_j	1	2	3	4	Row totals
1	2	1	2	4	20
2	1	3	4	3	9
3	1	3	2	5	14
Column totals	7	15	10	11	43

Minimize $f(X)$

(b)

S_i \ D_j	1	2	3	4	Row totals
1	5	1	3	3	34
2	3	3	5	4	15
3	6	4	4	3	12
4	4	−1	4	2	19
Column totals	21	25	17	17	80

Maximize $f(X)$

(c)

S_i \ D_j	1	2	3	4	5	Row totals
1	3	1	5	−2	1	22
2	6	3	4	1	5	13
3	2	0	2	2	4	37
Column totals	19	11	10	12	17	

Minimize $f(X)$

(d)

S_i \ D_j	1	2	3	4	5	Row totals
1	2	5	3	0	3	27
2	−1	1	3	1	5	25
3	5	3	1	2	4	20
4	0	2	3	5	4	13
Column totals	22	23	16	21	10	

Maximize $f(X)$

2. Solve Exercise 1 (c) and (d) using Vogel's method to obtain the initial solutions.

3. The following table shows all the necessary information on the available supply at each warehouse, the requirement of each market and the unit transportation cost from each warehouse to each market. The shipping clerk has worked out the following schedule from experience: 12 units from Warehouse A to Market II, 1 unit from A to III, 9 units from A to IV, 15 units from B to III, 7 units from C to I and 1 unit from C to III.

Warehouse \ Market	I	II	III	IV	Supply
A	5	2	4	3	22
B	4	8	1	6	15
C	4	6	7	5	8
Requirement	7	12	17	9	

(a) Check and see if the clerk has the optimal schedule.
(b) Find the optimal schedule and the minimal total shipping costs.
(c) If the clerk is approached by a carrier of route C-II, who offers to reduce his rate in the hope of getting some business, by how much must the rate be reduced before the clerk should consider giving him an order?

4. One of the popular applications of the transportation problem in industry is the scheduling of work on various machines so that total costs are minimized. For instance, assume that there are four orders to be filled. Order A calls for the production of 40,000 units of Part 106-2; order B, 25,000 units of Part 211-1; order C, 30,000 units of Part 89-12; and order D, 45,000 units of Part 101-8. Each of these parts may be turned out on machine x, y or z at rates and costs indicated by the tables below:

Hourly production rate of → on ↓	A	B	C	D	Hours available
Machine x	200	500	300	300	200
Machiue y	180	450	270	270	200
Machine z	160	400	240	240	250

Unit production cost of → on ↓	A	B	C	D
Machine x	\$0.025	\$0.014	\$0.015	\$0.015
Machine y	\$0.028	\$0.016	\$0.02	\$0.022
Machine z	\$0.03	\$0.016	\$0.02	\$0.02

The problem is: How should the four jobs be distributed among the three machines so that they can be completed at minimum cost?

As a straightforward mathematical formulation of the problem will show, this is normally not a transportation type of problem. However, one may realistically assume that the machines are related in terms of their relative production efficiency. In our case, the hourly production rates of the three machines are such that y may be regarded as 90% as efficient as x, and z only 80% as efficient. Once this relationship is established, the time spent on the different machines can be converted to an equivalent amount of time spent on a "standard" machine. The problem then becomes one of assigning "standard" machine time from various machines to various jobs, which is clearly a transportation problem.

(a) Solve the above machine-loading problem by the transportation method.

(b) Can it still be solved as a transportation problem if certain jobs cannot be done by certain machines? For instance, if C cannot be produced on z and D cannot be produced on y, how would you handle the problem?

5. Show that the row and column values (u_i, v_j) in the optimal solution of a transportation problem are actually the optimal solutions to its dual.

6. Show that the initial solution in Table 9.2 is associated with a set of vectors that comprise a basis.

7. Referring to Table 9.4a, write the non-basis vector associated with x_{13} (shipment from S_1 to D_3) as a linear combination of the basis vectors in the indicated loop. Prove its validity.

8. Referring to Table 9.4e, if the capacity of S_1 and the requirement of D_1 are each increased by 1, would the total shipping cost be greater or less? By how much? Explain.

9. Give an economic interpretation of the dual variables u_i and v_j. Of $u_i + v_j$.

10. Prove that in a transportation problem, $\sum_j \sum_i c_{ij}x_{ij} = \sum_i u_i a_i + \sum_j v_j b_j$.

11. Solve the following production scheduling problem by the method described in Section 9.7. What is the minimum cost?

Period	1	2	3	4	5	6
b_i	120	100	90	140	100	110
b_i'	30	40	20	40	30	30
a_j	110	160	100	100	120	200
k_i	6	6	8	5	7	5
k_i'	7	8	9	6	8	8

Here b_i is regular time capacity, b_i' is overtime capacity, a_j is sales requirement, k_i is regular time cost per unit and k_i' is overtime cost per unit. Inventory cost is assumed to be $\frac{1}{2}$ per unit per period.

12. In practice, the scheduling problem discussed in Section 9.7 can be solved in one pass (without iteration) by the following procedure.

(1) Schedule production from sales period 1 to sales period n in that order.

(2) Schedule production in each sales period by assigning as much as the available capacity permits to the lowest-cost cell first, then to cells with progressively higher costs (provided capacity is available) until the requirement of that period is completely satisfied.

(3) Repeat (2) for the next period until all requirements are scheduled.

(a) Try the procedure on Exercise 11.

(b) Prove its validity.

ADVANCED TOPICS IN LINEAR PROGRAMMING

10.1. Introduction.

In this chapter we shall study three of the more advanced topics in linear programming: the bounded-variable problem, integer programming and nonlinear programming. Our purpose is to familiarize the reader with the advancing frontier of linear programming knowledge and to provide him with a bridge to cross into territories that are not yet fully developed. It is hoped that the reader will be stimulated to strike out on his own, so to speak, and venture into the current literature where the latest progress is to be found.

In a bounded-variable problem, the variables are limited not only by the non-negativity constraints on the lower side but also by specific values on the upper side, called upper bounds. These upper bounds on individual variables do not present any new conceptual problem. Such a constraint $x_j \leq d_j$ may simply be regarded as a special case of the general structural constraint $a_{j1}x_1 + a_{j2}x_2 + \ldots + a_{jj}x_j + \ldots + a_{jn}x_n \leq d_j$, where $a_{j1}, a_{j2}, \ldots, a_{jn}$ are all equal to zero except a_{jj} which is equal to 1. The difficulty is rather a computational one since the additional upper-bound constraints will significantly enlarge the basis and the simplex tableau. To overcome this difficulty, we shall discuss a method of Charnes and Lemke,* which calls for a modification of the simplex decision and transformation rules rather than a change in the size of the tableau in solving the bounded-variable problem. This will form the subject matter of Section 10.2 in this chapter.

The integer programming problem is an ordinary linear programming problem with the additional condition that all or part of the variables must assume integer values. This type of problem is particularly important in business and industry because of the discreet nature of the variables involved in many decisions. For instance, in manufacturing, production is frequently scheduled in terms of batches, lots, or runs; in distribution, a shipment must involve a discrete number of trucks, barges, airplanes or freight cars. Although it is sometimes sufficient to have an approximate optimal solution which includes fractional values, the need for a method to determine the exact integer optimum remains both compelling to the practitioners and challenging to the theoreticians. Furthermore, some complex programming problems involving either nonlinear objective functions or nonlinear constraints are known to be transformable into linear programming counterparts with some of their variables constrained to be integers. The fixed-charge problem, problems with alternative constraint sets (dichotomies), or with conditional constraints, and many others all belong to this category.** It represented a significant step forward, therefore, when Gomory published his method of solving integer problems in 1958. In Section 10.3, we shall explain his method as well as make some general remarks on an alternative method proposed more recently by Land and Doig.

The third and last topic, to be discussed in Section 10.4, is nonlinear programming. This is a vast tract of virgin land, up to now only scantily charted and explored. By nonlinear programming we mean the broad class of programming problems which contain either a nonlinear objective function or a set of nonlinear constraints or both. The vastness of its scope and the frequent changes on its many fron-

* A. Charnes, and C. E. Lemke: "The Bounded Variables Problem," *ONR Research Memorandum,* No. 10, (Pittsburgh: Carnegie Institute of Technology, 1954).

** Cf. Dantzig, G. B., "On the Significance of Solving Linear Programming Problems with Some Integer Variables," *Econometrica,* XXVII (1960), No. 1, pp. 30–45.

tiers preclude a comprehensive treatment in this introductory text. However, we have already intimated that some of the nonlinear problems can be solved as integer programming problems. Therefore, in Sections 10.4 and 10.5 we shall only explain the general conceptual problems involved in nonlinear programming and discuss in some detail two specific nonlinear programming problems, (1) quadratic concave programming and (2) convex programming. These both call for the optimization of a nonlinear objective function (quadratic concave in one case and convex in the other) over a convex solution set defined by a system of linear inequalities.

10.2. The Bounded-Variable Problem.

Mathematically a bounded-variable problem is as follows:

$$
(10.2.1)\qquad
\begin{array}{lll}
\text{(a) Maximize } f(X) = & c_1x_1 + \dots + c_nx_n, & \\
\text{(b) subject to:} & a_{11}x_1 + \dots + a_{1n}x_n = b_1, & \\
 & \vdots & \\
 & a_{m1}x_1 + \dots + a_{mn}x_n = b_m, & \\
\text{(c)} & x_j \leq d_j, & j = 1, \dots, n, \\
\text{(d) and} & x_j \geq 0, & j = 1, \dots, n.
\end{array}
$$

The (a), (b) and (d) parts of (10.2.1) are, of course, the usual components of a straightforward linear programming problem. Part (c) which specifies the upper bounds (d_j) of x_j is new and will be referred to as the *upper-bound constraints*. By adding the necessary slack variables, $u_1, \dots, u_n$, the above problem can be rewritten as follows:

$$
(10.2.2)\qquad
\begin{array}{llllll}
\text{(a) Max } f(X) = & c_1x_1 + \dots + c_nx_n & & & & \\
\text{(b) subject to:} & a_{11}x_1 + \dots + a_{1n}x_n & & & = b_1, & \\
 & \vdots & & & \vdots & \\
 & a_{m1}x_1 + \dots + a_{mn}x_n & & & = b_m, & \\
\text{(c)} & x_1 & + u_1 & & = d_1, & \\
 & \ddots & \ddots & & \vdots & \\
 & x_n & & + u_n & = d_n, & \\
\text{(d) and} & x_j, u_j \geq 0, & & & j = 1, \dots n. &
\end{array}
$$

For future ease of reference, we will write (10.2.2) also in vector form as shown below:

(10.2.3) (a) $\text{Max } f(X) = c_1x_1 + \ldots + c_nx_n,$

(b) subject to:

$$\overline{P}_1x_1 + \ldots + \overline{P}_nx_n + \overline{Q}_1u_1 + \ldots + \overline{Q}_nu_n = \overline{P}_0,$$

(c) and $x_j, u_j \geq 0 \quad j = 1, \ldots, n,$

where

$$\overline{P}_j = \begin{bmatrix} a_{1j} \\ \cdot \\ \cdot \\ \cdot \\ a_{mj} \\ \hdashline 0 \\ \cdot \\ \cdot \\ \cdot \\ 1 \\ \cdot \\ \cdot \\ \cdot \\ 0 \end{bmatrix} = \begin{bmatrix} P_j \\ \hdashline I_j \end{bmatrix}, \quad \overline{Q}_j = \begin{bmatrix} 0 \\ \cdot \\ \cdot \\ \cdot \\ 0 \\ \hdashline 0 \\ \cdot \\ \cdot \\ \cdot \\ 1 \\ \cdot \\ \cdot \\ \cdot \\ 0 \end{bmatrix} = \begin{bmatrix} O \\ \hdashline I_j \end{bmatrix}, \quad j = 1, \ldots, n.$$

for x_j for u_j

$$\text{and } \overline{P}_0 = \begin{bmatrix} b_1 \\ \cdot \\ \cdot \\ \cdot \\ b_m \\ \hdashline d_1 \\ \cdot \\ \cdot \\ \cdot \\ d_n \end{bmatrix} = \begin{bmatrix} P_0 \\ \hdashline D \end{bmatrix}.$$

In the definitions above, $\overline{P}_j$ is given not only in terms of its elements, but also as a composite column vector composed of the two vectors P_j and I_j. The same is truc of $\overline{Q}_j$ and $\overline{P}_0$. It should be clear by inference that I_j is the jth column of the identity matrix I, and D is the column vector of the upper bounds d_1, d_2, etc.

Before we go into Charnes and Lemke's method of solving the bounded-variable problem, we must re-emphasize the point that it is a linear programming problem and can be solved by the standard simplex method. The Charnes and Lemke method is merely a simplified version of the simplex method and can always be explained in terms of the simplex. Therefore, it is very important to keep in mind a vivid impression of the structure of the bounded-variable problem as displayed in (10.2.2) and (10.2.3).

Let us now begin by analyzing the basis of a bounded-variable problem. Since there are altogether $m + n$ constraints (m structural and n upper-bound) there must be $m + n$ vectors in the basis. These basis vectors fall into three mutually exclusive and all inclusive groups with different implications as to the state their corresponding variables are in.

(a) Type 1 basis vectors— These include those $\overline{P}_j$ (or $\overline{Q}_j$) whose counterparts $\overline{Q}_j$ (or $\overline{P}_j$) are also in the basis. In other words, basis vectors of this type exist in pairs in the basis. We shall soon see that there are exactly m pairs of these vectors in the basis. In the meantime, it should be clear that since $\overline{P}_j$ and $\overline{Q}_j$ are both in the basis, x_j and u_j must both be positive, which means that x_j is positive and below its upper bound. That is to say, $0 < x_j < d_j$ as implied by (10.2.2)(c).

(b) Type 2 basis vectors— These include those $\overline{P}_j$ whose $\overline{Q}_j$ are not in the basis. This situation implies that x_j is positive but u_j is zero. According to (10.2.2)(c) again, $x_j = d_j$, which means that the variable x_j is at its upper-bound.

(c) Type 3 basis vectors— These include the $\overline{Q}_j$ in the basis whose corresponding $\overline{P}_j$ are not in the basis. This situation implies that $u_j > 0$ but $x_j = 0$. Therefore, $u_j = d_j$ according to (10.2.2)(c).

It is obvious that in any basic feasible solution, each x_j must be in one of the three states: $x_j = 0$, $0 < x_j < d_j$ and $x_j = d_j$. Since these states are reflected by the three types of basis vectors described above, this classification of basis vectors must be all-inclusive.

To understand why there are always m pairs of Type 1 basis vectors, we may reason in the following manner. First of all, if no x_j in the basis has reached its upper-bound, then the upper-bound constraints are in fact superfluous (non-operative) and the bounded-variable problem becomes an ordinary linear programming problem so far as that particular solution is concerned. Therefore, to satisfy the m structural constraints, there must be m $x_j > 0$ implying m $\overline{P}_j$

in the basis. To satisfy the n upper-bound constraints, there must be also n $\overline{Q}_j$ vectors in the basis, bringing the total number of basis vectors to the required number of $m + n$. Since all $\overline{Q}_j$ are in the basis, the m $\overline{P}_j$ in the basis must exist in pairs. Hence, there are m pairs of Type 1 vectors in such a basis. On the other hand, if there are r variables reaching their respective upper-bounds and s variables equal to zero, so the number of x variables which are positive but below their upper-bounds must be $n - (r + s)$. In order to satisfy the upper-bound constraints, this means that there must be $[s + n - (r + s)]$ $\overline{Q}_j$ vectors in the basis. By assumption we know that there are $[r + n - (r + s)]$ $\overline{P}_j$ vectors in the basis; therefore the total is $[r + n - (r + s)] + [s + n - (r + s)] = 2n - (r + s)$. But as pointed out earlier, the number of basis vector in a bounded-variable problem is $m + n$. Therefore, we have $2n - (r + s) = m + n$, which yields

$$r + s = n - m$$

Or

$$n - (r + s) = m$$

indicating that the number of x_j whose values are between 0 and d_j is m. Since each of these variables is represented by a pair of Type 1 basis vectors, there are at all times m pairs (or $2m$) of Type 1 vectors in the basis.

Going one step further, we may assume without loss of generality that the m pairs of Type 1 basis vectors are $\overline{P}_1, \ldots, \overline{P}_m$ and $\overline{Q}_1, \ldots, \overline{Q}_m$, the r type 2 basis vectors are $\overline{P}_{m+1}, \ldots, \overline{P}_{m+r}$, and the s $(= n - m - r)$ Type 3 basis vectors are $\overline{Q}_{m+r+1}, \ldots, \overline{Q}_n$. It is then convenient for our purpose here to express the bounded-variable problem described in (10.2.2) in terms of sub-matrices A_1, A_2, etc. as shown in Figure 10.1. As the reader will soon discover, such a reformulation is dictated by the composition of the basis assumed above expressly for the purpose of simplifying the description of the basis $\overline{B}$ as shown in (10.2.4). It will be noticed that the submatrix A_1 is square and normally non-singular. Hence, it is in fact thc basis B of the problem with its upper bounds disregarded. Moreover, it has an inverse B^{-1} which makes possible the inversion of the larger basis $\overline{B}$ as shown in (10.2.5). Recalling that $Y = B^{-1}A$ and

Vector Coverage	$\bar{P}_1 \leftrightarrow \bar{P}_m$	$\bar{P}_{m+1} \leftrightarrow \bar{P}_{m+r}$	$\bar{P}_{m+r+1} \leftrightarrow \bar{P}_n$	$\bar{Q}_1 \leftrightarrow \bar{Q}_m$	$\bar{Q}_{m+1} \leftrightarrow \bar{Q}_{m+r}$	$\bar{Q}_{m+r+1} \leftrightarrow \bar{Q}_n$	$\bar{P}_0$
Relation with Basis	In, Type 1	In, Type 2	Out	In, Type 1	Out	In, Type 3	
Row Dimension Column Dimension	m	r	$n-m-r$	m	r	$n-m-r$	1
m	A_1	A_2	A_3	O	O	O	P_0
m	I	O	O	I	O	O	D_1
r	O	I	O	O	I	O	D_2
$n-m-r$	O	O	I	O	O	I	D_3

where:

$$A_1 = \begin{bmatrix} a_{11} & \dots & a_{1m} \\ \cdot & & \cdot \\ \cdot & & \cdot \\ \cdot & & \cdot \\ a_{m1} & \dots & a_{mm} \end{bmatrix}, \quad A_2 = \begin{bmatrix} a_{1,m+1} & \dots & a_{1,m+r} \\ \cdot & & \cdot \\ \cdot & & \cdot \\ \cdot & & \cdot \\ a_{m,m+1} & \dots & a_{m,m+r} \end{bmatrix}, \quad A_3 = \begin{bmatrix} a_{1,m+r+1} & \dots & a_{1n} \\ \cdot & & \cdot \\ \cdot & & \cdot \\ \cdot & & \cdot \\ a_{m,m+r+1} & \dots & a_{mn} \end{bmatrix},$$

$$P_0 = \begin{bmatrix} b_1 \\ \cdot \\ \cdot \\ \cdot \\ b_m \end{bmatrix}, \quad D_1 = \begin{bmatrix} d_1 \\ \cdot \\ \cdot \\ \cdot \\ d_m \end{bmatrix}, \quad D_2 = \begin{bmatrix} d_{m+1} \\ \cdot \\ \cdot \\ \cdot \\ d_{m+r} \end{bmatrix}, \quad D_3 = \begin{bmatrix} d_{m+r+1} \\ \cdot \\ \cdot \\ \cdot \\ d_n \end{bmatrix}.$$

FIGURE 10.1

$$\text{(10.2.4)} \quad \bar{B} = \begin{bmatrix} A_1 & A_2 & O & O \\ I & O & I & O \\ O & I & O & O \\ O & O & O & I \end{bmatrix} = \begin{bmatrix} B & A_2 & O & O \\ I & O & I & O \\ O & I & O & O \\ O & O & O & I \end{bmatrix}.$$

$$\text{(10.2.5)} \quad \bar{B}^{-1} = \begin{bmatrix} B^{-1} & O & -B^{-1}A_2 & O \\ O & O & I & O \\ -B^{-1} & I & B^{-1}A_2 & O \\ O & O & O & I \end{bmatrix}$$

$$= \begin{bmatrix} B^{-1} & 0 & -Y_2 & 0 \\ 0 & 0 & I & 0 \\ -B^{-1} & I & Y_2 & 0 \\ 0 & 0 & 0 & I \end{bmatrix}, \text{ where } Y_2 = B^{-1}A_2.$$

$X = B^{-1}P_0 = \theta$ (cf. Chapter V Section 5.2), we finally obtain in matrix form the simplex tableau of a bounded-variable problem ($\overline{T}$) as shown in (10.2.6).

(10.2.6) $\overline{T} = \overline{B}^{-1}[\overline{P}_1 \ldots \overline{Q}_n \overline{P}_0]$

$$= \begin{bmatrix} B^{-1} & 0 & -Y_2 & 0 \\ 0 & 0 & I & 0 \\ -B^{-1} & I & Y_2 & 0 \\ 0 & 0 & 0 & I \end{bmatrix} \begin{bmatrix} B & A_2 & A_3 & 0 & 0 & 0 & P_0 \\ I & 0 & 0 & I & 0 & 0 & D_1 \\ 0 & I & 0 & 0 & I & 0 & D_2 \\ 0 & 0 & I & 0 & 0 & I & D_3 \end{bmatrix}$$

$$= \begin{bmatrix} I & 0 & Y_3 & 0 & -Y_2 & 0 & \phi \\ 0 & I & 0 & 0 & I & 0 & D_2 \\ 0 & 0 & -Y_3 & I & Y_2 & 0 & D_1 - \phi \\ 0 & 0 & I & 0 & 0 & I & D_3 \end{bmatrix},$$

$$\begin{aligned} \text{where } \phi &= B^{-1}P_0 - Y_2D_2 \\ &= \theta - Y_2D_2, \\ Y_3 &= B^{-1}A_3. \end{aligned}$$

In interpreting $\overline{T}$ the reader should bear in mind the identity of the submatrices defined in Fig. 10.1. For instance, the first column of $\overline{T}$ indicates that the vectors $\overline{P}_1, \ldots, \overline{P}_m$ are in the basis $\overline{B}$. Similarly the second, fourth and sixth columns of $\overline{T}$ comprise the rest of the basis $\overline{B}$. All of the non-basis vectors are then represented in the third and fifth columns with the former embracing all the non-basis $\overline{P}$ vectors ranging from $\overline{P}_{m+r+1}$ to $\overline{P}_n$ and the latter, all the non-basis $\overline{Q}$ vectors ranging from $\bar{Q}_{m+1}$ to $\bar{Q}_{m+r}$. For the purpose of determining the evaluations of these non-basis vectors, it is helpful to note the detailed contents of the third and fifth columns of $\overline{T}$ for a particular $\overline{P}_j$ and a particular $\overline{Q}_j$. Denoting them $\overline{Y}_{\overline{P}_j}$ and $\overline{Y}_{\overline{Q}_j}$ respectively (since they are the vectors of linear-combination coefficients for $\overline{P}_j$ and $\bar{Q}_j$), we infer from (10.2.6) and Fig. 10.1 that

$$\overline{Y}_{\overline{P}_i} = \begin{bmatrix} y_{1j} \\ \vdots \\ y_{mj} \\ 0 \\ \vdots \\ 0 \\ -y_{1j} \\ \vdots \\ -y_{mj} \\ 0 \\ \vdots \\ 1 \leftarrow \text{for } u_j \\ \vdots \\ 0 \end{bmatrix}, \qquad m + r + 1 \leq j \leq n;$$

$$\overline{Y}_{\overline{Q}_i} = \begin{bmatrix} -y_{1j} \\ \vdots \\ -y_{mj} \\ 0 \\ \vdots \\ 1 \leftarrow \text{for } x_j \\ \vdots \\ 0 \\ y_{1j} \\ \vdots \\ y_{mj} \\ 0 \\ \vdots \\ 0 \end{bmatrix}, \qquad m + 1 \leq j \leq m + r.$$

Bearing in mind further that c_j is the price coefficient of x_j and the price coefficient of u_j is 0, we thus arrive at the following evaluations of $\overline{P}_j$ and $\overline{Q}_j$ that are not in the basis.

$$(10.2.7) \quad z_{\overline{P}_j} - c_j = \left[\sum_{i=1}^{m} c_i y_{ij} + \sum_{i=m+1}^{m+r} c_i(0) + \sum_{i=1}^{m} 0(-y_{ij}) + 0(1)\right] - c_j$$

$$= \sum_{i=1}^{m} c_i y_{ij} - c_j = z_j - c_j$$

$$(10.2.8) \quad z_{\overline{Q}_j} - 0 = \left[\sum_{i=1}^{m} c_i(-y_{ij}) + c_j(1) + \sum_{i=1}^{m} 0(y_{ij}) + 0(0)\right] - 0$$

$$= -\sum_{i=1}^{m} c_i y_{ij} + c_j = -(z_j - c_j)$$

In other words, the evaluation of $\overline{P}_j$ is simply that of P_j and the evaluation of $\overline{Q}_j$ is simply the negative of that of P_j. This is the first indication that a bounded-variable problem may be treated within the much simpler framework of a "no upper-bound" problem with, of course, due attention to proper modifications of rules and procedure.

On the question of which basis vector should be replaced, i.e., Decision Rule I, there are three points to be considered, viz., (1) the continued feasibility of x_i in the basis, (2) the continued feasibility of u_i in the basis and (3) the upper-bound of the incoming variable x_k (or u_k) itself. Referring to (10.2.6) and concentrating on the third and sixth columns, one sees that feasibility of x_i is insured by selecting for replacement that basis vector which has the minimum of the ratios $\frac{\phi_i}{y_{ik}}$ for all $y_{ik} > 0$. Similarly, feasibility of u_i is insured by deciding in favor of the minimum of $\frac{d_i - \phi_i}{-y_{ik}}$ for all $y_{ik} < 0$. Finally, there is the upper-bound of x_k, namely d_k, to be satisfied. To satisfy all three requirements, the outgoing vector should be the one which yields the smallest of the three magnitudes. Specifically, if the incoming vector is a $\overline{P}$ vector, say $\overline{P}_k$ [in column 3 of (10.2.6)], then the outgoing vector is determined from

$$(.102.9) \qquad x_k = \phi_k' = \min\left\{\frac{\phi_i}{y_{ik}} \text{ for } y_{ik} > 0, \frac{d_i - \phi_i}{-y_{ik}} \text{ for } y_{ik} < 0, d_k\right\},$$

$$i = 1, \ldots, m.$$

On the other hand, if the incoming vector is a $\overline{Q}$ vector, say $\overline{Q}_k$ [in column 5 of (10.2.6)], then the outgoing vector is determined from

$$(10.2.10)\quad u_k = d_k - \phi'_k = \min\left\{\frac{\phi_i}{-y_{ik}} \text{ for } y_{ik} < 0, \frac{d_i - \phi_i}{y_{ik}} \text{ for } y_{ik} > 0, d_k\right\}, \quad i = 1, \ldots, m.$$

Thus, in a bounded-variable problem Decision Rule I is split into two parts. Which part actually applies in a particular iteration depends on whether the incoming vector is a $\overline{P}$ or a $\overline{Q}$ vector. It is noteworthy that most of the information required, e.g., y_{ik}, ϕ_i, is the same as that needed in an ordinary "no upper-bound" problem. This is another indication that the bounded-variable problem can be solved by the use of the regular simplex tableau without a substantial increase in size.

The full implication of the three-way comparison in (10.2.9) is that if the minimum is a ratio with index l in the first group $\left(\text{i.e.,} \frac{\phi_l}{y_{lk}}\right)$, then $\overline{P}_k$ replaces $\overline{P}_l$; if in the second group $\left(\text{i.e.,} \frac{d_l - \phi_l}{-y_{lk}}\right)$, then $\overline{P}_k$ replaces $\overline{Q}_l$; and finally, if the minimum is d_k, then $\overline{P}_k$ replaces $\overline{Q}_k$. The first type of exchange is one that we are accustomed to in usual linear programming problems and needs no further discussion. The second type of exchange, however, means that x_l is increased to its upper-bound value. The third type means that x_k itself jumps from zero to its upper bound in the solution. A similar interpretation may be given to (10.2.10) but will lead to different conclusions in each case. This is left to the reader.

We now come to the bulk of simplex computations, viz., the transformation of the tableau. Again referring to (10.2.6), we observe that most of the columns require no transformation at all because they represent vectors in both the old and the new $\overline{B}$. Moreover, inasmuch as the other columns, i.e., the third, fifth and sixth columns of $\overline{T}$, do require transformation, a great deal of effort can be saved by transforming only Y_2, Y_3, and ϕ, and obtaining the transformed values of $-Y_2$, $-Y_3$, and $D_1 - \phi$ simply by reversing the signs of the new Y_2, Y_3 and subtracting the new ϕ from the constant D_1.* This is still

* This statement is not intended to be mathematically complete: one can easily see that exception has to be allowed for the transformation of the incoming vector (either $\overline{P}_k$ or $\overline{Q}_k$).

another indication that a bounded-variable problem can be handled essentially as a problem without upper-bound constraints.

Having discussed the decision and transformation rules as applied to a bounded-variable problem, we are now ready to show how all the essential information needed for its solution can be compressed into a tableau comparable in size to the one used in a regular problem with no upper-bound constraints. Such a tableau is shown in Figure 10.2.* It is basically an $m \times n$ table with the m rows reserved for the m Type 1 basis vectors and the n columns for the linear-combination coefficients, $Y_1(= I)$, Y_2 and Y_3. All basis vectors of $\overline{B}$ are identified by the signs in the last row. If $\overline{P}_j$ is in $\overline{B}$, then the column P_j is given a plus sign in the last row. If $\overline{Q}_j$ is in $\overline{B}$, then the column P_j is given a minus sign. If both of them are in $\overline{B}$, then the column P_j is marked with both a plus and a minus such as in the case of the first m vectors which have been assumed to be Type 1. The column P_0 is split into two parts for the recording of ϕ_i and $d_i - \phi_i$ for $1 \leq i \leq m$ (Type 1 basis variables). No special space is provided for the solution values of Type 2 and Type 3 variables since in the former case $x_j = d_j$ $(m + 1 \leq j \leq m + r)$ and all upper bounds are displayed on the top of the tableau, and in the latter case, $x_j = 0$ $(m + r + 1 \leq j \leq n)$.

The following is a summary of rules and procedure for the solution of a bounded-variable problem by means of the special simplex tableau explained above. No further explanation will be given as they are either based directly on results already discussed or can be grasped by following a straightforward application of the standard simplex rules to the full tableau $(\overline{T})$ displayed in (10.2.6).

(a) For a maximizing (minimizing) problem, a solution is optimal if all $z_j - c_j \leq 0$ (≥ 0) in those columns marked plus, and all $z_j - c_j \geq 0$ (≤ 0) in those columns marked minus. The problem is then solved.

(b) If otherwise, select a vector to come into $\overline{B}$ either among those plus columns where $z_j - c_j > 0$ (< 0) or among those minus columns where $z_j - c_j < 0$ (> 0).

(c) If the incoming vector is among the plus columns, i.e., if it is $\overline{Q}_k$, determine the outgoing vector by means of (10.2.10). If the

---

* The format is essentially the one suggested by Charnes and Lemke, "Computational Theory of Linear Programming, I: The 'Bounded Variables' Problem," *ONR Research Memorandum*, No. 10, Carnegie Institute of Technology, (Pittsburgh, Pa., January 1954).

B		$d_1 \ldots\ldots\ldots d_m$	$d_{m+1} \ldots\ldots\ldots d_{m+r}$	$d_{m+r+1} \ldots\ldots\ldots d_n$	P_0	
		$c_1 \ldots\ldots\ldots c_m$	$c_{m+1} \ldots\ldots\ldots c_{m+r}$	$c_{m+r+1} \ldots\ldots\ldots c_n$		
c_i	P_i	$P_1 \ldots\ldots\ldots P_m$	$P_{m+1} \ldots\ldots\ldots P_{m+r}$	$P_{m+r+1} \ldots\ldots\ldots P_n$	x_i	u_i
c_1	P_1	$1 \ldots\ldots\ldots 0$	$y_{1,m+1} \ldots\ldots\ldots y_{1,m+r}$	$y_{1,m+r+1} \ldots\ldots\ldots y_{1n}$	ϕ_1	$d_1 - \phi_1$
$\vdots$	$\vdots$	$\vdots \quad\quad \vdots$	$\vdots \quad\quad \vdots$	$\vdots \quad\quad \vdots$	$\vdots$	$\vdots$
c_m	P_m	$0 \ldots\ldots\ldots 1$	$y_{m,m+1} \ldots\ldots\ldots y_{m,m+r}$	$y_{m,m+r+1} \ldots\ldots\ldots y_{mn}$	ϕ_m	$d_m - \phi_m$
$z_j - c_j$		$z_1 - c_1 \ldots z_m - c_m$	$z_{m+1} - c_{m+1} \ldots z_{m+r} - c_{m+r}$	$z_{m+r+1} - c_{m+r+1} \ldots z_n - c_n$	$f(X)$	
$\overline{B}$		$+\ - \ldots\ldots +\ -$	$+ \ldots\ldots\ldots +$	$- \ldots\ldots\ldots -$		

FIGURE 10.2

incoming vector is among the minus columns, i.e., if it is $\overline{P}_k$, then determine the outgoing vector by means of (10.2.9).

(d) After having selected the vectors for substitution, transform the tableau as follows:

(1) If $\overline{P}_k$ is replaced by $\overline{Q}_k$, change the sign of P_k to minus; add $y_{ik}d_k$ to ϕ_i and subtract it from $d_i - \phi_i$ under P_0 for $1 \leq i \leq m$. Also add $(z_k - c_k)d_k$ to $f(X)$. No other changes in the tableau are necessary.

(2) If $\overline{Q}_k$ is replaced by $\overline{P}_k$, change the sign of P_k to plus; subtract $y_{ik}d_k$ from ϕ_i and add it to $d_i - \phi_i$ under P_0 for $1 \leq i \leq m$. Also subtract $(z_k - c_k)d_k$ from $f(X)$. No other changes in the tableau are necessary.

(3) If $\overline{P}_l$ is replaced by either $\overline{P}_k$ or $\overline{Q}_k$, transform the whole tableau except the u_i part of P_0 and $f(X)$, by the standard simplex rules. Obtain the new u_i from the transformed x_i by subtracting the latter from d_i.* Remove the plus sign from P_l. Finally, depending on whether the incoming vector is $\overline{P}_k$ or $\overline{Q}_k$, (i) add a plus or minus sign to P_k, and (ii) subtract $\dfrac{\phi_l}{y_{lk}}(z_k - c_k)$ from or add $\dfrac{\phi_l}{-y_{lk}}(z_k - c_k)$ to $f(X)$ to obtain the new value of the objective function.

(4) If $\overline{Q}_l$ is replaced by either $\overline{P}_k$ or $\overline{Q}_k$, transform the whole tableau except the x_i part of P_0 and $f(X)$, by the standard simplex rules. Obtain the new x_i from the transformed u_i by subtracting the latter from d_i.* Remove the minus sign of P_l. Finally, depending on whether the incoming vector is $\overline{P}_k$ or $\overline{Q}_k$, (i) add plus or minus sign to P_k, and (ii) subtract $\dfrac{d_l - \phi_l}{-y_{lk}}(z_k - c_k)$ from or add $\dfrac{d_l - \phi_l}{y_{lk}}(z_k - c_k)$ to $f(X)$ to obtain the new value of the objective function.

This completes one iteration and we return to (a) for further improvements if possible.

Example 1. Minimize $f(X) = -x_2 - 3x_3$,

$$\begin{aligned} \text{subject to:}\quad & x_1 + x_2 + x_3 \leq 10, \\ & x_1 - 2x_3 \geq 0, \\ & 2x_2 - x_3 \leq 10, \\ & 0 \leq x_1 \leq 8, \quad 0 \leq x_2 \leq 4, \text{ and } x_3 \geq 0. \end{aligned}$$

* It must be borne in mind that in the new $\overline{B}$, $i = l$ is replaced by $i = k$. Hence, d_l is replaced by d_k. Moreover, in the event $\overline{P}_k$ replaces $\overline{Q}_l$, the value of x_k is obtained by reference to the u_i column but entered in the x_i column. The switch in column is reversed when $\overline{Q}_k$ replaces $\overline{P}_l$. Of course, if the replacement involves only $\overline{P}$ or only $\overline{Q}$, then there is no switching.

By multiplying both sides of the second constraint by -1, and then adding slack variables x_4, x_5, x_6 to each of the three constraints, we obtain the initial basic feasible solution shown in Table 10.1a. Since there are no upper bounds specified for x_3, x_4, x_5, x_6, we arbitrarily assume that they all have an upper bound of $+\infty$. Hence, the complete initial solution is: $x_4 = 10$, $x_5 = 0$, $x_6 = 10$, $u_1 = 8$, $u_2 = 4$, $u_3 = \infty$, $u_4 = \infty - 10$, $u_5 = \infty$, and $u_6 = \infty - 10$, where the u's are, of course, the upper-bound slacks of their corresponding x's. There are only 3 non-basis variables in this problem. These are in the case

Table 10.1a

d_j		8	4	∞	∞	∞	∞	P_0	
c_j		0	-1	-3	0	0	0		
c_i	P_i	P_1	P_2	P_3	P_4	P_5	P_6	x_i	u_i
0	P_4	1	1	1	1	0	0	10	$\infty - 10$
0	P_5	-1	0	(2)	0	1	0	(0)	∞
0	P_6	0	2	-1	0	0	1	10	$\infty - 10$
$z_j - c_j$		0	1	3	0	0	0	0	
$\bar{B}$		$-$	$-$	$-$	$+-$	$+-$	$+-$		

of the present initial solution, x_1, x_2 and x_3, which are all equal to zero.

Since the evaluations of P_2 and P_3 are both positive, either $\bar{P}_2$ or $\bar{P}_3$ may be put into $\bar{B}$ with a consequent decrease in $f(X)$ [cf. (a)

Table 10.1b

d_j		8	(4)	∞	∞	∞	∞	P_0	
c_j		0	-1	-3	0	0	0		
c_i	P_i	P_1	P_2	P_3	P_4	P_5	P_6	x_i	u_i
0	P_4	$\frac{3}{2}$	1	0	1	$-\frac{1}{2}$	0	10	$\infty - 10$
-3	P_3	$-\frac{1}{2}$	0	1	0	$\frac{1}{2}$	0	0	∞
0	P_6	$-\frac{1}{2}$	(2)	0	0	$\frac{1}{2}$	1	10	$\infty - 10$
$z_j - c_j$		$\frac{3}{2}$	1	0	0	$-\frac{3}{2}$	0	0	
$\bar{B}$		$-$	$-$	$+-$	$+-$	$-$	$+-$		

and (b) of rules and procedure for bounded-variable problem above]. Let us choose $\bar{P}_3$. This means that (10.2.9) has to be used in selecting the outgoing vector [cf. (c) above], which turns out to be $\bar{P}_5$. Thus, the first iteration calls for the replacement of $\bar{P}_5$ by $\bar{P}_3$. Now, since the outgoing vector is a $\bar{P}$ vector, according to (d)-(3) above, the ensuing transformation by standard simplex rules must exclude the u_i column. After the transformation is completed, the new values of u_i are then obtained from $d_i - \phi_i'$ (new values of x_i). The result is shown in Table 10.1b. In the next iteration, we elect to bring $\bar{P}_2$ into $\bar{B}$. But an application of the decision rule (10.2.9) indicates that d_2 is the limiting factor and hence $\bar{Q}_2$ has to be replaced by $\bar{P}_2$. According to (d)-(2) above then, we simply make appropriate changes in ϕ_i, $d_i - \phi_i$ and $f(X)$, and obtain Table 10.1c. Since $d_2 = 4$ and $z_2 - c_2 = 1$,

Table 10.1c

d_j		8	4	∞	∞	∞	∞	P_0	
c_j		0	-1	-3	0	0	0		
c_i	P_i	P_1	P_2	P_3	P_4	P_5	P_6	x_i	u_i
0	P_4	$\textcircled{\frac{3}{2}}$	1	0	1	$-\frac{1}{2}$	0	$\textcircled{6}$	$\infty - 6$
-3	P_3	$-\frac{1}{2}$	0	1	0	$\frac{1}{2}$	0	0	∞
0	P_6	$-\frac{1}{2}$	2	0	0	$\frac{1}{2}$	1	2	$\infty - 2$
$z_j - c_j$		$\frac{3}{2}$	1	0	0	$-\frac{3}{2}$	0	-4	
$\bar{B}$		$-$	$+$	$+ -$	$+ -$	$-$	$+ -$		

the new value of the objective function is $f(X') = f(X) - d_2(z_2 - c_2) = 0 - 4 \times 1 = -4$. At this point, the only favorable non-basis vector is $\bar{P}_1$, which according to (10.2.9) calls for the replacement of $\bar{P}_4$. This substitution leads to the new and final tableau (Table 10.1d) through standard transformation except for the u_i column. Since there is no more non-basis vector with a favorable evaluation, the solution contained in this table is an optimal solution, viz., $x_1 = 4$, $x_2 = 4\ (= d_2)$, $x_3 = 0$, $x_6 = 4$, $u_1 = 4$, $u_3 = \infty - 2$, $u_4 = u_5 = \infty$, $u_6 = \infty - 4$, and the optimal value of the objective function is $f(X) - \dfrac{\phi_4}{y_{41}}(z_1 - c_1) = -4 - 4\left(\dfrac{3}{2}\right) = -10.$

Table 10.1d

d_j		8	4	∞	∞	∞	∞	P_0	
c_j		0	-1	-3	0	0	0		
c_i	P_i	P_1	P_2	P_3	P_4	P_5	P_6	x_i	u_i
0	P_1		$\frac{2}{3}$		$\frac{2}{3}$	$-\frac{1}{3}$		4	4
-3	P_3		$\frac{1}{3}$		$\frac{1}{3}$	$\frac{1}{3}$		2	$\infty - 2$
0	P_6		$\frac{7}{3}$		$\frac{1}{3}$	$\frac{1}{3}$		4	$\infty - 4$
$z_j - c_j$		0	0	0	-1	-1	0	-10	
$\bar{B}$		$+\,-$	$+$	$+\,-$	$-$	$-$	$+\,-$		

Example 2. Maximize $f(X) = 4x_1 + 4x_2 + 3x_3$,

subject to:

$$\begin{aligned} -x_1 + 2x_2 + 3x_3 &\le 15, \\ -\ x_2 + \ x_3 &\le 4, \\ 2x_1 + x_2 - x_3 &\le 6, \\ x_1 - x_2 + 2x_3 &\le 10, \end{aligned}$$

and

$$0 \le x_1 \le 8,$$
$$0 \le x_2 \le 4,$$
$$0 \le x_3 \le 4.$$

The solution of this problem is laid out in Tables 10.2a through 10.2e, and should be largely self-explanatory. Attention is called, however, to Table 10.2c which indicates that a $\bar{P}$ vector ($\bar{P}_2$) should replace a $\bar{Q}$ vector ($\bar{Q}_3$). The value of the incoming variable x_2 according to (10.2.9) is the ratio $\dfrac{d_3 - \phi_3}{-y_{32}} = \dfrac{\frac{6}{5}}{-(-\frac{3}{5})} = 2$ which is entered in the

Table 10.2a

d_j		8	4	4	∞	∞	∞	∞	P_0	
c_j		4	4	3	0	0	0	0		
c_i	P_i	P_1	P_2	P_3	P_4	P_5	P_6	P_7	x_i	u_i
0	P_4	-1	2	3	1				15	$\infty - 15$
0	P_5	0	-1	1		1			4	$\infty - 4$
0	P_6	(2)	1	-1			1		(6)	$\infty - 6$
0	P_7	1	-1	2				1	10	$\infty - 10$
$z_j - c_j$		-4	-4	-3	0	0	0	0	0	
$\bar{B}$		$-$	$-$	$-$	$+\,-$	$+\,-$	$+\,-$	$+\,-$		

Table 10.2b

0	P_4		$\frac{5}{2}$	$\frac{5}{2}$	1		$\frac{1}{2}$		18	$\infty - 18$
0	P_5		-1	1		1	0		4	$\infty - 4$
4	P_1	1	$\frac{1}{2}$	$-\frac{1}{2}$			$\frac{1}{2}$		3	5
0	P_6		$-\frac{3}{2}$	($\frac{5}{2}$)			$-\frac{1}{2}$	1	(7)	$\infty - 7$
$z_j - c_j$		0	-2	-5	0	0	2	0	12	
$\overline{B}$		$+ -$	$-$	$-$	$+ -$	$+ -$	$-$	$+ -$		

Table 10.2c

0	P_4		4		1		1	-1	11	$\infty - 11$
0	P_5		$-\frac{2}{5}$			1	$\frac{1}{5}$	$-\frac{2}{5}$	$\frac{6}{5}$	$\infty - \frac{6}{5}$
4	P_1	1	$\frac{1}{5}$				$\frac{2}{5}$	$\frac{1}{5}$	$\frac{22}{5}$	$\frac{18}{5}$
3	P_3		($-\frac{3}{5}$)	1			$-\frac{1}{5}$	$\frac{2}{5}$	$\frac{14}{5}$	($\frac{6}{5}$)
$z_j - c_j$		0	-5	0	0	0	1	2	26	
$\overline{B}$		$+ -$	$-$	$+ -$	$+ -$	$+ -$	$-$	$-$		

Table 10.2d

0	P_4			$\frac{20}{3}$	1		$-\frac{1}{3}$	($\frac{5}{3}$)	(3)	$\infty - 3$
0	P_5			$-\frac{2}{3}$		1	$\frac{1}{3}$	$-\frac{2}{3}$	2	$\infty - 2$
4	P_1	1		$\frac{1}{3}$			$\frac{1}{3}$	$\frac{1}{3}$	4	4
4	P_2		1	$-\frac{5}{3}$			$\frac{1}{3}$	$-\frac{2}{3}$	2	2
$z_j - c_j$		0	0	$-\frac{25}{3}$	0	0	$\frac{8}{3}$	$-\frac{4}{3}$	36	
$\overline{B}$		$+ -$	$+ -$	$+$	$+ -$	$+ -$	$-$	$-$		

Table 10.2e

0	P_7			4	$\frac{3}{5}$		$-\frac{1}{5}$	1	$\frac{9}{5}$	$\infty - \frac{9}{5}$
0	P_5			2	$\frac{2}{5}$	1	$\frac{1}{5}$		$\frac{16}{5}$	$\infty - \frac{16}{5}$
4	P_1	1		-1	$-\frac{1}{5}$		$\frac{2}{5}$		$\frac{17}{5}$	$\frac{23}{5}$
4	P_2		1	1	$\frac{2}{5}$		$\frac{1}{5}$		$\frac{16}{5}$	$\frac{4}{5}$
$z_j - c_j$		0	0	-3	$\frac{4}{5}$		$\frac{12}{5}$		$\frac{192}{5}$	
$\overline{B}$		$+ -$	$+ -$	$+$	$-$	$+ -$	$-$	$+ -$		

x_i column of Table 10.2d. The new value of the objective function according to (d)-(4) is $f(X) - \dfrac{d_3 - \phi_3}{-y_{32}}(z_2 - c_2) = 26 - 2(-5) = 36$ as shown in Table 10.2d. The next iteration involves the substitution of $\overline{P}_4$ by $\overline{P}_7$. Hence, we follow the rules set down in (d)-(3) and obtain the optimal solution as shown in Table 10.2e with $x_1 = \frac{17}{5}$, $x_2 = \frac{16}{5}$, $x_3 = 4$ (indicated by the plus sign under P_3 column) and the maximal $f(X) = \frac{192}{5}$.

10.3. Integer Programming.

The difficulty in integer programming is primarily caused by our inability to write down explicitly the constraints necessary for restricting the solutions to integer values only. Let us illustrate this point by means of a simple 2-variable linear programming problem whose "continuous" solution is represented by the polygon ABCDEF in Figure 10.3. If this problem is further subjected to the requirement

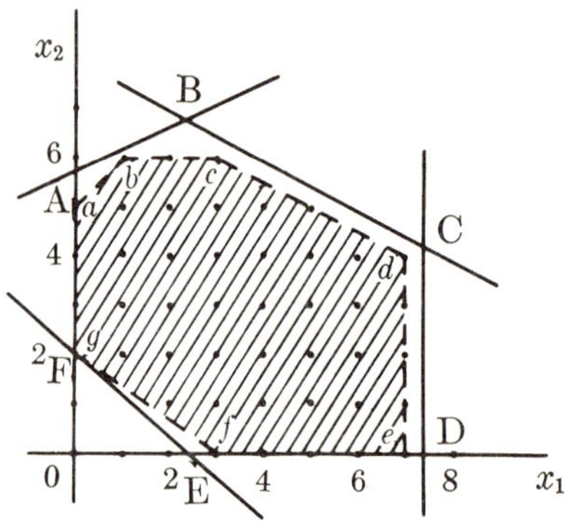

Figure 10.3

that the variables can assume integer values only, the acceptable solutions will be restricted to the indicated points within the polygon. Strictly speaking, these points do not constitute a convex set, hence the simplex method, which is based on the existence of a convex solution set, theoretically cannot be applied. Fortunately, however, all integer solutions can be covered by a minimal area abcdefg that is convex and a polygon. If this area is the minimal area that covers all integer solutions, then the extreme points of this convex polygon abcdefg must themselves be integer solutions. Thus, if we can construct a continuous linear programming problem whose solution set is the convex polygon abcdefg, then the optimal solution to that problem, obtained by the simplex method, must be the optimal integer solution to the original problem, whose solution set is the collection of integer points within the region ABCDEF. As intimated earlier, the real difficulty lies in discovering the constraints that define the smaller convex region abcdefg. We have at present two main proposals on the solution of integer programming problems. In chronological order, they are the Method of Integer Forms developed by Gomory* and the alternative method proposed by Land and Doig.** In this section, we will first present a broad description and comparison of the two methods, then a more detailed discussion of Gomory's method, which seems to have a wider acceptance at the present time.

The Method of Integer Forms starts off by using the simplex to obtain an optimal continuous solution. If this solution is not an integer solution, then a new constraint is constructed according to a certain rule and incorporated into the problem, and the new problem is then reoptimized. This process is repeated until, due to the nature of the new constraints added to the system, an optimal solution is found which is also an integer solution. We now have the optimal integer solution to the original problem. In the alternative approach of Land and Doig also, an optimal continuous solution is obtained first by the simplex method. If it is not an integer solution, one of the discrete variables is then chosen arbitrarily to be first integerized. This is accomplished by using parametric programming to determine the range of feasible values of this variable and noting the integer values within this range together with the corresponding values of the objective function. Next we fix this variable at the most desirable integer value in terms of the value of the objective function determined above, and proceed to find the range of feasible integer values of a second variable. We repeat this for the next best integer value of the first variable. From these two ranges of feasible integer values of the second variable, select particular integer values, which

---

* Gomory, R. E., "Outline of an Algorithm for Integer Solutions to Linear Programs," *Bulletin of the American Mathematical Society,* LXIV (1958), pp. 275–8.

** Land, A. H. and Doig, A. G., "An Automatic Method of Solving Discrete Programming Problems," *Econometrica,* XXVIII (1960), No. 3, pp. 497–521.

together with the predetermined integer values of the first variable, yield the higher values of the objective function. These combinations of integer values of the first two variables then provide the basis for integerizing a third variable, a fourth variable, and so on until all variables required to be integers are integerized. At every step the direction of further investigation is guided by reference to the value of the objective function yielded by partially integerized solutions obtained so far. In the end it is easy to find the best solution among the several fully integerized alternatives available. This then is the optimal integer solution to the problem.

From the brief descriptions of the two methods above, it should be clear that Gomory relies on reshaping the problem to "force" out the proper solution—whereas Land and Doig engage in a direct and systematic search for the optimum. The latter approach requires extensive and careful record-keeping in order to test exhaustively all integer solutions that are likely to develop into the optimal solution. Consequently, it would seem to be the more laborious of the two. However, as the process is one of progressive integerization, terminable at any point desired, it can be applied to the so-called mixed integer programming problems, i.e., problems where not all of the variables are required to be integers. On the other hand, since the Gomory method does not at any point require that any particular variable remain an integer, it is at present only applicable to problems where all variables are required to assume integer values.

Let us now concentrate on Gomory's Method of Integer Forms. We will deal only with the constraint system in the following discussion since obviously the objective function has no part in determining the nature of the solutions. Assume that the constraint system when converted into equations by adding slack variables, $x_{n+1}, \ldots, x_{n+m}$, is as shown in (10.3.1). Assume further that a_{ij} and

$$
\begin{array}{lllll}
 & a_{11}x_1 + \ldots + a_{1n}x_n + x_{n+1} & & & = b_1, \\
 & \cdot \qquad\qquad \cdot & \cdot & & \cdot \\
(10.3.1) & \cdot \qquad\qquad \cdot & \cdot & & \cdot \\
 & \cdot \qquad\qquad \cdot & \cdot & & \cdot \\
 & a_{m1}x_1 + \ldots + a_{mn}x_n & & + x_{n+m} & = b_m.
\end{array}
$$

b_i have all been converted to integer forms and that there exists an optimal solution consisting of basis variables $x_1, \ldots, x_m$. Let this optimal basis be B. It is now possible to write (10.3.1) in matrix form as follows:

$$AX = [B \;\; A']X = P_0, \tag{10.3.2}$$

where A is $m \times (n + m)$, B is $m \times m$, A' is $m \times n$, X is $(n + m) \times 1$, and P_0 is $m \times 1$.

Multiplying through by B^{-1}, we obtain $B^{-1}[B \;\; A']X = B^{-1}P_0$, or

$$[I \;\; Y]X = \theta, \tag{10.3.3}$$

where $Y = B^{-1}A'$ is an $m \times n$ matrix of linear combination coefficients y_{ij}; and $\theta = B^{-1}P_0$ is an $m \times 1$ vector of solution values, θ_i. All of these are contained in the simplex tableau representing the optimal solution, but for our present purpose it is more convenient to write (10.3.3) in the following index form:

$$x_i = \theta_i - \sum_{j=m+1}^{m+n} y_{ij}x_j, \qquad i = 1, \ldots, m. \tag{10.3.4}$$

Now, if all θ_i are integers, then the present optimal basic feasible solution is already an integer solution, i.e., $x_i = \theta_i$, $i = 1, \ldots, m$, and our problem is solved. However, if some θ_i are not integers, then according to our earlier discussion a new constraint, so structured as to admit all existing integer solutions while specifically barring the reappearance of the present non-integer optimal solution, will be added to the problem to create a new modified problem. This new problem will then be optimized as a continuous-variable problem, and the cycle is repeated until an optimal discrete solution is obtained.

But suppose that the problem has no integer solution. This can be detected immediately by noting that for a certain variable whose solution is not an integer, say $x_f = \theta_f$, all corresponding y_{fj} in (10.3.4) are, nevertheless, all integers. For we argue that if a discrete solution did exist, then the fractional part of θ_f must be offset by the fractional part of $\sum_{j=m+1}^{m+n} y_{fj}x_j$ in order that x_f be made an integer. This would not be possible, however, without some y_{fj} being nonintegers since x_j are by assumption all integers. We conclude, therefore, that if $x_f = \theta_f$ is not an integer while all y_{fj} are integers, then no integer solution is possible, and the solution of the problem comes to an end.

If, on the other hand, no such condition prevails and an integer solution is possible, we will proceed to construct the additional constraint (called hereafter the *Gomorian constraint*) called for in the method of integer forms. To do this, let us first define the term "non-negative fraction." A *non-negative fraction* is a fraction greater than or equal to zero but less than 1, which when *subtracted* from a given non-integer will convert it to an integer. For example, the non-negative fractions of the numbers 10.57 and -5.25 are 0.57 and 0.75 respectively.

Having done this, we can divide all non-integral θ_i and their y_{ij} in (10.3.4) into two parts: an integer (k) and a non-negative fraction (f). Thus,

$$\theta_i = k_i + f_i, \qquad i = 1, \ldots, m;$$
$$y_{ij} = k_{ij} + f_{ij}, \qquad i = 1, \ldots, m \quad \text{and} \quad j = m+1, \ldots, m+n.$$

Substituting these back into (10.3.4), we obtain

(10.3.5)

$$x_i = k_i + f_i - \sum_{j=m+1}^{m+n} (k_{ij} + f_{ij})x_j$$
$$= \left(k_i - \sum_{j=m+1}^{m+n} k_{ij}x_j\right) + \left(f_i - \sum_{j=m+1}^{m+n} f_{ij}x_j\right), \qquad i = 1, \ldots, m.$$

Now for any integer solution (which is already known to exist) the second part of the above expression, viz., $\left(f_i - \sum_{j=m+1}^{m+n} f_{ij}x_j\right)$ must be either zero or some negative integer. If it is neither zero nor an integer, the solution obviously cannot be an integer solution. However, it cannot be a positive integer because f_i, f_{ij} are non-negative fractions and x_j are also non-negative. Thus we arrive at the important conclusion that $f_i - \sum_{j=m+1}^{m+n} f_{ij}x_j \leq 0$. Introducing the *Gomorian slack variable* s_i, we have the following Gomorian constraint:

(10.3.6) $$s_i = -f_i + \sum_{j=m+1}^{m+n} f_{ij}x_j, \qquad \text{where} \quad s_i \geq 0.$$

This is a relationship that is satisfied by all existing (yet unknown) integer solutions, but not by the present optimal non-integer solution since in the latter case, x_j, $j = m+1, \ldots, m+n$ are non-basis variables and hence equal to zero, which forces s_i to become negative ($= -f_i$), rendering the whole solution infeasible. In other words, if such a constraint is incorporated into the original problem it will permanently bar from reappearance the optimal non-integer solution on which it is based. The problem with the new constraint is then re-optimized, and repeatedly modified and re-optimized until a discrete solution is obtained.

Computationally also the method of integer forms proves to be highly feasible. Since the Gomorian slack variable s_i has a price coefficient of zero and does not appear in any other constraints, it can be added to the original optimal basis without affecting the evaluations of the non-basis vectors. These evaluations are all unfavorable in this case, i.e., non-negative for a maximizing problem and non-positive for a minimizing one, since the original basis is an optimal basis. To re-optimize the new problem with $s_i = -f_i$, we therefore simply follow the dual simplex procedure discussed in Chapter VI Section 4.5.

It is probable that in an optimal non-discrete solution, more than one basis variable will assume non-integral values. In such cases a question arises as to which variable should be used to construct the new constraint. There is as yet no answer to this question based wholly on theoretical grounds. According to Gomory, limited computational experience indicates that a good rule-of-thumb is to use the variable with the largest non-negative fraction f_i.*

It will be noted that with each new Gomorian constraint, the dimension of the basis is increased by 1. Is there a limit to the number of such constraints that have to be added before an integer optimum is obtained? Fortunately, the answer is yes. For the effect of introducing a new Gomorian constraint is to reduce the optimality of the solution, i.e., to decrease $f(X)$ in case of a maximizing problem and to increase it in the opposite cases. This alone should prevent a return to the old non-discrete optimal solutions. So long as the successive re-optimized bases consist only of variables that are originally in the problem, their dimensions will have to remain as augmented by the addition of Gomorian constraints. If, however, a Gomorian slack variable added earlier to the problem should find its way back into the basis as a result of re-optimization, the corresponding Gomorian constraint could be dropped from the problem altogether since its presence in the basis serves only to satisfy a constraint that is not part of the original problem. In other words, the limiting influence

---

* Gomory, *op. cit.*, p. 278.

of the progressively inferior optimal solutions could be fully served by bases that include original variables only. Since there are only $m + n$ original variables in the problem, the maximal dimension of the basis need never be greater than $m + n + 1$. This is a factor that speaks well of the Method of Integer Forms.

Example 1. Find the optimal integer solution to the following problem:

$$\begin{aligned} \text{Maximize } f(X) &= 3x_2 \\ \text{subject to: } 3x_1 + 2x_2 &\leq 7 \\ x_1 - x_2 &\geq -2 \\ \text{and} \quad x_1, x_2 &\geq 0 \end{aligned}$$

After the usual initial changes and rearrangements, the problem is set up to yield the first simplex tableau (Table 10.3a), which shows the initial basic feasible solution in terms of the slack variables x_3 and x_4. It is then transformed by standard simplex techniques in two iterations to Table 10.3c which indicates that the optimal non-integer solution is $x_1 = \frac{3}{5}$, $x_2 = \frac{13}{5}$, $x_3 = x_4 = 0$, and $f(X) = \frac{39}{5}$.

Table 10.3a

c_i	P_i \ P_j	P_1	P_2	P_3	P_4	P_0
0	P_3	3	2	1	0	7
0	P_4	−1	①	0	1	2
z_j		0	0	0	0	
c_j		0	3	0	0	0
$z_j - c_j$		0	−3	0	0	

Table 10.3b

0	P_3	⑤	0	1	−2	3
3	P_2	−1	1	0	1	2
$z_j - c_j$		−3	0	0	3	6

Table 10.3c

0	P_1	1	0	$\frac{1}{5}$	$-\frac{2}{5}$	$\frac{3}{5}$
3	P_2	0	1	$\frac{1}{5}$	$\frac{3}{5}$	$\frac{13}{5}$
$z_j - c_j$		0	0	$\frac{3}{5}$	$\frac{9}{5}$	$\frac{39}{5}$

Next, we will construct a Gomorian constraint for x_2. From Table 10.3c we see that $f_2 = \frac{3}{5}$, $f_{23} = \frac{1}{5}$ and $f_{24} = \frac{3}{5}$. Hence, according to (10.3.6),

$$s_2 = -\tfrac{3}{5} + (\tfrac{1}{5}x_3 + \tfrac{3}{5}x_4)$$

Denoting the coefficient vector of s_2 by P_5, we obtain the tableau shown in Table 10.3d, which after one iteration by the dual simplex algorithm yields the optimal integer solution $x_1 = 0$, $x_2 = 2$, $x_3 = 3$, $x_4 = 0$ and $f(X) = 6$. It will be noticed that in the last iteration, the

Table 10.3d

c_i	P_i \ P_j	P_1	P_2	P_3	P_4	P_5	P_0
0	P_1	1	0	$\frac{1}{5}$	$-\frac{2}{5}$	0	$\frac{3}{5}$
3	P_2	0	1	$\frac{1}{5}$	$\frac{3}{5}$	0	$\frac{13}{5}$
0	P_5	0	0	$\boxed{-\frac{1}{5}}$	$-\frac{3}{5}$	1	$-\frac{3}{5}$
$z_j - c_j$		0	0	$\frac{3}{5}$	$\frac{9}{5}$	0	$\frac{39}{5}$

Table 10.3e

0	P_1	1	0	0	-1	1	0
3	P_2	0	1	0	0	1	2
0	P_3	0	0	1	3	5	3
$z_j - c_j$		0	0	0	0	3	6

ratio $\dfrac{z_j - c_j}{y_{lj}}$ [cf. (6.6.1)] is the same for P_3 and P_4. Therefore, if P_4 is put into the basis in place of P_5, the optimal integer solution will be $x_1 = 1$, $x_2 = 2$, $x_3 = 0$, $x_4 = 1$, with, of course, the same $f(X) = 6$.

The true significance of the Gomorian constraint in this problem can best be grasped by graphical representation as in Figure 10.4. From the first and second constraints in the problem, we obtain $x_3 = 7 - 3x_1 - 2x_2$ and $x_4 = 2 + x_1 - x_2$. Substituting these into the equation for the Gomorian constraint, we have

$$s_2 = -\tfrac{3}{5} + (\tfrac{1}{5}x_3 + \tfrac{3}{5}x_4) = -\tfrac{3}{5} + \tfrac{1}{5}(7 - 3x_1 - 2x_2) + \tfrac{3}{5}(2 + x_1 - x_2)$$
$$= 2 - x_2$$

The above is equivalent to a new constraint, viz., $x_2 \leq 2$, and this constraint, as is shown in Figure 10.4, eliminates the non-integer optimum A and establishes the new discrete optima B and C.

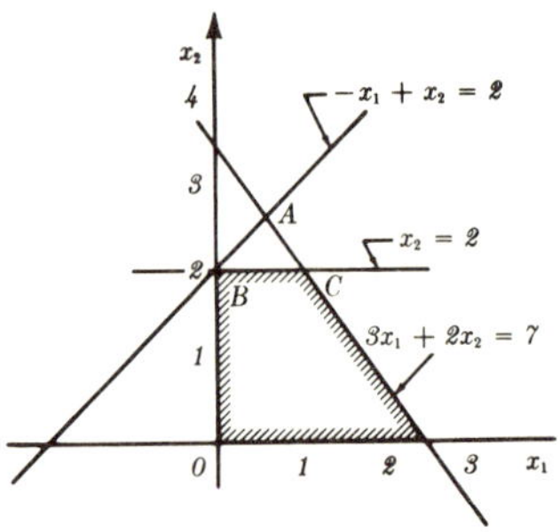

FIGURE 10.4

Example 2. In Chapter III we solved the simple production planning problem of determining the output levels of two products, A and B, so that existing production facilities would be most profitably utilized. An important simplifying assumption was the absence of fixed charges. If they had been included in the problem, the objective function would not have been linear and the problem could not have been solved by the standard simplex method. Realistically speaking, the production of any article usually carries with it a certain amount of fixed charges, whether large or small, that are dependent only upon the presence rather than the degree of productive activities. Such a problem, called the fixed-charge problem, is essentially one of optimizing a nonlinear objective function subject to a set of linear constraints. The following example demonstrates how it can be set up as an ordinary linear programming problem, with important aspects of integer and bounded-variable programming.

Without repeating the description in Chapter III Section 3.2, we may immediately set down the mathematical version of the original problem as follows:

$$\begin{aligned} \text{Maximize } f(X) &= 30x_1 + 20x_2, \\ \text{subject to:} \quad & 10x_1 + 6x_2 \leq 1000, \\ & 5x_1 + 4x_2 \leq 600, \\ \text{and} \quad & x_1, x_2 \geq 0. \end{aligned}$$

In the above formulation, x_1, x_2 stand for the outputs of A and B respectively and the two constraints represent the limitations imposed by available foundry and machine capacities. We shall assume further that there is a fixed cost of \$800 associated with the production of A and a fixed cost of \$500 associated with the production of B. This assumption in effect means that the total profit $f(x)$ will be diminished by \$800 (and/or \$500) if x_1 (and/or x_2) is greater than zero. On the other hand, $f(X)$ will not be so burdened if x_1 (and/or x_2) is zero. How can we design an objective function to reflect this behavior?

First, we could modify the objective function to read:

$$\text{Maximize } f(X) = 30x_1 + 20x_2 - 800x_3 - 500x_4,$$

where x_3 and x_4 are additional variables introduced to represent the state of production of A and B respectively. By the use of integer programming and additional upper-bound constraints for x_3 and x_4 we can limit their solution values to either 1 or 0. We interpret the solution x_3 (or x_4) $= 1$ to mean that there is production of A (or B), and the alternative solution x_3 (or x_4) $= 0$, that there is no production. Consistent with these interpretations, the objective function will then reflect through the terms $-800x_3$ and $-500x_4$ the presence or absence of fixed charges according to whether x_3 and x_4 are equal to 1 or 0 respectively. However, this apparently satisfactory scheme has hidden complications. The addition of x_3 and x_4 in the problem creates the possibility that the optimal solution may show (1) x_1 (or x_2) $= 0$ while x_3 (or x_4) $= 1$, or (2) x_1 (or x_2) > 0 while x_3 (or x_4) $= 0$. Both (1) and (2) represent inconsistent solutions in view of the definitions of x_3 and x_4. Therefore, both are unacceptable. But (1) is not really a problem because the price coefficients of x_1 and x_2 are far more favorable than those of x_3 and x_4. In a maximizing problem, x_1 and x_2 will always tend to enter the basis before x_3 and x_4. Therefore, we need only to concentrate on preventing the second type of inconsistent solutions from appearing. This can be accomplished by the inclusion of the following two additional constraints:

$$x_1 - d_1x_3 \leq 0,$$
$$x_2 - d_2x_4 \leq 0,$$

where d_1 and d_2 are arbitrary coefficients sufficiently large to permit the continued satisfaction of the constraints even for the largest possible values of x_1 and x_2. With these in the constraint system, it becomes impossible for x_1, x_2 to be in the basis (i.e., > 0) while x_3, and x_4 are not (i.e., $= 0$). Once x_3 is in, however, profit consideration in the form of a favorable evaluation of x_1 will inevitably bring x_1 into the basis also. The same argument holds also for x_4 and x_2. Thus the two types of inconsistent solution mentioned above are now successfully blocked from emergence. To obtain a reasonable estimate of the coefficients d_1 and d_2, we may use the figures representing maximal possible production of A and B by the foundry and machine departments. For product A, the foundry capacity permits a maximal possible production of $\frac{1000}{10} = 100$ units whereas the machine capacity allows $\frac{600}{5} = 120$. Hence, x_1 can never exceed 100, and it is adequate to let $d_1 = 100$. Similar computations will show that $d_2 = 150$. With these two coefficients determined, we may present the complete formulation of the fixed-charge problem as follows:

$$\begin{aligned} \text{Maximize } f(X) &= 30x_1 + 20x_2 - 800x_3 - 500x_4, \\ \text{subject to:} \quad & 10x_1 + 6x_2 \leq 1000, \\ & 5x_1 + 4x_2 \leq 600, \\ & x_1 - 100x_3 \leq 0, \\ & x_2 - 150x_4 \leq 0, \\ & x_3 \leq 1, \\ & x_4 \leq 1, \\ \text{and } x_j &= \text{non-negative integers}, \qquad j = 1, \ldots, 4. \end{aligned}$$

The actual solution of this specific problem is of little importance.*

---

* This problem may realistically be regarded as a pure integer programming problem even though there is no explicit requirement that the outputs of A and B be integers. Actually for problems as small as this, it is more practical to break up the problem into sub-problems: In the present example, (1) neither A nor B

The main purpose here is to demonstrate the flexibility and power of linear programming. By this example we are shown how it can be employed to solve problems of a *logical* nature as well as those that are strictly numerical in content. Unfortunately it is easier to teach linear programming than to teach the *use* of linear programming. Its use is largely an art and requires insight, intuition, creativity, and luck as well as experience and understanding of the subject matter. We can contribute here only illustrations and clear explanations of existing knowledge.

10.4. Quadratic Programming.

Quadratic programming is an important class of nonlinear programming problems. Specifically, it refers to the kind of problems which seek the optimal value of a quadratic (second degree) objective function whose variables are subject to the usual linear structural and non-negative constraints. These problems are frequently encountered in business and industry, sometimes as an accurate reflection of reality and sometimes as a necessary compromise between the conflicting needs for accuracy and simplicity in describing it. As an example, we may cite Dorfman's problem of revenue maximization by a monopolistic firm.* It is well known in economics that such a firm is confronted with negatively sloping rather than horizontal demand curves. If the assumption is made that $p_j = a_j - b_jx_j, j = 1, 2, \ldots, n$, where p_j is the price and x_j, the output of the jth product, and a_j, b_j, appropriate non-negative coefficients, then the total revenue $R = \sum_{j=1}^{n} p_jx_j$ becomes a quadratic function in x_j. The problem, together with the usually linear constraints regarding availability of resources employed, is thus a quadratic programming problem. In another example, Markowitz has considered the problem of selecting an investment portfolio which will yield a given expected total income with minimum total variance.** This again is a quadratic programming

---

is produced; (2) only A is produced; (3) only B is produced; (4) both A and B are produced; find the optimal solution to each of these sub-problems and select the best one as the final optimum.

* Robert Dorfman: *Application of Linear Programming to the Theory of the Firm* (Berkeley: University of California Press, 1951).

** Harry M. Markowitz: *Portfolio Selection, Monograph 16, Cowles Foundation for Research in Economics* (New York: John Wiley & Sons, Inc. 1959).

problem because total variance is $\sum_{i=1}^{n}\sum_{j=1}^{n} c_{ij}x_ix_j$ where x_i, x_j are the amount to be invested in securities i and j respectively, and c_{ij}, the co-variance in income between unit investment in i and j. In the simplest case, the constraints may consist of (1) $\sum_{j=1}^{n} a_jx_j = \text{I}$, where a_j is the expected yield of the jth security and I, a parameter denoting the total expected income desired, and (2) $\sum_{j=1}^{n} x_j = \text{M}$ where M is the amount of fund available for investment.

The algebraic form of a quadratic function in n variables is

$$f(X) = c_0 + c_1x_1 + \ldots + c_nx_n + c_{11}x_1^2 + c_{22}x_2^2 + \ldots + c_{nn}x_n^2 + 2c_{12}x_1x_2 + \ldots + 2c_{1n}x_1x_n + 2c_{23}x_2x_3 + \ldots + 2c_{2n}x_2x_n + \ldots + 2c_{n-1,\,n}x_{n-1}x_n. \tag{10.4.1}$$

However, if we split all terms of the form $2c_{ij}x_ix_j$ into $c_{ij}x_ix_j + c_{ji}x_jx_i$ with $c_{ij} = c_{ji}$, it is possible to write the quadratic function in the much simpler matrix form

$$f(X) = c_0 + cX + X^TCX, \tag{10.4.2}$$

where: X is the n-component column vector of variables, $x_1, \ldots, x_n$,
X^T is the transpose of X,
c is the row vector $[c_1, \ldots, c_n]$,
c_0 is a scalar, and

$$C = \begin{bmatrix} c_{11} & c_{12} & \cdots & c_{1n} \\ c_{21} & c_{22} & \cdots & c_{2n} \\ \vdots & \vdots & & \vdots \\ c_{n1} & c_{n2} & \cdots & c_{nn} \end{bmatrix} \text{ is an } n \times n \text{ symmetric matrix.}$$

From this expression, we can easily obtain its gradient vector*

$$\partial f(X) = \left[\frac{\partial f(X)}{\partial x_1}, \ldots, \frac{\partial f(X)}{\partial x_n}\right] = c + 2X^TC. \tag{10.4.3}$$

As will be seen later, this gradient vector which is a linear function of X, plays an important rôle in the general solution of the quadratic

* As discussed in a footnote in Section 3.2 of Chapter III, a gradient vector is a vector pointing at the direction of maximal increase in the value of $f(X)$.

programming problem. In the meantime let us look a little closer into the conceptual difficulties encountered in optimizing a quadratic function under linear constraints.

The fundamental difficulty with quadratic programming is that the quadratic function usually has an extremal value by itself whether or not there are any constraints.** As (10.4.3) implies, we cannot expect that continuous change in the values of the variables in the same directions will change the value of $f(X)$ consistently in one direction. Figuratively speaking, the surface of $f(X)$ is no longer a plane as in the linear case but is usually curved in the shape of a mountain or a cup. Consequently, in quadratic programming, it is a distinct possibility that the extreme value of the quadratic function may be reached before the variables are pushed to their boundary values determined by the associated constraints. The upshot of this is that the optimal solution may not be an extreme-point solution. In short, the simplex method may not work.

A function is said to be *concave* if, given two points X_1 and X_2 in the n-dimensional vector space

$$f[aX_1 + (1 - a)X_2] \geq af(X_1) + (1 - a)f(X_2) \qquad \text{where} \quad 0 \leq a \leq 1 \tag{10.4.4}$$

Similarly, it is *convex* if the inequality sign above is reversed. Now from (10.4.2) we have

$$\begin{aligned} af(X_1) &= a(c_0 + cX_1 + X_1^TCX_1), \\ (1 - a)f(X_2) &= (1 - a)(c_0 + cX_2 + X_2^TCX_2), \end{aligned}$$

and

$$\begin{aligned} f[aX_1 + (1 - a)X_2] = c_0 &+ c[aX_1 + (1 - a)X_2] \\ &+ [aX_1^T + (1 - a)X_2^T]C[aX_1 + (1 - a)X_2]. \end{aligned}$$

Therefore,

$$\begin{aligned} &f[aX_1 + (1 - a)X_2] - [af(X_1) + (1 - a)f(X_2)] \\ &\quad = [aX_1^T + (1 - a)X_2^T]C[aX_1 + (1 - a)X_2] \\ &\qquad\qquad - [aX_1^TCX_1 + (1 - a)X_2^TCX_2] \\ &\quad = a^2X_1^TCX_1 + 2a(1 - a)X_1^TCX_2 + (1 - a)^2X_2^TCX_2 \\ &\qquad\qquad - aX_1^TCX_1 - (1 - a)X_2^TCX_2 \end{aligned}$$

** The existence of an extremum (maximum or minimum) depends upon whether the quadratic function is definite (negative or positive respectively). *Vide* G. Debreu, "Definite and Semidefinite Quadratic Forms," *Econometrica*, XX (1952), pp. 295–300, and Wm. F. Osgood, *Advanced Calculus* (New York: The Macmillan Co., 1958), pp. 172–179.

$$= -a(1 - a)X_1^TCX_1 + 2a(1 - a)X_1^TCX_2 - a(1 - a)X_2^TCX_2$$
$$= -a(1 - a)[X_1^TCX_1 - 2X_1^TCX_2 + X_2^TCX_2]$$
$$= -a(1 - a)[(X_1^T - X_2^T)C(X_1 - X_2)].$$

It is clear that if (10.4.4) is to hold, the above expression must be non-negative. Since the range of "a" is from 0 to 1, $a(1 - a)$ is always non-negative and consequently $(X_1^T - X_2^T)C(X_1 - X_2)$ must be non-positive. This is equivalent to saying in proper mathematical language that the quadratic form $(X_1^T - X_2^T)C(X_1 - X_2)$ must be negative semi-definite. The condition for the concavity of a quadratic function, therefore, is the same as the condition for the negative semi-definiteness of a quadratic form, namely that the successively higher-ordered determinants of C along the main diagonal, viz. $|c_{11}|$, $\begin{vmatrix} c_{11} & c_{12} \\ c_{21} & c_{22} \end{vmatrix}$, etc., must be alternatively non-positive and non-negative. By similar procedure we can show that a quadratic function is convex if it is positive semi-definite, i.e., if the successively higher-ordered determinants of C are all non-negative.*

Let us be more specific and look at a simple quadratic function such as $f(X) = 10 + 20x_1 + 10x_2 - 2x_1^2 - x_2^2$. Written in matrix form, this function will have $c = [20 \ 10]$ and $C = \begin{bmatrix} -2 & 0 \\ 0 & -1 \end{bmatrix}$. Since $|c_{11}| = -2$ and $\begin{vmatrix} c_{11} & c_{12} \\ c_{21} & c_{22} \end{vmatrix} = 2$, $f(X)$ is concave. Our next question is: What are the difficulties in maximizing or minimizing this function under linear constraints?

To answer this question, we draw (roughly) the level curves of $f(X)$ on the x_1-x_2 plane as shown in Figure 10.7. In (a), an arbitrary system of constraints is drawn such that the maximum (M) of $f(X)$ (which incidentally is at $x_1 = x_2 = 5$) lies outside of the feasible solution region whereas in (b) it is inside. If we are to maximize $f(X)$, the optimal solution $X_{\max}$ in (a) is a point on the line segment between two extreme points of the convex solution set and in (b) is M itself inside the set. In general, therefore, the standard simplex method which takes us from extreme point to extreme point will not give us the correct answer unless by coincidence $X_{\max}$ happens to be an extreme point also. Now, consider the problem of minimizing $f(X)$. It can be seen that in both (a) and (b), the optimal solution $X_{\min}$ is

* Osgood, *op. cit.*, pp. 172–179.

indeed an extreme-point solution. Unfortunately, however, the simplex method is still unsuitable. The reason is that it permits us to evaluate (by computing $z_j - c_j$) the adjacent extreme points only. But, if $f(X)$ is quadratic and we are at the extreme point A [Figure 10.7 (b)], the evaluations of its adjacent extreme points B and E will both indicate higher $f(X)$, and prevent us from going over the hump, so to speak, to reach the true minimal feasible solution O. The danger of using the simplex method in minimizing a concave quadratic function over a set of linear constraints therefore is that it may lead us to a local minimum instead of the true or global minimum. With these observations, we may now conclude the following regarding the prob-

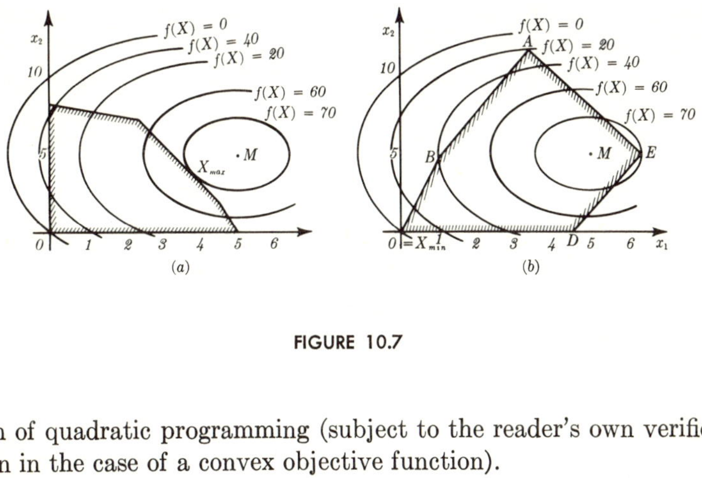

FIGURE 10.7

lem of quadratic programming (subject to the reader's own verification in the case of a convex objective function).

If the problem is to maximize a concave quadratic function (or minimize a convex quadratic function) over a set of linear constraints, there is always a unique maximal (minimal) solution, and it is generally not an extreme point of the solution set. On the other hand, if the problem is to minimize a concave quadratic function (or maximize a convex quadratic function), there may very likely be more than one extreme-point solution which is minimal (maximal) in the *local* sense, i.e., relative to the solutions in its immediate neighborhood. The problem is not solved until all local minima (maxima) are compared and the minimum of the minima, or *global* minimum (maximum of the maxima, or global maximum) is determined.

Thus, it is seen that each of the two types of quadratic programming problem has its own problems. For various reasons the first type seems to have attracted more attention in the past than the second. The remainder of this section will be addressed to the question of finding the single optimum that is not a basic feasible solution. So far as locating the global optimum among a group of local optima is concerned, we shall have to be content with the unhappy observation that no satisfactory systematic procedure has yet been proposed. Although no real difficulty exists when the problem is small enough to make feasible complete evaluation of all extreme-point solutions, in large problems the true optimal solution remains tinged with uncertainty.

Of the methods that have been published for the maximization of concave quadratic function under linear constraints, the best-known are those of Wolfe, Beale, Houthakker and Markowitz.* We shall not discuss all of them here, nor attempt to rank them in order of merit. We shall merely explain one of them in sufficient detail that the reader can actually put it to use. The selection of Wolfe's method is based on two basic considerations: (1) it has an interesting and somewhat sophisticated mathematical approach and more importantly, (2) it has a close affinity with the simplex method, which should be well known to the reader by now.

The heart of Wolfe's method is a theorem due to Kuhn and Tucker,** which states that in maximizing any function $f(X)$ over any set of constraints $f_i(X) = 0$, $i = 1, \ldots, m$, (either linear or nonlinear), a solution X_0 is a local maximum if the hyperplane passing through X_0 and normal to the gradient vector of $f(X)$ at X_0, is locally a supporting hyperplane of the constraint set, i.e., if it touches the constraint set in the vicinity of X_0 only at X_0. Furthermore, as we

---

* See M. Frank and Philip Wolfe, "An Algorithm for Quadratic Programming," *Naval Research Logistics Quarterly*, III (1956), No. 1 and 2; E. M. L. Beale, "On Quadratic Programming," *Ibid.*, VI (1959), No. 3; Harry Markowitz, "The Optimization of a Quadratic Function Subject to Linear Constraints," *Ibid.*, III (1956), No. 1 and 2; H. S. Houthakker, "The Capacity Method of Quadratic Programming," *Econometrica*, XXVIII (1961), No. 1.

** Cf. H. W. Kuhn and A. W. Tucker, "Nonlinear Programming," *Proceedings of the Second Berkeley Symposium of Mathematical Statistics and Probability*, pp. 481–492, 1951.

have discussed previously, if $f(X)$ is concave quadratic and $f_i(X)$ are linear, then X_0 is also the unique global maximum desired. In such standard quadratic programming problems, Wolfe noted that the projection of X_0 along the gradient vector of $f(X)$ at X_0 must be the greatest of all similar projections from other points contained in the convex constraint set. In other words, if our problem is

$$\text{Maximize } f(X) = c_0 + CX + X^TCX, \tag{10.4.2}$$

$$\text{subject to: } \quad AX \leq b,$$

(10.4.5) $$\text{and} \quad X \geq O,$$

where: $f(X)$ is concave (or equivalently, C is negative semi-definite), A is an $m \times n$ matrix, and b is an $m \times 1$ column vector,

and if X_0 is the maximal solution, then the gradient vector of $f(X)$ at X_0 according to (10.4.3) is $\partial f(X_0) = \left[\frac{\partial f(X)}{\partial x_1}, \ldots, \frac{\partial f(X)}{\partial x_n}\right]_{X=X_0}$, and the projection of X_0 along the gradient vector is

$$\partial f(X_0)X_0 = \max \{\partial f(X_0)X \mid AX \leq b, X \geq O\}.\text{*} \tag{10.4.6}$$

Let us pause and illustrate the above by a simple quadratic problem involving 3 constraints and 2 variables only. The concave quadratic function $f(X)$ is shown in Figure 10.8 as a series of level curves, $f(X) = d_1$, $f(X) = d_2$, etc. until $f(X) =$ maximum at M. The constraints are represented by the shaded convex area with extreme-point solutions O, K, H, X_1 and G. By construction X_0 is found to be the maximal solution as it touches the highest level curve of all the points in the convex solution set. Since the gradient vector points to the direction of maximal increase of $f(X)$, we know the gradient vector at X_0, or $\partial f(X_0)$ with components $\frac{\partial f(X_0)}{\partial x_1}$ and $\frac{\partial f(X_0)}{\partial x_2}$, as a vector perpendicular to the tangent of $f(X) = d_4$ at X_0. The gradient vector forms an angle ϕ with the x_1-axis, which is defined by

$$\sin \phi = \frac{\dfrac{\partial f(X_0)}{\partial x_2}}{h} \quad \text{and} \quad \cos \phi = \frac{\dfrac{\partial f(X_0)}{\partial x_1}}{h},$$

$$\text{where } h = \sqrt{\left[\frac{\partial f(X_0)}{\partial x_1}\right]^2 + \left[\frac{\partial f(X_0)}{\partial x_2}\right]^2}.$$

---

* The sign | is read "subject to the conditions."

Now the projection of X_0 (or vector $\overrightarrow{OX_0}$) along the gradient vector is the vector $\overrightarrow{AX_0}$ whose length is

$$\overline{AX_0} = \overline{AB} + \overline{BX_0} = \overline{OC} \cos \phi + \overline{CX_0} \sin \phi$$

Since OC and CX_0 are the x_1, x_2 coordinates of X_0, therefore

$$\overline{AX_0} = \frac{1}{h}\, \partial f(X_0)X_0$$

Similarly, the projection of any point X in the convex region is $\frac{1}{h}\, \partial f(X_0)X$. The projection of X_1, for instance, is $\overline{AD} = \frac{1}{h}\, \partial f(X_0)X_1$ which is smaller than $\overline{AX_0}$. The reader may try any other point in the

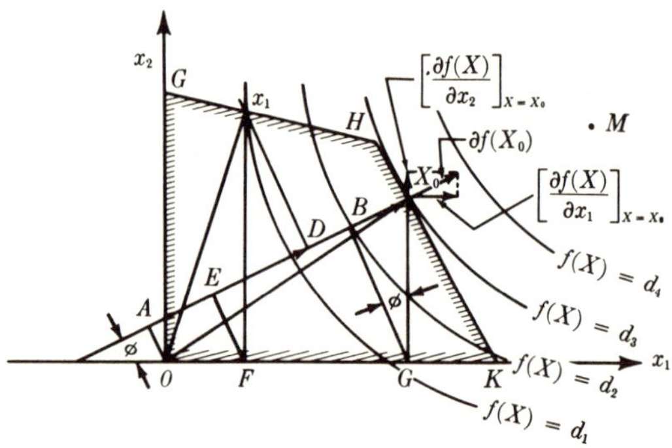

Figure 10.8

convex region defined by the linear constraint set and find that its projection is no greater than $\overline{AX_0}$. Thus the validity of (10.4.6) is verified.

Resuming our discussion of Wolfe's method, we note that (10.4.6) is actually a linear programming problem with $\partial f(X_0)X$ as the linear objective function and $AX \leq b$ and $X \geq O$ as the linear constraints. According to the duality theorem discussed in Chapter VIII, there is a dual problem, viz., minimize Wb subject to the constraints $WA \geq \partial f(X_0)$ and $W \geq O$, such that (1) $Wb \geq \partial f(X_0)X$, and (2) min $Wb = \max f(X_0)X$. Consequently, we know

(10.4.7)

$$\text{Max}\ \{\partial f(X_0)X - Wb \mid AX \leq b, WA \geq \partial f(X_0), X \geq O, W \geq O\} = 0.$$

This conclusion is interesting but unfortunately does not provide us with a method to solve for X_0 since $\partial f(X_0)$ is still unknown. Trying to overcome this obstacle, Wolfe investigated a similar function, i.e., $\partial f(X)X - Wb$ with X and W subjected to the constraints $AX \leq b$, $X \geq O$, $WA \geq \partial f(X)$ and $W \geq O$. By substituting $WA \geq \partial f(X)$ into the function, he noticed that

$$\partial f(X)X - Wb \leq WAX - Wb = W(AX - b).$$

Since $AX \leq b$,

$$\partial f(X)X - Wb \leq O, \tag{10.4.8}$$
$$\text{subject to: } AX \leq b,\ X \geq O,\ WA \geq \partial f(X) \text{ and } W \geq O.$$

This is an important discovery for if (10.4.8) is true, then (10.4.7) and (10.4.8) are equivalent and X_0 is not only the maximizing solution to the former but also to the latter. Thereupon, Frank and Wolfe defined as an auxiliary problem of the original quadratic programming problem (10.4.5) the following:

$$\text{Maximize } g(X, W) = \partial f(X)X - Wb,$$

$$\text{(10.4.9)} \quad \begin{array}{lll} \text{subject to:} & AX \leq b, & WA \geq \partial f(X), \\ \text{and} & X \geq O, & W \geq O. \end{array}$$

It is understood that the maximal value of $g(X, W)$ is 0 and that the vector which brings this about is the desired optimum X_0 which also maximizes the original quadratic function $f(X)$ in (10.4.5).

This auxiliary problem, however, is not a linear programming problem since $\partial f(X)$ in $g(X, W)$ is not constant. To solve it, let us first convert the constraints into equations by adding the nonnegative slack vector X_s to $AX \leq b$ and subtracting the nonnegative slack vector W_s from $WA \geq \partial f(X)$. Thus

$$\text{(10.4.10)} \quad \begin{aligned} AX + X_s &= b, \\ WA - W_s &= \partial f(X). \end{aligned}$$

Substituting the latter into $g(X, W)$, we have

$$\begin{aligned} g(X, W) = f(X)X - Wb &= (WA - W_s)X - Wb \\ &= W(AX - b) - W_sX = -WX_s - W_sX. \end{aligned}$$

This in turn can be written as

(10.4.11) $g(X, W) = -\frac{1}{2}\tilde{Z}Z,$ where $\tilde{Z} = [W_s \; W \; X_s^T \; X^T]$ and

$$Z = \begin{bmatrix} X \\ X_s \\ W^T \\ W_s^T \end{bmatrix}.$$

Recalling that $\partial f(X) = c + 2X^TC$ (10.4.3), the constraints in (10.4.10) may now also be written in terms of Z. The following conversions may be easily verified:

$$\text{(10.4.12)} \qquad \begin{array}{lll} [\;\;A \;\; I \;\; O \;\;\;\; O\;\;]Z = b & \text{from} & AX + X_s = b, \\ [-2C \; O \; A^T \; -I\;]Z = c^T & \text{from} & WA - W_s = \partial f(X). \end{array}$$

Now, combining (10.4.11) and (10.4.12), we have the following restatement of the auxiliary problem given in (10.4.9):

$$\text{(10.4.13)} \qquad \begin{array}{l} \text{Maximize } -\frac{1}{2}\tilde{Z}Z, \\ \text{subject to: } TZ = d \text{ and } Z \geq O, \\ \text{where } \quad T = \begin{bmatrix} A \;\; I \;\; O \;\;\;\; O \\ -2C \; O \; A^T \; -I \end{bmatrix}, \quad d = \begin{bmatrix} b \\ c^T \end{bmatrix}, \\ \text{and } \quad \tilde{Z}, Z \text{ as defined in (10.4.11).} \end{array}$$

Since it is already known that the maximal value of $g(X, W)$ is 0, according to (10.4.11), the maximal value of $-\frac{1}{2}\tilde{Z}Z$ is also 0. Our next logical question is then: What kind of vector must Z be in order to yield this desired optimal value?

To answer this question, we note that Z is constrained to be non-negative, and that $\tilde{Z}$ is nothing but a rearrangement of the components of Z. Since Z has $2(m + n)$ components, $-\frac{1}{2}\tilde{Z}Z$ is always the negative sum of $2(m + n)$ positive terms and will not vanish unless each of these terms is 0. This obviously cannot happen unless each non-zero component in Z is matched by a corresponding zero component in $\tilde{Z}$. Therefore, the optimal Z must not have more than $m + n$ non-zero components. Noticing that there are $m + n$ constraints in the constraint system $TZ = d$, we conclude that the optimal solution to (10.4.13) is a basic feasible solution (which may or may not be degenerate). In order to satisfy the primal constraints, viz., $[A \;\; I \;\; O \;\; O]Z = b$, this optimal solution must have at least m positive x-components, which implies that it could be a non-basic feasible solution to the original quadratic programming problem (10.4.5). Frank and Wolfe thus have succeeded in finding a way which allows a non-basic feasible solution to a quadratic programming problem to be represented as a basic feasible solution to an enlarged but equivalent problem. Since the simplex method works with basic feasible solutions, this transformation is an important step towards using the simplex method to solve quadratic programming problems.

But the enlarged problem is not yet a linear programming problem because its objective function, $-\frac{1}{2}\tilde{Z}Z$, is clearly nonlinear. This difficulty, however, was skillfully overcome by a two-phase iterative procedure which as proven by Frank and Wolfe, insures convergence.

Phase I—Obtain a feasible solution to (10.4.13). If no feasible solution exists, disregard the dual constraints and obtain a feasible solution of only the primal constraints $AX \leq b$. If one exists, then the constraint set is unbounded above and hence, maximal $f(X) = \infty$. If one still cannot be found, then the quadratic problem has no solution, showing that it has inconsistency among its constraints.

Phase II—If a feasible solution, say V, is found for (10.4.13), then let $\tilde{Z} = \tilde{V}$ and construct the linear programming problem

$$\text{(10.4.14)} \qquad \text{Maximize } -\tfrac{1}{2}\tilde{V}Z \quad \text{subject to: } TZ = d \quad \text{and } Z \geq O.$$

Use the simplex method to obtain progressively better basic feasible solutions $Z_1, Z_2 \ldots$ until Z_h is found which satisfies either of the following two conditions:

(a) $\tilde{Z}_h Z_h = 0$, which indicates that Z_h is the desired optimum and the quadratic programming problem is solved;

(b) $-\tilde{V}Z_h \geq -\frac{1}{2}\tilde{V}V$ for the first time,* whereupon a new feasible solution V' is obtained thus,

$$\text{(10.4.15)} \qquad V' = (1 - a)V + aZ_h\,,^{**}$$

$$\text{where} \quad a = \min\left\{\frac{\tilde{V}(V - Z_h)}{(\tilde{V} - \tilde{Z}_h)(V - Z_h)}, 1\right\},$$

and by letting $\tilde{Z} = \tilde{V}'$, Phase II is begun anew.

---

* Frank and Wolfe, *op. cit.*, p. 102, showed in their proof of convregence that the linear programming problem (10.4.14) has an optimal value that is not less than $-\frac{1}{2}\tilde{V}V$ due to the concave nature of $g(X, W)$. Determination of the exact optimum of (10.4.14) may or may not be worth the additional time required.

** The objective of (10.4.15) is to find a new feasible solution V' on the line segment $\overline{VZ_h}$, which has the highest value of $g(X, W)$. Thus, substituting for Z, $V' = (1 - a)V + aZ_h$ into $g(X, W) = -\frac{1}{2}\tilde{Z}Z$, we have

$$-\tfrac{1}{2}[(1 - a)\tilde{V} + a\tilde{Z}_h]\,[(1 - a)V + aZ_h] = -\tfrac{1}{2}[(1 - a)^2\tilde{V}V + 2a(1 - a)\tilde{V}Z_h + a^2\tilde{Z}_hZ_h].$$

Example. We will now illustrate the Frank-Wolfe algorithm by a simple numerical problem.

$$\begin{aligned} \text{Maximize } f(X) &= 10 + 20x_1 + 10x_2 - 2x_1^2 - x_2^2, \\ \text{subject to:} \quad & 2x_1 + x_2 \leq 10, \\ & x_1 + x_2 \leq 7, \\ \text{and} \quad & x_1, x_2 \geq 0. \end{aligned}$$

First we must determine the nature of the quadratic objective function. This is accomplished by testing the matrix C, which, by the definition given in (10.4.1) and (10.4.2), is for this problem $\begin{bmatrix} -2 & 0 \\ 0 & -1 \end{bmatrix}$. Its first-ordered and second-ordered determinants therefore have the values -2 and 2 respectively. Thus the function is negative definite, or equivalently, concave, and a unique maximal solution is possible, provided that the constraints are not inconsistent.

Secondly, we shall write down the constraints of the auxiliary problem in accordance with (10.4.13). These are

$$(10.4.16) \quad \begin{aligned} 2x_1 + x_2 + x_3 \qquad\qquad\qquad\qquad\qquad &= 10, \\ x_1 + x_2 \qquad + x_4 \qquad\qquad\qquad\qquad &= 7, \\ 4x_1 \qquad\qquad\qquad + 2w_1 + w_2 - w_3 \qquad &= 20, \\ 2x_2 \qquad\qquad + w_1 + w_2 \qquad - w_4 &= 10. \end{aligned}$$

Following Phase I of the Frank-Wolfe algorithm, we observe that there is a feasible solution to the above constraint system. This is: $x_3 = 10$, $x_4 = 7$, $w_1 = 10$, and the rest all equal to zero. We are now ready to begin Phase II.

In Phase II we use the above feasible solution, viz., $\tilde{V} = [w_3 \; w_4 \; w_1 \; w_2 \; x_3 \; x_4 \; x_1 \; x_2] = [0 \; 0 \; 10 \; 0 \; 10 \; 7 \; 0 \; 0]$, to construct the linear objective function in accordance with (10.4.14). Thus

---

Differentiating the above with respect to 'a,' and setting the resulting derivative to zero, we obtain

$$\frac{dg(X, W)}{da} = 0 = -(1 - a)\tilde{V}V + (1 - 2a)\tilde{V}Z_h + a\tilde{Z}_hZ_h,$$

which then yields the expression of "a" in (10.4.15). The value so determined can never be less than 0, since $-\tilde{V}Z_h \geq -\frac{1}{2}\tilde{V}V$, but may exceed 1. In the latter case, V' would be infeasible as Z_h is an extreme-point solution. For this reason, the value of a is given in (10.4.15) as the smaller of the two magnitudes. Because of this feature of determining V' in the algorithm, Frank and Wolfe called their method the "gradient and interpolation" method.

$$(10.4.17) \qquad -\tfrac{1}{2}\tilde{V}Z = -\tfrac{1}{2}(w_3x_1 + w_4x_2 + w_1x_3 + w_2x_4 + x_3w_1 + x_4w_2 + x_1w_3 + x_2w_4) = -5x_3 - 5w_1 - \tfrac{7}{2}w_2$$

and our linear programming problem consists in maximizing (10.4.17) subject to (10.4.16) and the non-negativity constraints.

In the interest of clarity, the process of solving the linear programming problem is shown in detail in the following sequence of tableaus. The initial feasible solution obtained in Phase I suggests that the initial basic feasible solution for Phase II could be $x_3 = 10$, $x_4 = 7$, $w_1 = 10$ and $w_4 = 0$. Using these as basis variables, we obtain the following basis and its inverse:

$$B = \begin{matrix} P_3 & P_4 & P_5 & P_8 \\ \end{matrix} \begin{bmatrix} 1 & 0 & 0 & 0 \\ 0 & 1 & 0 & 0 \\ 0 & 0 & 2 & 0 \\ 0 & 0 & 1 & -1 \end{bmatrix} \qquad B^{-1} = \begin{bmatrix} 1 & 0 & 0 & 0 \\ 0 & 1 & 0 & 0 \\ 0 & 0 & \frac{1}{2} & 0 \\ 0 & 0 & \frac{1}{2} & -1 \end{bmatrix}$$

Thus, the simplex tableau corresponding to the initial basic feasible solution is (cf. Chapter V, Section 5.2 regarding its construction),

	P_1	P_2	P_3	P_4	P_5	P_6	P_7	P_8	P_0
-5 P_3	2	1	1						10
0 P_4	1	1		1					7
-5 P_5	2				1	$\frac{1}{2}$	$-\frac{1}{2}$		10
-0 P_8	②	-2				$-\frac{1}{2}$	$-\frac{1}{2}$	1	0
z_j	-20	-5	-5	0	-5	$-\frac{5}{2}$	$\frac{5}{2}$	0	
c_j	0	0	-5	0	-5	$-\frac{7}{2}$	0	0	
$z_j - c_j$	-20	-5	0	0	0	1	$\frac{5}{2}$	0	-100

Substituting P_1 for P_8 in the basis, we obtain the second tableau as follows:

-5 P_3		③	1			$\frac{1}{2}$	$\frac{1}{2}$	-1	10
0 P_4		2		1		$\frac{1}{4}$	$\frac{1}{4}$	$-\frac{1}{2}$	7
-5 P_5		2			1	1		-1	10
0 P_1	1	-1				$-\frac{1}{4}$	$-\frac{1}{4}$	$\frac{1}{2}$	0
$z_j - c_j$	0	-25	0	0	0	-4	$-\frac{5}{2}$	10	-100

The corresponding basic feasible solution is $Z_1^T = [0\ 0\ 10\ 7\ 10\ 0\ 0\ 0]$. Hence, $\tilde{Z}_1 Z_1 \neq 0$, and $-\tilde{V}Z_1\ (= -200)$ is smaller than $-\frac{1}{2}\tilde{V}V$ $(= -100)$. These test results indicate that the solution of the linear programming problem should continue and the following third tableau is partly completed to identify the next basic feasible solution

	P_1	P_2	P_3	P_4	P_5	P_6	P_7	P_8	P_0
0 P_2		1							$\frac{10}{3}$
0 P_4				1					$\frac{1}{3}$
−5 P_5					1				$\frac{10}{3}$
0 P_1	1								$\frac{10}{3}$
$z_j - c_j$									$-\frac{50}{3}$

which is $Z_2^T = [\frac{10}{3}\ \frac{10}{3}\ 0\ \frac{1}{3}\ \frac{10}{3}\ 0\ 0\ 0]$. Since $\tilde{Z}_2 Z_2 = 0$, the problem is solved, and the maximizing solution is Z_2, or $x_1 = x_2 = \frac{10}{3}$, $x_4 = \frac{1}{3}$ with $f(X) = 76\frac{2}{3}$. If desired, the validity of this answer can be quickly checked graphically. An analytical confirmation would be to ascertain whether the level curve $f(X) = 76\frac{2}{3}$ is tangent to the convex constraint set at $x_1 = x_2 = \frac{10}{3}$. That this solution is a point on the boundary and not a basic feasible solution to the original quadratic programming problem is evidenced by the fact that $x_3 = 0$ and $x_4 = \frac{1}{3}$.

10.5. Convex Programming by Piece-wise Linear Approximation.

In this section we shall consider a different method of solving programming problems whose objective functions are nonlinear, separable and convex (or concave). This means an objective function which can be expressed as a linear combination of several separate one-variable functions which are all convex and of which some or all are nonlinear. An example of this type of function may be taken from almost any manufacturing firm. The cost of direct labor, for instance, is a linear function of the level of production only to the extent that the total amount of work is within the limit of the regular work week. When the limit is exceeded, premium rates for extra shifts or bonuses for overtime work cause the cost curve to assume a steeper climb as shown in Figure 10.9 (a). Thus, the entire direct labor cost curve becomes nonlinear, being a string of connected linear segments with increasing slopes. Another example may be found in the behavior of machine repair-and-maintenance cost. Up to a certain level of production, this cost may remain quite constant. But as this level is surpassed, it may rise rapidly with further increase in production resulting in nonlinearity. This is illustrated in Figure 10.9 (b). For the sake of argument, we assume that these are the only two costs

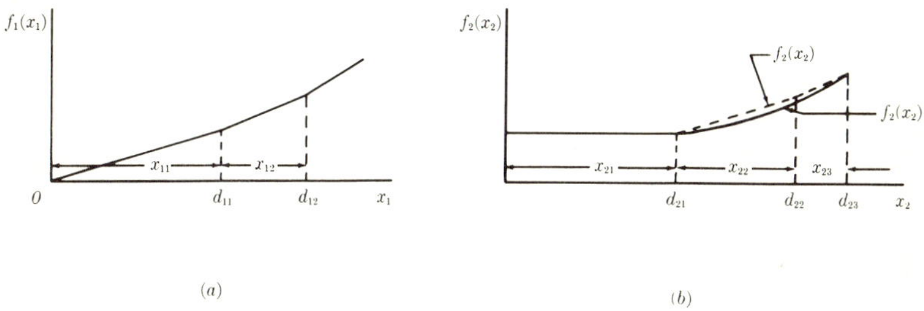

Figure 10.9

relevant to the problem. Since they are both convex, the total cost would be nonlinear, but separable and itself also *convex*. It would then have a *unique minimum* which can be approximated by the piece-wise linear approximation method discussed below. Obviously if the function is nonlinear, separable and *concave*, then a *unique maximum* exists and the same approximation method can be applied with equal success.

Suppose that we are to solve the following problem:

$$\text{Minimize } f(X) = \sum_{j=1}^{n} f_j(x_j), \qquad \text{where } f_j(x_j) \text{ are convex,}$$

(10.5.1)
$$\text{subject to:} \quad \sum_{j=1}^{n} a_{ij}x_j \geq b_i \qquad i = 1, \ldots, m.$$

$$\text{and} \qquad x_j \geq 0 \qquad j = 1, \ldots, n.$$

We shall first reduce those $f_j(x_j)$ that are nonlinear to their piece-wise linear approximations by fitting linear segments to the original curve as indicated in Figure 10.9 (b). This approximation can be carried out to any degree of accuracy by dividing the curve into a suitable number of intervals for linear representation. Let us now assume that each $f_j(x_j)$ is divided into p_j intervals marked by $x_j = 0,\ d_{j1},\ d_{j2},\ \ldots,\ d_{jp_j}$, and that the slopes of the linear segments occupying these intervals are correspondingly, $s_{j1}, \ldots, s_{jp_j}$. Since $f_j(x_j)$ is convex by assumption, it follows that $s_{j1} < s_{j2} < \ldots < s_{jp_j}$, which makes it possible for us to substitute

$$x_j = \sum_{t=1}^{p_j} x_{jt}, \quad \text{where} \quad 0 \le x_{jt} \le d_{jt} - d_{j,t-1}, \tag{10.5.2}$$

into (10.5.1) and obtain the following equivalent problem:

$$\begin{aligned}
&\text{Minimize } \hat{f}(X) = \sum_{j=1}^{n} \sum_{t=1}^{p_j} s_{jt} x_{jt} + \sum_{j=1}^{n} f_j(0), \\
(10.5.3) \qquad &\text{subject to:} \quad \sum_{j=1}^{n} a_{ij} \left(\sum_{t=1}^{p_j} x_{jt} \right) \ge b_i, \qquad i = 1, \ldots, m, \\
&\qquad x_{jt} \le d_{jt} - d_{j,t-1}, \qquad j = 1, \ldots, n, \quad t = 1, \ldots, p_j, \\
&\text{and} \qquad x_{jt} \ge 0, \qquad j = 1, \ldots, n, \quad t = 1, \ldots, p_j.
\end{aligned}$$

We now have a bounded-variable problem with m constraints and $\sum_{j=1}^{n} p_j$ variables. It can be solved by the procedure discussed in Section 10.2 of this Chapter.

One important theoretical point needs to be discussed and resolved. Referring to (10.5.3), we notice that the constraints impose no condition whatsoever on the order of appearance in the optimal solution of the several components (x_{jt}) of a given variable x_j. That is to say, it is conceivable that in the optimal solution some higher-ordered component, e.g., x_{j4}, might assume a positive value while its lower-ordered counterpart, e.g., x_{j2}, remained zero. And yet, such a solution would be totally meaningless and inadmissible because the components have no meaning in themselves except as *connected* parts of the original variable x_j (10.5.2). Fortunately this difficulty does not arise in the case of (10.5.3), for two reasons. One is that the different components of the same variable all have the same coefficient vector (P_j) and, hence, only one of them can be in the basis B [not $\overline{B}$;

see Section 10.2, particularly (10.2.6) and Figure 10.2]. Secondly, since their coefficient vectors are all P_j, their z_j ($= B^{-1}P_j$) are also the same. This renders the evaluation ($z_j - c_j$) of the components strictly a matter of comparing their respective price coefficients, viz., the s_{jt}. Since it has already been pointed out that $s_{j,t-1} < s_{jt}$ for the entire range of x_j, therefore $x_{j,t-1}$ is always preferred to x_{jt} in a minimizing solution. These two properties together insure that no x_{jt} will be positive unless $x_{j,t-1} = d_{jt} - d_{j,t-1}$. Hence, the solution is always meaningful.

The above discussion clearly suggests that without the assumption of the convexity of $f_j(x_j)$, the bounded-variable method cannot be applied. Charnes and Lemke, however, have developed a technique* which solves problems having nonlinear, non-convex but separable objective functions. It makes use of the same piece-wise linear approximation approach but defines auxiliary variables and additional constraints differently so as to prevent the occurrence of "disconnected representation" of original variables in the final optimal solution. It does not insure that the optimum obtained is the global optimum,** hence will not be discussed in detail here.

In conclusion, we note that success in nonlinear programming is still strictly limited as of this date. There is as yet no proven solution to a general nonlinear programming problem which can effectively distinguish between local and global optima and guide us unerringly to the latter by means of some pre-determined rules and procedure. The removal of this obstacle will require major research efforts in the years to come.

EXERCISES

1. Solve Exercise 1 (e) of Chapter V with the additional upper-bound constraints: $x_1 \leq 7$, $x_2 \leq 3$, and $x_3 \leq 4$.

2. Solve Exercise 1 (g) of Chapter *V* with the additional upper-bound constraint, $x_1 \leq 40$.

---

* See A. Charnes and C. E. Lemke, "Minimization of Non-linear Separable Convex Functionals, Computational Theory of Linear Programming—IV," *ONR Research Memorandum No. 16*, (Pittsburgh: Carnegie Institute of Technology, 1954).

** See A. Charnes and W. W. Cooper, "Non-linear Power of Adjacent Extreme-Point Method in Linear Programming," *Econometrica*, XXV (1957), No. 1.

3. Determine the optimal integer solution of Exercise 2, Chapter I.

4. Determine the optimal integer solution of Exercise 7, Chapter V for the production of X, Y and Z.

5. Solve the following quadratic programming problem:

$$
\begin{aligned}
\text{Maximize } f(X) &= 20x_1 - 10x_2 - 3x_1^2 - 2x_2^2, \\
\text{subject to:} \quad & 2x_1 - x_2 \le 6, \\
& -x_1 + x_2 \le 10, \\
& 2x_1 + 3x_2 \ge 8, \\
\text{and} \quad & x_1, x_2 \ge 0.
\end{aligned}
$$

BIBLIOGRAPHY

1. Beale, E. M. L., "An Algorithm for Solving the Transportation Problem, When the Shipping Cost over Each Route is Convex," *Naval Research Logistics Quarterly,* Vol. 6 (1959), pp. 43–56.
2. Beale, E. M. L., "On Quadratic Programming," *Naval Research Logistics Quarterly,* Vol. 6, No. 3 (1959), pp. 227–243.
3. Bennion, E. G., *Elementary Mathematics of Linear Programming and Game Theory,* Michigan: Michigan State University, 1960.
4. Boulding, K. E., and W. A. Spivey, *Linear Programming and the Theory of the Firm,* New York: The Macmillan Company, 1960.
5. Bowman, E. H., "Production Scheduling by the Transportation Method of Linear Programming," *Operations Research,* Vol. 4 (1956), No. 1, pp. 100–3.
6. Charnes, A., "Optimality and Degeneracy in Linear Programming," *Econometrica,* XX (1952), No. 2.
7. Charnes, A., and W. W. Cooper, "Generalizations of the Warehouse Model," *Operations Research,* Vol. 6 (1955), pp. 131–172.
8. Charnes, A., and W. W. Cooper, "The Stepping Stone Method of Explaining Linear Programming Calculations in Transportation Problems," *Management Science,* Vol. 1 (1954–5).
9. Charnes, A., and W. W. Cooper, "Non-linear Power of Adjacent Extreme Point Method in Linear Programming," *Econometrica,* XXV (1957), No. 1, pp. 132–153.
10. Charnes, A., and W. W. Cooper, "Management Models and Industrial Applications of Linear Programming," *Management Science,* Vol. 4 (1957), No. 1, pp. 38–92.

11. Charnes, A., and W. W. Cooper, *Management Models and Industrial Applications of Linear Programming,* New York: John Wiley & Sons, Inc., 1961, Vols. I and II.
12. Charnes, A., W. W. Cooper, and A. Henderson, *Introduction to Linear Programming,* New York: John Wiley & Sons, Inc., 1953.
13. Charnes, A., W. W. Cooper, and B. Mellon, "Blending Aviation Gasolines—A Study in Programming Interdependent Activities in an Integrated Oil Company," *Econometrica,* XX (1952).
14. Charnes, A., and C. E. Lemke, *The Bounded Variables Problem,* Office of Naval Research, Research Memorandum 10, Pennsylvania: GSIA, Carnegie Institute of Technology, 1954.
15. Charnes, A., and C. E. Lemke, *Minimization of Non-linear Separable Functions,* Pennsylvania: GSIA, Carnegie Institute of Technology, 1954.
16. Dantzig, G. B., "Maximization of a Linear Function of Variables Subject to Linear Inequalities," Chapter XXI of Koopmans: *Activity Analysis of Production and Allocation* (*see* item 45 in Bibliography).
17. Dantzig, G. B., "Application of the Simplex Method to a Transportation Problem," Chapter XXXII of Koopmans: *ibid.*
18. Dantzig, G. B., "Linear Programming under Uncertainty," *Management Science,* Vol. 1 (1955), pp. 197–206.
19. Dantzig, G. B., "Upper Bounds, Secondary Constraints and Block Triangularity in Linear Programming," *Econometrica,* XXIII (1955), pp. 174–183.
20. Dantzig, G. B., "On the Significance of Solving Linear Programming Problems with Some Integer Variables," *Econometrica,* XXVIII (1960), No. 1, pp. 30–45.
21. Dantzig, G. B., L. R. Ford, Jr., and D. R. Fulkerson, "A Primal-Dual Algorithm for Linear Programs," Ann. Math. Studies, No. 38 (*Linear Inequalities and Related Systems*), N.J.: Princeton University Press, 1956.
22. Dantzig, G. B., A. Orden, and P. Wolfe, "The Generalized Simplex Method for Minimizing a Linear Form under Linear Inequality Restraints," *Pacific J. Math.,* Vol. 5 (1955), pp. 183–195.
23. Debreu, G., "Definite and Semidefinite Quadratic Forms," *Econometrica,* XX (1952), pp. 295–300.
24. Dorfman, R., "Mathematical or 'Linear' Programming," *American Economic Review,* XLIII (1953), No. 5.
25. Dorfman, R., P. A. Samuelson, and R. Solow, *Linear Programming and Economic Analysis,* New York: McGraw-Hill Book Company, Inc., 1958.

26. Egervary, E, "On Combinatorial Properties of Matrices," *Matematika es Fizikai Lapok,* Vol. 38 (1931); translated as "On Combinatorial Properties of Matrices," by H. W. Kuhn, Office of Naval Research Logistics Project Report, Department of Mathematics, Princeton University, 1953.

27. Ferguson, A. R., and G. B. Dantzig, "The Allocation of Aircraft to Routes—An Example of Linear Programming under Uncertain Demand," *Management Science,* Vol. 3 (1956), pp. 45–73.

28. Ford, L. R., Jr., and D. R. Fulkerson, "Solving the Transportation Problem," *Management Science,* Vol. 3 (1956), pp. 24–32.

29. Ford, L. R., Jr., and D. R. Fulkerson, "A Primal-Dual Algorithm for the Capacitated Hitchcock Problem," *Naval Research Logistics Quarterly,* Vol. 4 (1957), pp. 47–54.

30. Gale, D., H. W. Kuhn, and A. W. Tucker, "Linear Programming and the Theory of Games," Chapter XIX of Koopmans (*see* item 45 in Bibliography).

31. Gass, S. I., *Linear Programming,* New York: McGraw-Hill Book Company, Inc., 1958.

32. Gass, S. I., and T. L. Saaty, "Parametric Objective Function. Part II. Generalization," *Journal of the Operations Research Society of America,* Vol. 3 (1955).

33. Gass, S. I., and T. L. Saaty, "The Computational Algorithm for the Parametric Objective Function," *Naval Research Logistics Quarterly,* Vol. 2 (1955).

34. Garvin, W. W., *Introduction to Linear Programming,* New York: McGraw-Hill Book Company, Inc., 1960.

35. Hadley, G., *Linear Algebra,* Reading, Mass.: Addison-Wesley Publishing Company, Inc., 1961.

36. Hadley, G., *Linear Programming,* Reading, Mass.: Addison-Wesley Publishing Company, Inc., 1962.

37. Henderson, A., and R. Schlaifer, "Mathematical Programming," *Harvard Business Review,* Vol. 32 (1954) May–June.

38. Henderson, J. M., *The Efficiency of the Coal Industry,* Cambridge, Mass.: Harvard University Press, 1958.

39. Hirsch, W. M., and G. B. Dantzig, *The Fixed Charge Problem,* Rand Research Memorandum RM-1383, Santa Monica, Calif.: The Rand Corporation, 1954.

40. Hoffman, A. J., and W. W. Jacobs, "Smooth Patterns of Production," *Management Science,* Vol. 1 (1954–5).

41. Hoffman, A. J., M. Mannos, D. Sokolowsky, and N. Wiegmann, "Computational Experience in Solving Linear Programs," *Journal of the Society for Industrial and Applied Mathematics,* Vol. 1 (1953).

42. Houthakker, H. S., "The Capacity Method of Quadratic Programming," *Econometrica*, XXVIII (1960), No. 1, pp. 62–87.

43. Jacobs, W. W., "The Caterer Problem," *Naval Research Logistics Quarterly*, Vol. 1 (1954).

44. Karush, W., "A Theorem in Convex Programming," *Naval Research Logistics Quarterly*, Vol. 6 (1959), pp. 245–260.

45. Koopmans, T. C. (ed.), *Activity Analysis of Production and Allocation*, Cowles Commission Monograph 13, New York: John Wiley & Sons, Inc., 1951.

46. Kuhn, H. W., "The Hungarian Method for the Assignment Problem," *Naval Research Logistics Quarterly*, Vol. 2 (1955), pp. 83–97.

47. Kuhn, H. W., and A. W. Tucker, *Linear Inequalities and Related Systems*, Annals of Mathematics Studies **38**, Princeton, N.J.: Princeton University Press, 1956.

48. Kuhn, H. W., and A. W. Tucker, "Nonlinear Programming," in Proceedings of the Second Symposium on Mathematical Statistics and Probability (Berkeley, Calif.: University of California Press, 1951).

49. Lemke, C. E., "The Dual Method of Solving the Linear Programming Problem," *Naval Research Logistics Quarterly*, Vol. 1 (1954), No. 1.

50. Manne, A. S., "A Note on the Modigliani-Hohn Production Smoothing Model," *Management Science*, Vol. 3 (1957), No. 4, pp. 371–380.

51. Markowitz, H. M., "The Optimization of a Quadratic Function Subject to Linear Constraints," *Naval Research Logistics Quarterly*, Vol. 3 (1956), Nos. 1 and 2, pp. 111–133.

52. Markowitz, H. M., Portfolio Selection, Monograph 16, Cowles Foundation for Research in Economics, New York: John Wiley & Sons, Inc., 1959.

53. Markowitz, H. M., and A. S. Manne, "On the Solution of Discrete Programming Problems," *Econometrica*, Vol. XXV (1957), No. 1.

54. Metzger, R. W., *Elementary Mathematical Programming*, New York: John Wiley & Sons, Inc., 1958.

55. Orchard-Hays, W., "Background, Development and Extensions of the Revised Simplex Method," Rand Corporation Report RM-1433, 1954.

56. Orden, A., "The Transhipment Problem," *Management Science*, Vol. 2 (1956), pp. 276–285.

57. Saaty, T. L., and S. I. Gass, "The Parametric Objective Function. Part I." *Journal of the Operations Research Society of America*, Vol. 2 (1954), pp. 316–9.

58. Schwartz, Jacob, *Introduction to Matrices and Vectors*, New York: McGraw-Hill Book Company, Inc., 1961.

59. Tompkins, C. B., "Some Methods of Computational Attack on Pro-

gramming Problems, Other than the Simplex Method," *Naval Research Logistics Quarterly,* Vol. 4 (1957), pp. 95–6.

60. Tucker, A. W., "Linear and Non-linear Programming," *Operations Research,* Vol. 5 (1957), No. 2.
61. Vajda, S., *The Theory of Games and Linear Programming,* New York: John Wiley & Sons, Inc., 1956.
62. Wagner, H. M., "A Comparison of the Original and Revised Simplex," *Operations Research,* Vol. 5 (1957), pp. 361–9.
63. Wagner, H. M., "A Linear Programming Solution to Dynamic Leontief Type Models," *Management Science,* Vol. 3 (1957), pp. 234–253.
64. Waugh, F. V., "The Minimum-cost Dairy Feed," *Journal of Farm Economics,* August, 1951.
65. Wegner, P., "A Non-linear Extension of the Simplex Method," *Management Science,* Vol. 7 (1961), pp. 43–56.
66. Weiss, M. J., *Higher Algebra,* New York: John Wiley & Sons, Inc., 1956.
67. Wolfe, Philip, "The Simplex Method for Quadratic Programming," *Econometrica,* XXVII (1959), No. 3, pp. 382–399.
68. Yale University School Mathematics Study Group, *Introduction to Matrix Algebra, Mathematics for High School,* Revised ed., New Haven, Conn.: Yale University Press, 1961.

INDEX